CMS/CAIMS Books in Mathematics

Volume 3

CMS/CAIMS Books in Mathematics is a collection of monographs and graduate level textbooks published in cooperation jointly with the Canadian Mathematical Society- Societé mathématique du Canada and the Canadian Applied and Industrial Mathematics Society-Societé Canadienne de Mathématiques Appliquées et Industrielles. This series offers authors the joint advantage of publishing with two major mathematical societies and with a leading academic publishing company. The series is edited by Karl Dilcher, Frithjof Lutscher, Nilima Nigam, and Keith Taylor. The series publishes high-impact works across the breadth of mathematics and its applications. Books in this series will appeal to all mathematicians, students and established researchers. The series replaces the CMS Books in Mathematics series that successfully published over 45 volumes in 20 years.

CMS
SMC

CAIMS
SCMAI

Kira Adaricheva • Jennifer Hyndman •
J. B. Nation • Joy N. Nishida

A Primer of Subquasivariety Lattices

Springer

Kira Adaricheva
Department of Mathematics
Hofstra University
Hempstead, NY, USA

Jennifer Hyndman
Department of Mathematics and Statistics
University of Northern British Columbia
Prince George, BC, Canada

J. B. Nation
Department of Mathematics
University of Hawaii
Honolulu, HI, USA

Joy N. Nishida
Department of Mathematics
University of Hawaii at Manoa
Honolulu, HI, USA

ISSN 2730-650X ISSN 2730-6518 (electronic)
CMS/CAIMS Books in Mathematics
ISBN 978-3-030-98090-0 ISBN 978-3-030-98088-7 (eBook)
https://doi.org/10.1007/978-3-030-98088-7

Mathematics Subject Classification: 08C15, 08B15, 06B15

This Springer imprint is published by the registered company Springer Nature Switzerland AG
The registered company address is: Gewerbestrasse 11, 6330 Cham, Switzerland

Preface

Praise no day until evening, no wife before her cremation, no sword till tested, no maid before marriage, no ice till crossed, no ale till it's drunk. – Viking proverb from *Hávamál*

Скоро сказка сказывается, да не скоро дело делается. – Русская пословица

It being cold in New York, colder in British Columbia, and colder still in Kazakhstan, the authors met in a series of seminars in Hawai'i. The topic was the connection between lattices $\mathrm{L_q}(\mathcal{K})$ of subquasivarieties of a quasivariety $\mathcal{K}$ and lattices $\mathrm{S_p}(\mathbf{L}, H)$ of H-closed algebraic subsets of an algebraic lattice $\mathbf{L}$ with a monoid of operators H. Our original intention was to produce a short memo describing this connection and some of its consequences. The method involving H-closed algebraic subsets is a refinement of older approaches, and gives new information about subquasivariety lattices, so the reward seemed to justify the effort, besides providing an excuse to meet in a warm clime.

Alas, the project got out of hand, continually raising more questions. We started off with a short itinerary, but then followed where the road took us, with some scenic stops along the way. We hope that you will enjoy the trip and be inspired to your own explorations.

По существу, научная работа в математике - коллективная работа. – Андрей Николаевич Колмогоров

Adventure is not in the guidebook and beauty is not on the map. The best one can hope for is to be able to persuade some people to do some traveling on their own. – Bjarni Jónsson

Hempstead, NY, USA — Kira Adaricheva
Prince George, BC, Canada — Jennifer Hyndman
Honolulu, HI, USA — J. B. Nation
Honolulu, HI, USA — Joy N. Nishida

Contents

Chapter 1
Introduction

Из Шопенгауэра (пер. Страхова) я прочёл тоже только первую половину первой страницы (заплатив 3 руб.): но на ней-то первою строкою и стоит это: «Мир есть моё представление».
— Вот это хорошо, — подумал я по-обломовски. — «Представим», что дальше читать очень трудно и вообще для меня, собственно, не нужно. – Василий Розанов, Уединенное

An old problem of Birkhoff and Mal'cev asks for characterizations of lattices of subvarieties and lattices of subquasivarieties. We focus here on subquasivarieties. There are two parts to the question:

(1) Describe properties that hold in subquasivariety lattices,
(2) Represent lattices satisfying some given properties as subquasivariety lattices.

The hope is that the twain will meet in the middle to yield a characterization.

This problem was tackled by Viktor Gorbunov and his Siberian school, with much success. Progress up until 2004 is recorded in Gorbunov's book on quasivarieties [52] and the survey [1]. But more recent results have cast the problem in a different light, so that it now seems appropriate to reconstruct the situation from scratch.

The first part of this monograph (Chaps. 1 and 2) develops the basic theory of subquasivariety lattices in the very general context of implicational classes of structures with operations and/or relations, in a language that may or may not include equality. This version is based on results of the authors in [20, 64, 92]; cf. Hoehnke [60]. The main result represents the lattice $\mathrm{L_q}(\mathcal{K})$ of subquasivarieties of a quasivariety $\mathcal{K}$ as the lattice $\mathrm{S_p}(\mathbf{S}, H)$ of H-closed algebraic subsets of an algebraic lattice $\mathbf{S}$ with a monoid H of operators.

One can use the representation to find restrictions on the equational closure operator on a lattice of subquasivarieties $\mathrm{L_q}(\mathcal{K})$, strengthening results from [14, 16, 20]. The second part of the book (Chaps. 3 and 4) discusses some of these restrictions.

K. Adaricheva et al., *A Primer of Subquasivariety Lattices*, CMS/CAIMS Books in Mathematics 3, https://doi.org/10.1007/978-3-030-98088-7_1

The third part (Chaps. 5–8) presents a method that allows us to represent some pairs $(\mathbf{L}, \gamma)$ consisting of a finite lattice and an equaclosure operator on it as a lattice of subquasivarieties $\mathrm{L_q}(\mathcal{K})$ and its equational closure operator. The method mimics [91] to initially obtain a representation $\mathbf{L} \cong \mathrm{L_q}(\mathcal{K}_0)$ with a quasivariety $\mathcal{K}_0$ in a language without equality, which can sometimes be converted to a quasivariety $\mathcal{K}_1$ in a language with equality.

The fourth and final part (Chap. 9) discusses representations of dually algebraic distributive lattices.

Many aspects of quasivarieties are not covered in this book. The monograph [63] focuses on locally finite quasivarieties, and its Appendix contains a survey of manifold results on quasivarieties in general. The original problem, raised independently by Birkhoff and Mal'cev, was to describe all subquasivariety lattices. It has become increasingly clear, however, that subquasivariety lattices have a rich inner structure; perhaps Anvar Nurakunov's description of them as "unreasonable" is closer to the mark [96]. In particular, small algebras can too easily generate a Q-universal quasivariety. It is doubtful whether even the class of finite subquasivariety lattices admits a tractable characterization. The complexity of subquasivariety lattices is discussed in Sect. 1.5.

1.1 An Overview

Our results on subquasivariety lattices include the following.

A complete lattice $\mathbf{L}$ is isomorphic to the lattice $\mathrm{L_q}(\mathcal{K})$ of all subquasivarieties of a quasivariety $\mathcal{K}$ (in a language that may or may not contain equality) if and only if $\mathbf{L}$ is isomorphic to the lattice $\mathrm{S_p}(\mathbf{S}, H)$ of all H-closed algebraic subsets of an algebraic lattice $\mathbf{S}$, where H is a monoid of operators on $\mathbf{S}$. (Theorems 2.36, 2.43, 2.54)

The representation of a subquasivariety lattice as $\mathrm{S_p}(\mathbf{S}, H)$ yields new restrictions on the equational closure operator on $\mathrm{L_q}(\mathcal{K})$. (Theorems 3.14, 3.26, 3.32)

Let us consider pairs $(\mathbf{L}, \gamma)$ consisting of a finite lattice and a weak equaclosure operator on it, *with the additional property that every join irreducible element of* $\mathbf{L}$ *is the least element of its* γ*-class.*

(a) If such a pair $(\mathbf{L}, \gamma)$ can be represented as $(\mathrm{S_p}(\mathbf{S}, H), \Gamma)$, where $\mathbf{S}$ is an algebraic lattice and Γ is the natural weak equaclosure operator on lattices of algebraic subsets, then we can take $\mathbf{S} = \gamma^d(\mathbf{L})$. (Theorem 5.13)
(b) If for such a pair the lattice $\gamma(\mathbf{L})$ of closed sets is a chain, then $(\mathbf{L}, \gamma)$ can be represented as $(\mathrm{S_p}(\mathbf{S}, H), \Gamma)$. (Corollary 5.9)
(c) There is an algorithm to determine whether such a pair $(\mathbf{L}, \gamma)$ can be represented as $(\mathrm{S_p}(\mathbf{S}, H), \Gamma)$. (Sect. 5.2)

If $\mathbf{S}$ is an algebraic lattice, and H is a monoid of operators on $\mathbf{S}$ with the property that, for every $h \in H$, $h(x) = 1_\mathbf{S}$ only if $x = 1_\mathbf{S}$, and $1_\mathbf{S}$ is compact, then $\mathrm{S_p}(\mathbf{S}, H)$

is isomorphic to $L_q(\mathcal{K})$ for a quasivariety of structures in a language with equality. (Theorem 7.7)

If $\mathbf{L}$ is a complete lattice such that $\mathbf{L} \cong S_p(\mathbf{S}, H)$ for some algebraic lattice $\mathbf{S}$ and monoid H of operators on $\mathbf{S}$, then the linear sum $\mathbf{1} + \mathbf{L}$ is a subquasivariety lattice: $\mathbf{1} + \mathbf{L} \cong L_q(\mathcal{K})$ for a quasivariety with equality. (Corollary 7.10)

Any distributive, dually algebraic lattice can be represented as $S_p(\mathbf{S}, H)$ with $\mathbf{S}$ an algebraic lattice and H a monoid of operators. (Theorem 9.1)

Combining the last two ideas yields the result that every dually algebraic, distributive lattice $\mathbf{D}$, the lattice $\mathbf{1} + \mathbf{D}$ is isomorphic to $L_q(\mathcal{K})$ for a quasivariety of structures with equality. (Theorem 9.12)

If $\mathbf{D}$ is a distributive lattice that is both algebraic and dually algebraic, and the least element of $\mathbf{D}$ is dually compact, then $\mathbf{D}$ can be represented as $L_q(\mathcal{K})$ for some quasivariety $\mathcal{K}$ with equality (Theorem 9.18).

A chain $\mathbf{C}$ is isomorphic to $L_q(\mathcal{K})$ for some quasivariety $\mathcal{K}$ of structures with equality if and only if $\mathbf{C}$ is dually algebraic and has an atom. (Corollary 9.27)

We have included many examples to illustrate the results and encourage the reader to construct her own.

1.2 Lattices, Theories, and Models

Let us remind the reader of some basic notions that we will use.

A *complete lattice* is an ordered set $\mathbf{L}$ in which every subset $A \subseteq L$ has a join (least upper bound), denoted $\bigvee A$, and a meet (greatest lower bound), denoted $\bigwedge A$. That is, there is an element $m = \bigvee A$ such that $x \geq m$ iff $x \geq a$ for all $a \in A$, and dually for meets. Taking $A = L$ or $A = \varnothing$, a complete lattice has a greatest element, denoted $1_{\mathbf{L}}$, and a least element, denoted $0_{\mathbf{L}}$.

An element c in a complete lattice is *compact* if whenever $c \leq \bigvee A$ for some $A \subseteq L$, then $c \leq \bigvee F$ for some finite subset $F \subseteq A$. The lattice $\mathbf{L}$ is *algebraic* if every element of L is a join of compact elements, i.e., for every $x \in L$ it holds that $x = \bigvee\{c \in L : c \leq x \text{ and } c \text{ is compact}\}$.

Algebraic lattices have some strong structural properties.

- Every element of an algebraic lattice is a meet of completely meet irreducible elements.
- An algebraic lattice is *weakly atomic:* if $a < b$, then there exist c, d such that $a \leq c \prec d \leq b$.
- An algebraic lattice is *upper continuous:* if D is an up-directed set and $a \in L$, then $a \wedge \bigvee D = \bigvee_{d \in D}(a \wedge d)$.

Good references include [31, 33, 57, 90]. Moreover, algebraic lattices play a special role in universal algebra; see [28, 31, 56, 87]. But for us, they are crucial because of the following very general theorem.

A collection of first-order sentences (in some fixed language) is a *theory* if it is closed under deduction. The Compactness Theorem of logic says that if a first-order sentence can be deduced from a collection A of sentences, then it can be deduced from a finite subset $F \subseteq A$.

Theorem 1.1 *Let $\mathcal{U}$ be a first-order theory. The collection of all theories $\mathcal{T}$ containing $\mathcal{U}$ forms an algebraic lattice* $\mathrm{L}(\mathcal{U})$*. The compact theories are just the theories that are finitely based with respect to $\mathcal{U}$, i.e., generated by $\mathcal{U} \cup \mathcal{F}$ for some finite set of sentences $\mathcal{F}$.*

We are generally concerned, not with all first-order theories, but all theories of a certain type: equational or quasi-equational theories. Each of these forms a complete sublattice of $\mathrm{L}(\mathcal{U})$, making them algebraic lattices in their own right.

The *models* of a theory are all structures that satisfy every sentence in the theory. Models are dual to theories: the more sentences in a theory, the fewer models it has. Thus the lattice of all model classes will be dual to the class of all theories of a given kind. In particular, for equational classes and quasi-equational classes, the lattices $\mathrm{L_v}(\mathcal{U})$ and $\mathrm{L_q}(\mathcal{U})$ are dual to the corresponding lattices of theories. Thus $\mathrm{L_v}(\mathcal{U})$ and $\mathrm{L_q}(\mathcal{U})$ are *dually algebraic* lattices.

The reader casting for a familiar example might remember that the lattice of group varieties is dually isomorphic to the lattice of fully invariant subgroups of a countably generated free group $\mathrm{FG}(\omega)$. In the latter, fully invariant subgroups correspond to sets of equations $w \approx 1$. These are the equational theories of groups, and the fully invariant subgroups are a complete sublattice of the lattice of normal subgroups of $\mathrm{FG}(\omega)$, and that lattice is algebraic. The lattice of group varieties is dually isomorphic to the lattice of fully invariant subgroups, making $\mathrm{L_v}(\mathcal{G})$ dually algebraic.

1.3 A Review of Classical Subvariety Lattices

While our main topic is quasivarieties, to help us set the stage let us begin with a summary of varieties of algebras and their subvariety lattices.

A *variety*, or *equational class*, is the collection of all algebras of a fixed type that satisfy some set of equations. For example, the variety $\mathcal{A}$ of abelian groups consists of all algebras with one binary operation, one unary operation, and one constant (nullary operation) satisfying the equations

$$\begin{aligned} x(yz) &\approx (xy)z \\ x1 &\approx x \\ xx^{-1} &\approx 1 \\ xy &\approx yx. \end{aligned}$$

Birkhoff showed that a class $\mathcal{K}$ of similar algebras is a variety if and only if it is closed under homomorphic images, subalgebras, and direct products [24], in notation, $\mathbb{HSP}(\mathcal{K}) = \mathcal{K}$. More generally, if $\mathcal{K}$ is not already a variety, the least variety containing $\mathcal{K}$ is $\mathbb{HSP}(\mathcal{K})$. The key step in Birkhoff's argument is the construction, for any variety $\mathcal{V}$ and set X, of the $\mathcal{V}$-free algebra generated by X, denoted $\mathbf{F}_{\mathcal{V}}(X)$.

When a variety $\mathcal{U}$ is contained in a variety $\mathcal{V}$, we say that $\mathcal{U}$ is a *subvariety* of $\mathcal{V}$ and write $\mathcal{U} \leq \mathcal{V}$. For example, the variety $\mathcal{A}_n$ of all abelian groups of exponent n is a subvariety of $\mathcal{A}$. The algebras in $\mathcal{A}_n$ satisfy the equations for $\mathcal{A}$, plus the additional equation

$$x^n \approx 1.$$

The collection of all subvarieties of a variety $\mathcal{V}$, ordered by inclusion, forms a dually algebraic, complete lattice denoted $\mathrm{L_v}(\mathcal{V})$.

Corresponding to each variety $\mathcal{V}$ is the set of all equations satisfied by all the algebras in $\mathcal{V}$, called the *equational theory of* $\mathcal{V}$. For example, the equational theory of abelian groups would contain all equations that are consequences of the 4-equation basis, including things such as $xy^2z \approx yzyx$.

If we fix a variety $\mathcal{V}$, then the collection of all equational theories containing the theory of $\mathcal{V}$, ordered by containment, forms an algebraic lattice denoted by $\mathrm{ETh}(\mathcal{V})$. Indeed, $\mathrm{ETh}(\mathcal{V})$ is dually isomorphic to $\mathrm{L_v}(\mathcal{V})$, since more equations means fewer models. By the compactness theorem of logic, equational theories that are finitely based relative to $\mathcal{V}$ are the compact elements of $\mathrm{ETh}(\mathcal{V})$. Dually then, varieties that are finitely based relative to $\mathcal{V}$ are the dually compact members of $\mathrm{L_v}(\mathcal{V})$.

A congruence relation φ on an algebra $\mathbf{A}$ is *fully invariant* if for every endomorphism ε of $\mathbf{A}$, and every pair $a, b \in A$, we have that $a \mathrel{\varphi} b$ implies $\varepsilon a \mathrel{\varphi} \varepsilon b$. The fully invariant congruences of $\mathbf{A}$ form a complete sublattice of Con $\mathbf{A}$, denoted FiCon $\mathbf{A}$. As a complete sublattice of an algebraic lattice, FiCon $\mathbf{A}$ is itself algebraic.

The connection between lattices of equational theories and fully invariant congruences is that, for any variety $\mathcal{V}$, the lattice of theories $\mathrm{ETh}(\mathcal{V})$ is isomorphic to FiCon $\mathbf{F}_{\mathcal{V}}(\omega)$, where $\mathbf{F}_{\mathcal{V}}(\omega)$ is the countably generated $\mathcal{V}$-free algebra; see Theorem 2.54(1). This explains why $\mathrm{ETh}(\mathcal{V})$ is algebraic and $\mathrm{L_q}(\mathcal{V})$ is dually algebraic, besides giving us a concrete structure for dealing with equational theories.

Only a few restrictions are known on lattices of subvarieties $\mathrm{L_v}(\mathcal{V})$. They are dually algebraic, and since the least element corresponds to the variety $x \approx y$, it is dually compact. Lampe proved that subvariety lattices satisfy a form of join semidistributivity at 0, the so-called Zipper Condition [78]:

$$\text{if } a_i \vee c = z \text{ for all } i \in I \text{ and } \bigwedge_{i \in I} a_i = 0, \text{ then } c = z.$$

A similar but stronger condition was found by Erné [41] and Tardos (independently), which was refined yet further by Lampe [79]. These results show that the structure of lattices of subvarieties is quite constrained at the bottom. On the other hand, Pigozzi and Tardos proved that every dually algebraic lattice with a completely meet irreducible least element 0, that is, every linear sum $\mathbf{L} = \mathbf{1} + \mathbf{K}$ with $\mathbf{K}$ dually algebraic, is isomorphic to a lattice of subvarieties $\mathrm{L_v}(\mathcal{V})$ [99]. This strengthens earlier results of Ježek [65].

Nurakunov proved that a lattice is isomorphic to a lattice of subvarieties if and only if it is dually isomorphic to congruence lattice of a monoid with a right zero and two unary operations satisfying certain properties [95]. This extends earlier work of McKenzie [86] and Newrly [94].

Traditionally, varieties are studied for algebras in a language that includes equality as the only relation. If we expand the language to include other relations, so that the type may contain both operations and relations, then the analogue of an equational class is a class of relational structures that satisfy a set of atomic formulas, which could be of the form either $s \approx t$, where s and t are terms, or $R(\mathbf{u})$ for some relation symbol, where $\mathbf{u}$ is a k-tuple of terms. In this case, $x \approx y$ need not be the least theory in the lattice of atomic theories. Thus, for future consideration, we pose the following problems.

(1) Describe lattices of atomic theories for relational structures in languages that include equality as a congruence relation.
(2) Describe lattices of atomic theories for relational structures in languages that may not include equality as a congruence relation.

Lattices of atomic theories for relational structures in languages that *do not* include equality are the topic of Appendix A.2.

1.4 A Review of Classical Subquasivariety Lattices

Quasivarieties are usually studied for relational structures, which may have both operations and relations. Thus a typical structure has the form $\mathbf{S} = \langle S, \mathcal{F}, \mathcal{R} \rangle$ where $\mathcal{F}$ is a set of operations on S and $\mathcal{R}$ is a set of relations on S. It is traditional to assume that equality is in the language, in which case we might as well assume that $\approx$ is one of the relations, and that it is a congruence relation, i.e., reflexive, symmetric, transitive, and compatible with each $f \in \mathcal{F}$ and each $R \in \mathcal{R}$. Later, in Chap. 2, we will discard that assumption.

Again, for structures, atomic formulas can be of the form either $\mathbf{s} \approx \mathbf{t}$, where s and t are terms, or $R(\mathbf{u})$ for some relation symbol, where $\mathbf{u}$ is a k-tuple of terms.

A *quasi-equation* is a sentence of the form $\beta_1 \mathbin{\&} \cdots \mathbin{\&} \beta_{n-1} \rightarrow \beta_n$ with each β_j an atomic formula. We allow $n = 1$, so that equations are quasi-equations with the antecedent empty.

A *quasivariety* is the collection of all relational structures of a fixed type that satisfy some set of quasi-equations. For example, to say that a group has no element of order 5, you can use the quasi-equation

$$x^5 \approx 1 \rightarrow x \approx 1. \tag{ϱ}$$

The class of all algebras satisfying the group axioms, which are equations, and the quasi-equation (ϱ) is a quasivariety.

The *join semidistributive* law for lattices is another familiar quasi-equation:

$$x \vee y \approx x \vee z \rightarrow x \vee y \approx x \vee (y \wedge z). \tag{$\mathrm{SD}_\vee$}$$

The quasivariety of join semidistributive lattices is determined by the lattice axioms, which are equations, and ($\mathrm{SD}_\vee$).

When the index set I is finite and the constant 0 is in the language, the Zipper Condition is a quasi-equation:

$$x_1 \vee y \approx z \,\&\, \dots \,\&\, x_n \vee y \approx z \,\&\, x_1 \wedge \dots \wedge x_n \approx 0 \;\rightarrow\; y \approx z.$$

The classic characterization of quasivarieties uses the class operators: for a collection $\mathcal{X}$ of structures of the same type,

- $\mathbb{H}(\mathcal{X})$ denotes all homomorphic images of structures in $\mathcal{X}$;
- $\mathbb{S}(\mathcal{X})$ denotes all isomorphic copies of substructures of structures in $\mathcal{X}$;
- $\mathbb{P}(\mathcal{X})$ denotes all direct products of structures in $\mathcal{X}$;
- $\mathbb{U}(\mathcal{X})$ denotes all ultraproducts of structures in $\mathcal{X}$;
- $\mathbb{R}(\mathcal{X})$ denotes all reduced products of structures in $\mathcal{X}$.

Note that $\mathbb{P}(\mathcal{X})$ and $\mathbb{R}(\mathcal{X})$ include the empty product, which is a 1-element structure with all possible relations of the type holding. The classes $\mathbb{H}(\mathcal{X})$, $\mathbb{S}(\mathcal{X})$, etc. are taken to be closed under isomorphism, by convention.

Theorem 1.2 *The following are equivalent for a class $\mathcal{Q}$ of structures of the same type.*

(1) *$\mathcal{Q}$ is a quasivariety.*
(2) *$\mathcal{Q}$ is closed under $\mathbb{S}$, $\mathbb{P}$ and $\mathbb{U}$.*
(3) *$\mathcal{Q} = \mathbb{SPU}(\mathcal{Q})$.*
(4) *$\mathcal{Q} = \mathbb{SR}(\mathcal{Q})$.*

Let us denote the quasivariety generated by $\mathcal{X}$ as $\mathbb{Q}(\mathcal{X})$.

Corollary 1.3 *If $\mathcal{X}$ is a collection of structures of the same type, then*

$$\mathbb{Q}(\mathcal{X}) = \mathbb{SR}(\mathcal{X}) = \mathbb{SPU}(\mathcal{X}).$$

Corollary 1.4 *If $\mathcal{X}$ is a finite set of finite structures, then $\mathbb{Q}(\mathcal{X}) = \mathbb{SP}(\mathcal{X})$.*

The second corollary is because any ultraproduct of a finite set of finite structures is isomorphic to one of them.

Theorem 1.2 is Corollary 2.3.4 in Gorbunov's book [52], based on Mal'cev [83], and incorporating ideas of Łoś (see [29]), Mal'cev [82], Horn [61], Chang and Morel [30] and Frayne, Morel and Scott [43, 44]. There is a complete proof in Chapter V of Burris and Sankappanavar [28]; see Theorem V.2.25. Note that the equality $\mathbb{SR}(\mathcal{X}) = \mathbb{SPU}(\mathcal{X})$ just reflects the fact that any filter on a Boolean algebra (in this case the Boolean algebra $\mathbf{2}^I$, where I is the index set for a direct product) is an intersection of ultrafilters; see Lemma 2.8 (1).

When a quasivariety $\mathcal{Q}$ is contained in a quasivariety $\mathcal{K}$, we say that $\mathcal{Q}$ is a *subquasivariety* of $\mathcal{K}$, denoted $\mathcal{Q} \leq \mathcal{K}$. The collection of all subquasivarieties of $\mathcal{K}$, ordered by inclusion, forms a dually algebraic, complete lattice denoted $\mathrm{L}_\mathrm{q}(\mathcal{K})$. Our main goal in this book is to investigate subquasivariety lattices.

Analogous to the correspondence between varieties and equational theories, there is a duality between quasivarieties and implicational theories. The *implicational theory*, or sometimes *quasi-equational theory*, of a quasivariety $\mathcal{K}$ consists of all quasi-equations satisfied by all the structures in $\mathcal{K}$. This will include all consequences of the quasi-equations in a basis for $\mathcal{K}$. For example, if a quasivariety of groups satisfies $x^5 \approx 1 \rightarrow x \approx 1$, then it also satisfies $x^{10} \approx 1 \rightarrow x^2 \approx 1$.

If we fix a quasivariety $\mathcal{K}$, then the collection of all implicational theories containing the theory of $\mathcal{K}$, ordered by containment, forms an algebraic lattice denoted by $\mathrm{ITh}(\mathcal{K})$. As with varieties, the lattice of theories $\mathrm{ITh}(\mathcal{K})$ is dually isomorphic to $\mathrm{L_q}(\mathcal{K})$.

Again, implicational theories that are finitely based relative to $\mathcal{K}$ are the compact elements of $\mathrm{ITh}(\mathcal{K})$. Dually, quasivarieties that are finitely based relative to $\mathcal{K}$ are the dually compact members of $\mathrm{L_q}(\mathcal{K})$.

Being closed under substructures and direct products, quasivarieties have *free structures*. That is, given a quasivariety $\mathcal{K}$ and a set X, there is a structure $\mathbf{F} = \mathbf{F}_{\mathcal{K}}(X)$ such that $\mathbf{F} \in \mathcal{K}$, $\mathbf{F}$ is generated by X, and any map $h_0 : X \rightarrow S$, where $\mathbf{S}$ is a structure in $\mathcal{K}$, can be extended to a homomorphism.

But not every homomorphic image of a structure in a quasivariety $\mathcal{K}$ need be in $\mathcal{K}$; this applies to free structures are well. The set of all congruences θ on a structure $\mathcal{S}$ such that $\mathbf{S}/\theta$ is in $\mathcal{K}$ forms a complete meet subsemilattice of $\mathrm{Con}\,\mathbf{S}$, denoted $\mathrm{Con}_{\mathcal{K}}\,\mathbf{S}$. Though $\mathrm{Con}_{\mathcal{K}}\,\mathbf{S}$ need not be closed under arbitrary joins in $\mathrm{Con}\,\mathbf{S}$, it is closed under directed joins, and hence is an algebraic subset of $\mathrm{Con}\,\mathbf{S}$. That makes $\mathrm{Con}_{\mathcal{K}}\,\mathbf{S}$ an algebraic lattice in its own right.

The lattice $\mathrm{L_q}(\mathcal{K})$ of subquasivarieties of $\mathcal{K}$ is isomorphic to a lattice $\mathrm{S_p}(\mathbf{S}, H)$ of H-closed algebraic subsets, where $\mathbf{S}$ is an algebraic lattice and H a monoid of operators on $\mathbf{S}$ (see Theorem 2.54(2)). Lattices of algebraic subsets are described in Sect. 1.6. In this representation, $\mathbf{S}$ is the lattice of $\mathcal{K}$-congruences of the countably generated $\mathcal{K}$-free structure $\mathrm{Con}_{\mathcal{K}}\,\mathbf{F}_{\mathcal{K}}(\omega)$, and the operators are induced by the endormorphisms of $\mathbf{F}_{\mathcal{K}}(\omega)$. Note that each endomorphism ε of a free structure $\mathbf{F}_{\mathcal{K}}(X)$ is determined by the values of $\varepsilon(x)$ for $x \in X$, that is, a substitution of terms for variables. Moreover, the correspondence between subquasivarieties and H-closed algebraic subsets is such that a subquasivariety $\mathcal{Q} \leq \mathcal{K}$ maps to the algebraic subset $Q = \{\theta \in \mathrm{Con}_{\mathcal{K}}(\mathbf{F}) : \mathbf{F}/\theta \in \mathcal{Q}\}$.

Dually, the lattice $\mathrm{ITh}(\mathcal{K})$ of implicational theories containing the theory of $\mathcal{K}$ is isomorphic to the congruence lattice of a semilattice with operators, $\mathrm{ITh}(\mathcal{K}) \cong \mathrm{Con}(\mathbf{K}, \vee, 0, \widehat{H})$ where $\mathbf{K}$ is the semilattice of compact $\mathcal{K}$-congruences of $\mathbf{F}_{\mathcal{K}}(\omega)$ and the mappings in $\widehat{H}$ are the restrictions of those in H to compact elements (see Theorem 2.36). Unfortunately, the implicational theories are not congruence classes, but classes of another type of relation called *dongruences*, as described in [20].

Summarizing, for any quasivariety $\mathcal{K}$,

$$\mathrm{L_q}(\mathcal{K}) \cong^d \mathrm{ITh}(\mathcal{K}) \cong \mathrm{Con}(\mathbf{K}, \vee, 0, \widehat{H}) \cong^d \mathrm{S_p}(\mathbf{S}, H),$$

where $\mathbf{S} = \mathrm{Con}_{\mathcal{K}}(\mathbf{F}_{\mathcal{K}}(\omega))$, $\mathbf{K}$ is the semilattice of compact elements of $\mathbf{S}$, the operators in H are induced by endomorphisms of $\mathbf{S}$, and $\widehat{H}$ is their restriction to $\mathbf{K}$. The details will come in Chap. 2.

The use of operators in the above approach, based on Adaricheva and Nation [20], refines the representation of Gorbunov and Tumanov for subquasivariety lattices as $\mathrm{L}_\mathrm{q}(\mathcal{K}) \cong \mathrm{S}_\mathrm{p}(\mathbf{S}, \varepsilon)$ the lattice of ε-closed algebraic subsets for a distributive quasi-order ε [52, 54]. The operator restriction is strictly stronger: if $\mathbf{S}$ is a finite meet semilattice with 1, then any sublattice of $\mathrm{Sub}(\mathbf{S}, \wedge, 1)$ can be represented as $\mathrm{Sub}(\mathbf{S}, \wedge, 1, \varepsilon)$, but not necessarily as $\mathrm{Sub}(\mathbf{S}, \wedge, 1, H)$; see Section 4-3 of [22]. Nonetheless, we still occasionally find it worthwhile to use distributive quasi-orders, as in Sect. 4.3.

Some of the basic structural properties of subquasivariety lattices are consequences of the representation $\mathrm{L}_\mathrm{q}(\mathcal{K}) \cong \mathrm{S}_\mathrm{p}(\mathbf{S}, H)$. In view of Theorem 1.21, we have the following.

Theorem 1.5 *For any quasivariety $\mathcal{K}$, the lattice $\mathrm{L}_\mathrm{q}(\mathcal{K})$ is dually algebraic and join semidistributive.*

It follows from the representation of subquasivariety lattices as $\mathrm{S}_\mathrm{p}(\mathbf{S}, H)$ that $\mathrm{L}_\mathrm{q}(\mathcal{K})$ supports an equaclosure operator (see Chap. 3), but that is backwards. We should begin with the *natural equaclosure operator on* $\mathrm{L}_\mathrm{q}(\mathcal{K})$.

Following Wiesław Dziobiak [39], for each subquasivariety $\mathcal{Q}$ of a quasivariety $\mathcal{K}$, let

$$\Gamma(\mathcal{Q}) = \mathcal{K} \cap \mathbb{HSP}(\mathcal{Q}) = \mathcal{K} \cap \mathbb{H}(\mathcal{Q}).$$

Clearly Γ is a closure operator on the lattice $\mathrm{L}_\mathrm{q}(\mathcal{K})$, assigning to each $\mathcal{Q}$ its closure under homomorphic images that lie in $\mathcal{K}$.

Since a structure in $\mathcal{Q}$ may have homomorphic images that are not in $\mathcal{K}$, we should be careful in describing $\Gamma(\mathcal{Q})$. A *relative subvariety* is a subclass of $\mathcal{K}$ determined within $\mathcal{K}$ by a set of atomic formulas. The closure $\Gamma(\mathcal{Q}) = \mathcal{K} \cap \mathbb{H}(\mathcal{Q})$ of a subquasivariety is a relative subvariety of $\mathcal{K}$; it need not be a variety with respect to some larger quasivariety $\mathcal{H} > \mathcal{K}$.

The gist of Chap. 3 is that the equaclosure operator Γ imposes strong restrictions on what lattices can be lattices of subquasivarieties. We defer the details and history until then.

The big difference between lattices of subquasivarieties $\mathrm{L}_\mathrm{q}(\mathcal{K})$ and lattices of algebraic sets $\mathrm{S}_\mathrm{p}(\mathbf{S}, H)$, where $\mathbf{S}$ is any algebraic lattice, is the role played by the variety $x \approx y$ of 1-element structures. In a language with relation symbols, this need not be the least subquasivariety. The variety $x \approx y$ is, of course, dually compact, being finitely based.

Recall that a lattice is *atomic* if it has a least element 0 and for every $x > 0$ there exists an atom a such that $x \geq a > 0$. Gorbunov used the quasivariety $x \approx y$ to prove that if the language of $\mathcal{K}$ has an equality relation, then $\mathrm{L}_\mathrm{q}(\mathcal{K})$ is atomic [52]. This is Theorem 2.60 below.

The special role of $x \approx y$ is reflected in property (I8) of equaclosure operators. Property (I8) holds for the natural equaclosure operator on $\mathrm{L}_\mathrm{q}(\mathcal{K})$, but need not hold for the natural weak equaclosure operator on $\mathrm{S}_\mathrm{p}(\mathbf{S}, H)$. This is discussed in Sect. 3.1.

All of which raises the question: *What would life be like without equality?* And that is exactly where we are headed, starting in Chap. 2.

As we proceed, it is useful to have a couple of sources of generic examples of subquasivariety lattices.

Theorem 1.6

(1) *Every finite distributive lattice is isomorphic to* $\mathrm{L_q}(\mathcal{K})$ *for some quasivariety* $\mathcal{K}$.

(2) *If* $\mathbf{S}$ *is a finite meet semilattice with* 1, *then the subalgebra lattice* $\mathrm{Sub}(\mathbf{S}, \wedge, 1)$ *is isomorphic to* $\mathrm{L_q}(\mathcal{K})$ *for some quasivariety* $\mathcal{K}$.

Part (1) is due to Tumanov [112], while part (2) is from Gorbunov and Tumanov [53]. The lattices in part (2) are atomistic (every element is a join of atoms), and Sect. 2.7 gives a fuller version of what is known for atomistic lattices.

In the course of our investigation, we will often deal with quasivarieties generated by finitely many, finite structures. These quasivarieties are, of course, locally finite. There are special methods for working with locally finite quasivarieties, which are the topic of the book [63]. The Appendix of [63] also contains a survey of results about quasivarieties of particular types of algebras.

1.5 It's Complicated: The Complexity of Subquasivariety Lattices

The original statement of the Birkhoff-Mal'cev problem: *Characterize lattices of subquasivarieties* is too general to be practical. It is useful for perspective to list variations of the problem. These necessarily involve one or more parts of one of the following:

- A quasivariety $\mathcal{K}$ of structures and the lattice of subquasivarieties $\mathrm{L_q}(\mathcal{K})$ and the natural equaclosure operator Γ on $\mathrm{L_q}(\mathcal{K})$,
- An algebraic lattice $\mathbf{S}$ and a monoid H of operators on $\mathbf{S}$ and the lattice $\mathrm{S_p}(\mathbf{S}, H)$ of H-closed algebraic subsets of $\mathbf{S}$ and the natural weak equaclosure operator Γ on $\mathrm{S_p}(\mathbf{S}, H)$,
- A dually algebraic, join semidistributive lattice $\mathbf{L}$ and a (weak) equaclosure operator γ on $\mathbf{L}$.

These terms will be defined in the course of the book. Here are some possible versions of the Birkhoff-Mal'cev problem.

- Given a quasivariety $\mathcal{K}$, find $\mathrm{L_q}(\mathcal{K})$.
- Find lattice properties that hold in every subquasivariety lattice $\mathrm{L_q}(\mathcal{K})$.
- Find properties of $\mathrm{L_q}(\mathcal{K})$ with restrictions on $\mathcal{K}$, e.g., $\mathcal{K}$ *is locally finite* or $\mathcal{K}$ *is a quasivariety of algebras.*
- Find lattice properties that hold in every lattice of algebraic subsets $\mathrm{S_p}(\mathbf{S}, H)$.
- Given a lattice $\mathbf{L}$, when is $\mathbf{L} \cong \mathrm{L_q}(\mathcal{K})$ for some quasivariety $\mathcal{K}$?

- Given a lattice $\mathbf{L}$, when is $\mathbf{L} \cong \mathrm{S_p}(\mathbf{S}, H)$ for some $\mathbf{S}$ and H?
- Given a lattice $\mathbf{L}$ with an equaclosure operator, when can $(\mathbf{L}, \gamma)$ be represented as $(\mathrm{L_q}(\mathcal{K}), \Gamma)$ for some quasivariety $\mathcal{K}$?
- Given a lattice $\mathbf{L}$ with a weak equaclosure operator, when can $(\mathbf{L}, \gamma)$ be represented as $(\mathrm{S_p}(\mathbf{S}, H), \Gamma)$ for some $\mathbf{S}$ and H?
- Given a lattice $\mathbf{L}$, when does it admit a (weak) equaclosure operator γ?

In fact, none of these questions has a straightforward answer. The best we can hope for is partial solutions, by putting extra restrictions on $\mathcal{K}$ or $\mathbf{L}$. In the rest of this section, let us review some of the complications in characterizing subquasivariety lattices.

A quasivariety $\mathcal{K}$ is said to be *Q-universal* if, for every quasivariety $\mathcal{Q}$ of finite type, $\mathrm{L_q}(\mathcal{Q})$ is a homomorphic image of a sublattice of $\mathrm{L_q}(\mathcal{K})$. This notion was introduced by Sapir [104], who showed that certain quasivarieties of semigroups are Q-universal. Moreover, if $\mathcal{K}$ is a Q-universal quasivariety, then

- $|\mathrm{L_q}(\mathcal{K})| = 2^{\aleph_0}$,
- the free lattice $\mathrm{FL}(\omega)$ is a sublattice of $\mathrm{L_q}(\mathcal{K})$, whence, in particular, $\mathrm{L_q}(\mathcal{K})$ satisfies no lattice identity.

It turns out that Q-universal quasivarieties are ubiquitous. Various ways to show that a given quasivariety is Q-universal are in Adams and Dziobiak [3, 4], Gorbunov [52], and Sapir [104]. Every quasivariety $\mathcal{K}$ that is known to be Q-universal has not only the free lattice, but its ideal lattice $\mathcal{I}(\mathrm{FL}(\omega))$, as a sublattice of $\mathrm{L_q}(\mathcal{K})$, and that property is sufficient to guarantee Q-universality [3]. For good discussions of Q-universal quasivarieties, see Adams and Dziobiak [6] or Adams *et al.* [1].

Let us look at some typical examples of subquasivariety lattices.

Bands Bands are idempotent semigroups, satisfying $x(yz) \approx (xy)z$ and $x^2 \approx x$. Simple examples include the 2-element left-zero semigroup $\mathbf{L}$, satisfying $xy \approx x$, the 2-element right-zero semigroup $\mathbf{R}$, satisfying $xy \approx y$, and the 2-element semilattice $\mathbf{I}$, satisfying $xy \approx yx$. To each of $\mathbf{L}$ and $\mathbf{R}$ we can add a new zero element, forming $\mathbf{L}^0$ and $\mathbf{R}^0$. Note that $\mathbf{I} \leq \mathbf{L}^0$ and $\mathbf{I} \leq \mathbf{R}^0$. *Normal bands* are the subvariety, $\mathcal{N}$, of semigroups satisfying $xyzt \approx xzyt$. Shafaat [109] showed that the semigroups we have listed are precisely all the quasicritical normal bands. Moreover, each of the bands $\mathbf{L}, \mathbf{L}^0, \mathbf{I}, \mathbf{R}, \mathbf{R}^0$ generates a quasivariety that is join prime in $\mathrm{L_q}(\mathcal{B})$; see, e.g., Corollaries 2.33 and 3.5 of [63] for join prime locally finite quasivarieties. Hence the lattice $\mathrm{L_q}(\mathcal{N})$ of normal bands is a small distributive ideal of $\mathrm{L_q}(\mathcal{B})$, as shown in Fig. 1.1.

But the variety of all bands is Q-universal (Adams and Dziobiak [5]), so $\mathrm{L_q}(\mathcal{B})$ has the cardinality of the continuum and satisfies no lattice identity. In fact, the subvariety $\mathcal{LN}$ of *left normal bands* defined by $xyxz \approx xyz$ and the subvariety $\mathcal{RN}$ of *right normal bands* defined by $xzyz \approx xyz$ are both in the interval $[\mathcal{N}, \mathcal{B}]$ and Q-universal [1].

Modular Lattices The variety of modular lattices exhibits similar behavior. Let $\mathcal{M}$ denote the variety of modular lattices. The lattices $\mathbf{M}_k$ with k atoms, and $\mathbf{M}_{3,3}$ and $\mathbf{M}^+_{3,3}$, are drawn in Figs. 1.2 and 1.3, respectively. Section 2.4 of [63] describes the ideal $\downarrow(\mathbb{Q}(\mathbf{M}^+_{3,3}) \vee \mathbb{Q}(\mathbf{M}_\omega))$ of $\mathrm{L_q}(\mathcal{M})$, which is countable and distributive, containing

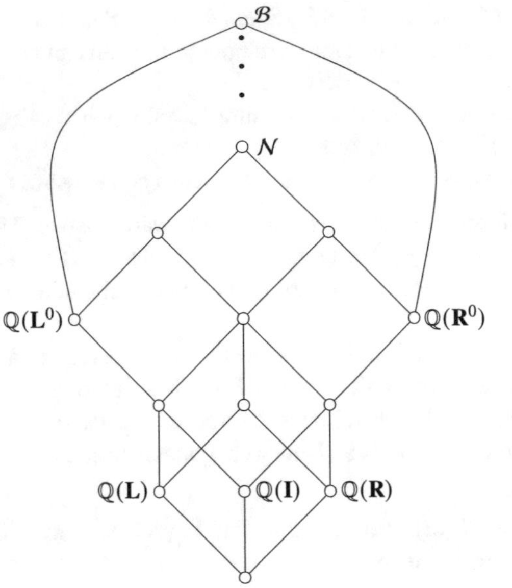

Fig. 1.1 The lattice of subquasivarieties of bands. Here $\mathcal{B}$ denotes bands, $\mathcal{N}$ normal bands. The variety $\mathcal{B}$ is Q-universal, so $\mathrm{L_q}(\mathcal{B})$ is uncountable and satisfies no lattice identity. The ideal $\downarrow \mathcal{N}$, though, is finite and distributive

only the subquasivarieties indicated schematically in Fig. 1.4. The argument is based on Jónsson [68] and Grätzer [55].

On the other hand, Grätzer and Lakser showed that in $\mathrm{L_q}(\mathcal{M})$ the interval $[\mathbb{Q}(\mathbf{M}^+_{3,3}), \mathbb{Q}(\mathbf{M}_{3,3})]$ contains $2^{\aleph_0}$ subquasivarieties [58]. A refinement of their proof shows that $\mathbb{Q}(\mathbf{M}_{3,3})$ is Q-universal, due to Adams and Dziobiak [3, 38].

But there is another wrinkle: adding constants to the language changes which subsets of a structure are substructures, and hence which classes of structures are quasivarieties. Consider what happens when we enlarge the type of lattices to include 0 and 1 as constants. *If a variety $\mathcal{K}$ of* $(0, 1)$*-lattices contains a finite simple nondistributive* $(0, 1)$*-lattice, then $\mathcal{K}$ is Q-universal* (Adams and Dziobiak [4]). In particular, $\mathbf{M}_3$, regarded as a $(0, 1)$-lattice, generates a variety that is a Q-universal quasivariety, whereas without the constants $\mathrm{L_q}(\mathbb{V}(\mathbf{M}_3))$ is a 3-element chain.

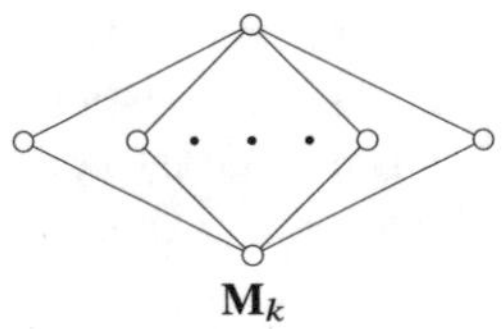

Fig. 1.2 The lattice $\mathbf{M}_k$ with k atoms

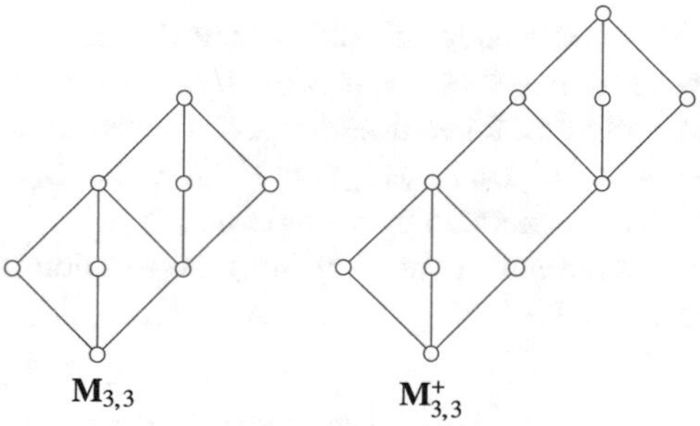

Fig. 1.3 The lattices $\mathbf{M}_{3,3}$ and $\mathbf{M}_{3,3}^{+}$

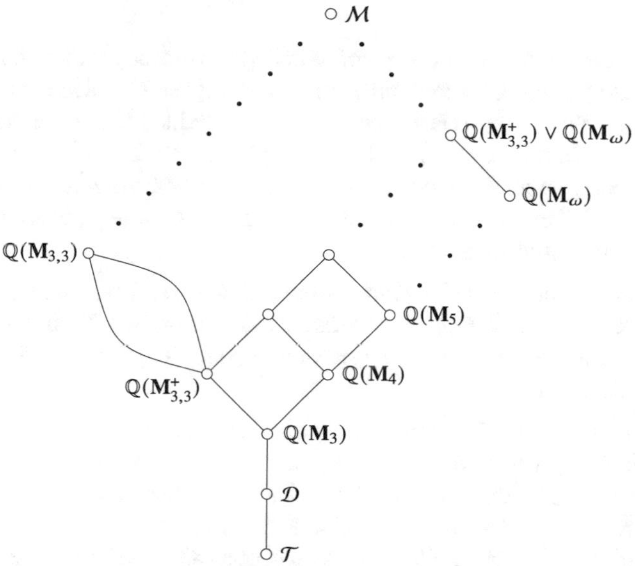

Fig. 1.4 The lattice $\mathrm{L_q}(\mathcal{M})$ of quasivarieties of modular lattices. Here $\mathcal{T}$ denotes all 1-element lattices, $\mathcal{D}$ denotes distributive lattices. The interval $[\mathbb{Q}(\mathbf{M}_{3,3}^{+}), \mathbb{Q}(\mathbf{M}_{3,3})]$ contains uncountably many subquasivarieties, while the ideal $\downarrow(\mathbb{Q}(\mathbf{M}_{3,3}^{+}) \vee \mathbb{Q}(\mathbf{M}_{\omega}))$ is countable and distributive as it contains exactly those quasivarieties shown

Lattices Turning to lattices in general, a different story emerges, with lots of small distributive subquasivariety lattices. A quasivariety Q is said to be *primitive* if every subquasivariety $\mathcal{K} \leq Q$ is equational relative to Q. As we shall see, many small lattices generate a variety $\mathbb{V}(\mathbf{L})$ that is a primitive quasivariety.

We begin with a general notion. A structure $\mathbf{A}$ is said to be *weakly projective* in a class $\mathcal{K}$ if whenever $\mathbf{A}$ is a homomorphic image of a structure $\mathbf{B} \in \mathcal{K}$, then $\mathbf{A}$ embeds in $\mathbf{B}$. The characterization of primitive, locally finite quasivarieties is due to Gorbunov [50, 52] and Slavík [111], independently.

Theorem 1.7 *A locally finite quasivariety Q of finite type is primitive if and only if every finite Q-subdirectly irreducible structure $\mathbf{A} \in Q$ is weakly projective in Q_{fin}, the class of finite structures in Q. Moreover, if Q is primitive, $\mathrm{L_q}(Q)$ is distributive.*

For lattices, we have two wonderful tools at our disposal. The first is Jónsson's Lemma: *If* $\mathbf{L}$ *is a finite lattice, then the subdirectly irreducible lattices in* $\mathbb{V}(\mathbf{L})$ *are contained in* $\mathbb{HS}(\mathbf{L})$. Note that for lattices, or more generally any class of algebras with idempotent elements, $\mathbb{V}(\{\mathbf{L}_1, \ldots, \mathbf{L}_m\}) = \mathbb{V}(\mathbf{L}_1 \times \cdots \times \mathbf{L}_m)$. Thus every finitely generated lattice variety is generated by a single lattice.

The second tool is Whitman's condition from the solution of the word problem for free lattices [115]:

$$\text{if } s = \bigwedge_{i=1}^{m} s_i \leq \bigvee_{j=1}^{n} t_j = t, \text{ then either } s_i \leq t \text{ for some } i, \qquad \text{(W)}$$
$$\text{or } s \leq t_j \text{ for some } j.$$

Jónsson, McKenzie, and Kostinsky combined (W) with bounded homomorphisms to characterize finitely generated projective lattices; see Chapter II of [46] for the arguments and discussion. Davey and Sands [34] added the observation that for lattices with no infinite chains, all homomorphisms are bounded, to yield: *every finite lattice satisfying* (W) *is projective in the class of finite lattices.* Thus we can apply Theorem 1.7 to obtain a sufficient criterion for the variety generated by a finite lattice to be a primitive quasivariety.

Theorem 1.8 *Assume that* $\mathbf{L}$ *is a finite lattice with the property that every subdirectly irreducible lattice in* $\mathbb{HS}(\mathbf{L})$ *satisfies Whitman's condition* (W). *Then the variety* $\mathbb{V}(\mathbf{L})$ *is a primitive quasivariety, i.e., every subquasivariety is equational. Thus* $\mathrm{L_q}(\mathbb{V}(\mathbf{L}))$ *is finite and distributive.*

For modular lattices, Theorem 1.8 applies only to the lattices $\mathbf{M}_k$ (including by extension $\mathbf{M}_\omega$), while $\mathbf{M}_{3,3}$ fails (W). But many nonmodular lattices have the property, for example, the pentagon $\mathbf{N}_5$ and all 16 lattices that generate a cover of $\mathbb{V}(\mathbf{N}_5)$. Figure 1.5 indicates a couple of sequences of lattices, all of which generate primitive varieties. Uncountably many such sequences of primitive lattice varieties are constructed in Jipsen and Nation [66]. They also found primitive varieties $\mathbb{V}(\mathbf{L})$ generated by lattices that fail (W).

The bottom of the lattice $\mathrm{L_q}(\mathcal{L})$ of lattice varieties is very thin. There is one atom (distributive lattices) and 2 varieties of height 2 ($\mathbb{V}(\mathbf{M}_3)$ and $\mathbb{V}(\mathbf{N}_5)$), 18 varieties of height 3, etc., until eventually the number of elements of a given height becomes infinite. For a recent summary, see Jipsen and Rose [67]. The lattice $\mathrm{L_q}(\mathbb{V}(\mathbf{L}))$ for a primitive lattice variety inherits these restrictions: the variety $\mathcal{D}$ of distributive lattices is the unique atom, the second level contains $\mathbb{V}(\mathbf{M}_3)$ or $\mathbb{V}(\mathbf{N}_5)$ or both, and so forth. Note that the third level has at most 17 varieties, not including $\mathbb{V}(M_{3,3})$.

Varieties of 2-Unary Algebras *versus* 1-Binary Algebras In the course of writing [63], the authors decided to test the algorithms therein on the variety $\mathcal{T}_3$ generated by a more-or-less randomly chosen 3-element, 2-unary algebra $\mathbf{T}_3$. A familiar pattern emerged: a segment at the bottom of $\mathrm{L_q}(\mathcal{T}_3)$ contained quasivarieties with only finitely many subquasivarieties, but $\mathbb{Q}(\mathbf{T}_3)$ itself turned out to be Q-universal. The

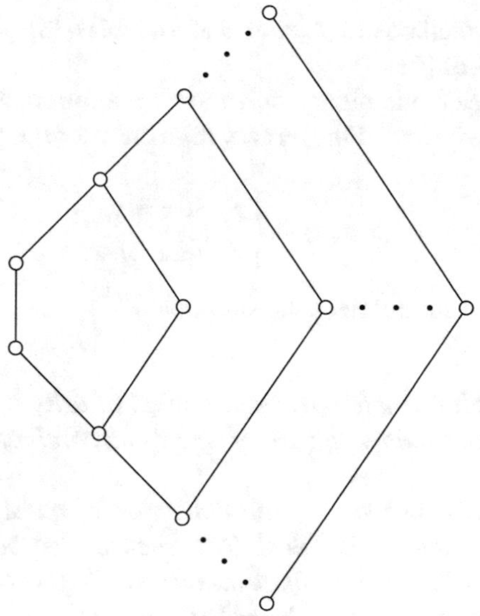

$\mathbf{N}_k$

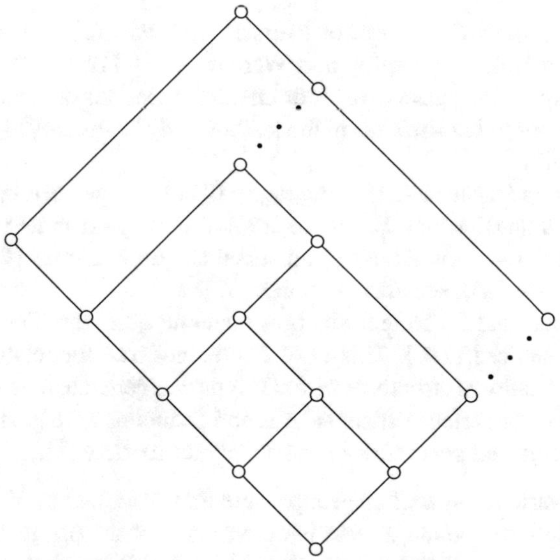

$\mathbf{J}_k$

Fig. 1.5 Lattices $\mathbf{N}_k$ and $\mathbf{J}_k$ which generate primitive lattice varieties

proof again used the methods of Adams and Dziobiak [3]. A detailed analysis of $L_q(\mathcal{T}_3)$ is in Chapter 5 of [63].

Algebras with at least one binary operation are a different matter. Recall that a finite algebra **A** is *quasiprimal* if the ternary discriminator $t(x, y, z)$ is a term function on **A**, where

$$t(x, y, z) = \begin{cases} z & \text{if } x = y, \\ x & \text{otherwise.} \end{cases}$$

The following theorem summarizes the situation.

Theorem 1.9

(1) *Fix a type τ with a single operation symbol of arity $k > 1$. As $n \to \infty$, the probability that a random algebra of type τ and cardinality $\leq n$ is quasiprimal goes to* 1.

(2) *Let $C = \{\mathbf{A}_1, \ldots, \mathbf{A}_n\}$ be a finite collection of quasiprimal algebras contained in a common arithmetical (congruence distributive and congruence permutable) variety. Then the finite members of $\mathbb{V}(C)$ are finite direct products of subalgebras of $\mathbf{A}_1, \ldots, \mathbf{A}_n$. Hence $\mathbb{V}(C) = \mathbb{Q}(C)$, and $L_q(\mathbb{Q}(C)) = L_v(\mathbb{V}(C))$ is a finite distributive lattice.*

Part (1) is a renowned theorem of Murskiĭ [88]. Part (2) is due to Keimel and Werner [72], and Bulman-Fleming and Werner [27]. For the classical theory of primal and quasiprimal algebras, we recommend the sections on *semisimple varieties* and *directly representable varieties* in the textbooks by Bergman [23] and Burris and Sankappanavar [28].

Note that adding constants to the language will not change the property of whether a finite algebra is quasiprimal. But there are non-finitely-generated varieties with a ternary discriminator term that are Q-universal as quasivarieties (Blanco, Campercholi, and Vaggione [25], see also Dziobiak [37]).

Shafaat showed that 2-element algebras generate quasivarieties with 2-element subquasivariety lattices [110]. This is of course not true for relational structures: even 1-element relational structures of finite type can generate a quasivariety with a large (finite) subquasivariety lattice. Adams and Dziobiak were the first to show that a 3-element algebra can generate a Q-universal quasivariety [2].

Groups Quasivarieties of abelian groups were first described by Vinogradov [113]. Every quasivariety of abelian groups is determined by the prime-power or infinite cyclic groups it contains. For a quasivariety $Q \leq \mathcal{A}$, this is either

(1) $\{\mathbf{Z}_q : q \in S\}$ for a finite hereditary set S of prime powers, or
(2) $\{\mathbf{Z}\} \cup \{\mathbf{Z}_q : q \in T\}$ for an arbitrary hereditary set T of prime powers,

where $\mathbf{Z}_q$ denotes the cyclic group of order q. Consequently, $L_q(\mathcal{A})$ is a distributive lattice of cardinality $2^{\aleph_0}$.

In contrast, Nurakunov has shown that the quasivariety generated by a finite abelian group with an extra constant can be Q-universal [97].

We strongly suspect that the variety of all nilpotent groups of class 2 is Q-universal but do not even know whether this is true for the variety of all groups.

Thus far we have seen that subquasivariety lattices can range from being tractable to complex, with the only surprise being how easy it is for the quasivariety generated by a finite structure to be $\mathcal{Q}$-universal. There are also results which say directly that describing subquasivariety lattices is hard.

Kearnes and Nation showed that the class of lattices that can be embedded into subquasivariety lattices is not first-order axiomatizable [71]. (The result there is for lattices embeddable into congruence lattices of semilattices, but in view of [20, 64] and Theorems 2.36 and 2.64, that is equivalent to being embedded into the dual of some $\mathrm{L_q}(\mathcal{K})$.)

A lattice **L** is called *reasonable* if the set of all finite sublattices of **L** is decidable, i.e., if there is an algorithm deciding whether a finite lattice is a sublattice of **L**. Lattices that are not reasonable are *unreasonable*. In [96], Nurakunov showed that for any signature that contains at least one non-constant operation, there is a quasivariety $\mathcal{R}$ with this signature such that $\mathrm{L_q}(\mathcal{R})$ is unreasonable. The same result is true if "decidable" is replaced by "computable" (recursively enumerable).

Schwidefsky relates unreasonableness to $\mathcal{Q}$-universality by showing that if a quasivariety $\mathcal{K}$ of finite type satisfies a certain generalization of the Adams–Dziobiak properties (P1)–(P4) for $\mathcal{Q}$-universality, then there is a subquasivariety $\mathcal{Q} \leq \mathcal{K}$ such that $\mathrm{L_q}(\mathcal{Q})$ is unreasonable [107]. But the connection is in one direction only: Lutsak proved that for each non-constant finite signature there are continuum many quasivarieties of that signature such that $\mathrm{L_q}(\mathcal{K})$ is unreasonable, but $\mathcal{K}$ is not $\mathcal{Q}$-universal [81].

Further undecidability results for quasivariety lattices are in Kravchenko, Nurakunov, and Schwidefsky [73–77].

We began this section by stating that the original statement of the Birkhoff-Mal'cev problem is "too general to be practical." *Quixotic* might be a more apt description. This will not deter us from seeing what we can find!

1.6 A Review of Lattices of Algebraic Sets

Recall that a subset X of an ordered set is *up-directed* if for every $x, y \in X$ there exists $z \in X$ such that $z \geq x$ and $z \geq y$. In a complete lattice, a *directed join* is the join of a nonempty up-directed subset.

An *algebraic subset* of a complete lattice **S** is a subset closed under arbitrary meets (including $\bigwedge \varnothing = 1_{\mathbf{S}}$) and nonempty directed joins. The intersection of a collection of algebraic subsets of **S** is again an algebraic subset, so the set of all algebraic subsets of **S** forms a complete lattice, denoted $\mathrm{S_p}(\mathbf{S})$. Here we provide a brief summary of the properties of lattices of algebraic subsets. A more thorough exposition is given in Section 4-1 of [22], and further properties will be developed in the course of this book.

We are primarily interested in the case when **S** is algebraic, though for some purposes upper continuity would suffice.

A fundamental fact is that an algebraic subset of an algebraic lattice is an algebraic lattice in its own right. Even though we do not use the statement directly, we include its proof, because it is classic, and because one of the authors remembers having trouble with the exercise 50 years ago in graduate school.

Theorem 1.10 *If A is an algebraic subset of an algebraic lattice* **S**, *then A with the order inherited from* **S** *is an algebraic lattice.*

Proof Since A is a complete meet subsemilattice of $\mathbf{S}$, it is a complete lattice. Joins in A may be different, with in general $\bigvee_A X \geq \bigvee_{\mathbf{S}} X$ for $X \subseteq A$, but for directed sets they are the same.

For $x \in \mathbf{S}$, define $\alpha(x) = \bigwedge(\uparrow x \cap A)$, so that $x \leq \alpha(x) \in A$, and when $a \in A$ we have $x \leq a$ iff $\alpha(x) \leq \alpha(a)$.

Let us show that if c is compact in $\mathbf{S}$, then $\alpha(c)$ is compact in A. Assume $\alpha(c) \leq \bigvee_A B$ for a subset $B \subseteq A$. Then

$$\begin{aligned} c \leq \alpha(c) &\leq \bigvee_A B \\ &= \bigvee_A \{\bigvee_A F : F \subseteq B \text{ is finite}\} \\ &= \bigvee_{\mathbf{S}} \{\bigvee_A F : F \subseteq B \text{ is finite}\} \end{aligned}$$

which implies that $c \leq \bigvee_A G$ for some finite $G \subseteq B$, since c is compact and the right-hand side is directed. As $c \leq \bigvee_A G \in A$ we get $\alpha(c) \leq \bigvee_A G$, and conclude that $\alpha(c)$ is compact in A.

Let $a \in A$. In $\mathbf{S}$ we have $a = \bigvee_{\mathbf{S}}\{c \in S : c \text{ is compact and } c \leq a\}$. As $c \leq a$ implies $c \leq \alpha(c) \leq a$, it follows that $a = \bigvee_A\{\alpha(c) : c \text{ is compact and } c \leq a\}$. Thus every $a \in A$ is a join of elements compact in A. □

Before adding operators, it is useful to write down the properties of $\mathrm{S_p}(\mathbf{S})$, the lattice of algebraic subsets of an algebraic lattice $\mathbf{S}$. These go back to Gorbunov [51]; see also [11, 20].

Theorem 1.11 *Let* **S** *be an algebraic lattice. The lattice* $\mathrm{S_p}(\mathbf{S})$ *is*

(1) *atomistic,*
(2) *dually algebraic,*
(3) *join semidistributive,*
(4) *and it supports a weak equaclosure operator.*

Item (1) reflects the fact that, in the absence of operators, $\{a, 1\}$ is an algebraic set for any $a \in S$. Properties (2)–(4) hold more generally for $\mathrm{S_p}(\mathbf{S}, H)$: (2) is proved in Theorem 2.71, (3) is a consequence of Theorem 2.59 (the Jónsson-Kiefer Property), and (4) is Theorem 3.6. Still, these properties do not totally characterize lattices of algebraic subsets, as evidenced by the following result of Gorbunov and Tumanov [53].

Theorem 1.12 *A Boolean lattice* **B** *is isomorphic to* $\mathrm{S_p}(\mathbf{S})$ *for an algebraic lattice* **S** *if and only if* $\mathbf{B} \cong \mathbf{2}^\kappa$ *for some* $\kappa \leq \aleph_0$.

Indeed, for k finite, $\mathbf{2}^k$ is isomorphic to $\mathrm{S_p}(\mathbf{k}+\mathbf{1})$ for a $k+1$-element chain, while $\mathbf{2}^{\aleph_0}$ is isomorphic to $\mathrm{S_p}(\omega+1)$. The proof that these are the only Boolean lattices $\mathrm{S_p}(\mathbf{S})$ is given with Theorem 9.29.

Now we turn our attention to algebraic lattices with operators. An *operator* on an algebraic lattice is a map $h : S \to S$ that preserves arbitrary meets (including $\bigwedge \varnothing = 1_S$) and nonempty directed joins. (These are sometimes called *continuous operators*.) Clearly the composition of operators is an operator, and so is the identity map.

If h is an operator on **S**, an algebraic subset $A \subseteq S$ is said to be *h-closed* if $h(a) \in A$ for all $a \in A$. Likewise, for a set H of operators, A is said to be *H-closed* if it is h-closed for every $h \in H$.

Lemma 1.13 *Let* H_0 *be a set of operators on a complete lattice* **S**. *Let* H *be the monoid of operators generated by* H_0 *under composition, with the identity map. An algebraic subset* $A \subseteq S$ *is* H*-closed if and only if it is* h*-closed for every* $h \in H_0$.

Thus we usually speak of a *monoid H* of operators on **S**, while in practice it suffices to identify a set of operators H_0 that will generate the monoid H. By convention, *no operators* means $H_0 = \varnothing$, in which case H consists of only the identity map.

Likewise, define the meet of operators by $(g \wedge h)(x) = g(x) \wedge h(x)$. Finite meets of operators are operators (but not necessarily infinite meets nor directed joins), and if an algebraic subset A is H_0 closed, then it is H_1 closed for the meet semilattice H_1 generated by H_0. We return to this topic in Sect. 3.2.

If **S** is an algebraic lattice and H a monoid of operators on **S**, then $\mathrm{S_p}(\mathbf{S}, H)$ denotes the lattice of all H-closed algebraic subsets of **S**, i.e., algebraic subsets A of **S** such that $h(a) \in A$ for all $a \in A$ and $h \in H$. Clearly the least H-closed algebraic subset is the singleton $\{1\}$, and the largest one is all of S. Meets in the lattice $\mathrm{S_p}(\mathbf{S}, H)$ are just set intersection: $\bigwedge_i X_i = \bigcap_i X_i$ for H-closed algebraic subsets.

We will see in Theorem 1.21 that $\mathrm{S_p}(\mathbf{S}, H)$ is dually algebraic, join semidistributive, and supports a weak equaclosure operator. It need not be atomistic, as $\{1, a\}$ may not be H-closed.

Note that if **S** is finite, then $\mathrm{S_p}(\mathbf{S}, H) = \mathrm{Sub}(\mathbf{S}, \wedge, 1, H)$, the subalgebra lattice of a meet semilattice with a largest element 1 (as a constant) and a set of unary operations that preserve $\wedge$ and 1.

Lattices of algebraic subsets are intimately related to congruence lattices of semilattices. Freese and Nation [48] proved that for a finite semilattice **K** with 0 and 1 (which is, therefore, a lattice), $\mathrm{Con}(\mathbf{K}, \vee, 0) \cong^d \mathrm{Sub}(\mathbf{K}, \wedge, 1)$. Fajtlowicz and J. Schmidt [42] gave the extension to the infinite case: If **K** is a join semilattice with 0, then $\mathrm{Con}(\mathbf{K}, \vee, 0) \cong^d \mathrm{S_p}(\mathcal{I}(\mathbf{K}))$. The lattice of ideals $\mathcal{I}(\mathbf{K})$ is just the algebraic lattice with **K** as its semilattice of compact elements. The ideas involved here are implicit in E. T. Schmidt's slightly earlier work on congruence lattices [105, 106].

In view of the connection to subquasivariety lattices, we want an extension to congruence lattices of semilattices with operators, found in Hyndman, Nation, and

Nishida [64]. Let $\mathbf{K}$ be a join semilattice with 0, and let G be a monoid of maps $g : K \longrightarrow K$ that preserve $\vee$ and 0, i.e., 0-semilattice endomorphisms. Let $\mathbf{S}$ be the algebraic lattice $\mathcal{I}(\mathbf{K})$ of (nonempty) ideals of $\mathbf{K}$. For each $g \in G$ we can define an operator $\overline{g} : S \longrightarrow S$, called the *adjoint* of g, by

$$\begin{aligned}\overline{g}(I) &= \{k \in K : g(k) \in I\}\\ &= g^{-1}(I)\end{aligned}$$

for an ideal I of $\mathbf{K}$. A routine check shows that $\overline{g}$ is indeed an operator. Then, for the set of maps $\overline{G} = \{\overline{g} : g \in G\}$, a more involved argument from [64] yields

$$\mathrm{Con}(\mathbf{K}, \vee, 0, G) \cong^d \mathrm{S_p}(\mathbf{S}, \overline{G}).$$

Looking ahead to Chap. 6, it is also possible to start on the other side. Let $\mathbf{S}$ be an algebraic lattice with a monoid of operators H. Let $\mathbf{K}$ be its semilattice of compact elements, so that $\mathbf{S} \cong \mathcal{I}(\mathbf{K})$. For each $h \in H$, define the *adjoint* $h' : S \longrightarrow S$ by

$$h'(s) = \bigwedge\{x \in S : h(x) \geq s\}.$$

It turns out that each h' maps K to K, preserving finite joins and 0. Setting $H' = \{h' : h \in H\}$, we have

$$\mathrm{Con}(\mathbf{K}, \vee, 0, H') \cong^d \mathrm{S_p}(\mathbf{S}, H).$$

We refer the reader to [64] for the detailed proofs.

For future reference, let us summarize the preceding discussion.

Theorem 1.14

(1) *Given a join semilattice with operators* $(\mathbf{S}, \vee, 0, G)$, *for* $\overline{G}$ *as above,*

$$\mathrm{Con}(\mathbf{K}, \vee, 0, G) \cong^d \mathrm{S_p}(\mathcal{I}(\mathbf{K}), \overline{G}).$$

(2) *Given an algebraic lattice* $\mathbf{S}$ *with a monoid* H *of operators, and* $\mathbf{K}$ *its semilattice of compact elements, for* H' *as above*

$$\mathrm{Con}(\mathbf{K}, \vee, 0, H') \cong^d \mathrm{S_p}(\mathbf{S}, H).$$

A complete lattice $\mathbf{L}$ is *upper continuous* if whenever $x \in L$ and $\{d_i : i \in I\}$ is an up-directed set in L, then $x \wedge \bigvee_i d_i = \bigvee_i (x \wedge d_i)$. Algebraic lattices are upper continuous, and many facts about algebraic subsets of algebraic lattices can be extended to the case when $\mathbf{S}$ is just upper continuous, including the following one. Gorbunov showed that finite joins in $\mathrm{S_p}(\mathbf{S}, H)$ have a simple description [51].

Theorem 1.15 *If* $\mathbf{S}$ *is a complete, upper continuous lattice, and the sets* $X_1, \ldots, X_n$ *are in* $\mathrm{S_p}(\mathbf{S}, H)$, *then*

$$X_1 \vee \cdots \vee X_n = \{x_1 \wedge \cdots \wedge x_n : x_j \in X_j \textit{ for } 1 \leq j \leq n\}.$$

This is important for us, so let us include the proof, following Lemma 4-1.12 of [22].

Proof Let us show that if X, Y are H-closed algebraic subsets of $\mathbf{S}$, then $X \vee Y = \{x \wedge y : x \in X, y \in Y\}$. Indeed, since $1 \in X \cap Y$, we have $X, Y \subseteq S_{XY} = \{x \wedge y : x \in X, y \in Y\}$, and hence $S_{XY} \subseteq X \vee Y$. It remains to show that S_{XY} is an H-closed algebraic subset. Obviously, S_{XY} is closed with respect to $\bigwedge$ and operators $h \in H$.

So consider an up-directed subset $d_i \in S_{XY}$, $i \in I$. Then each $d_i = x_i' \wedge y_i'$ for some $x_i' \in X$, $y_i' \in Y$. The sets $\{x_i' : i \in I\}$ and $\{y_i' : i \in I\}$ need not be directed. To fix this, define $x_i = \bigwedge\{x \in X : d_i \leq x\}$ and $y_i = \bigwedge\{y \in Y : d_i \leq y\}$. Then $x_i \in X$, $y_i \in Y$ and $d_i = x_i \wedge y_i$. Besides, $\{x_i : i \in I\}$ is up-directed in X, whence $x = \bigvee_I x_i$ is in X. Similarly, $y = \bigvee_I y_i \in Y$. Due to the upper continuity of $\mathbf{S}$, one has

$$x \wedge y = x \wedge \bigvee_I y_i = \bigvee_I (x \wedge y_i) = \bigvee_{i \in I} (\bigvee_{j \in I} x_j \wedge y_i) = \bigvee_{i \in I} \bigvee_{j \in I} x_j \wedge y_i \,.$$

Note that $d_i \leq d_k$ implies $x_i \leq x_k$ and $y_i \leq y_k$. Hence for every pair $x_j \wedge y_i$ and $x_p \wedge y_q$ one can find $k \in I$ such that $x_k \wedge y_k \geq (x_i \wedge y_j), (x_p \wedge y_q)$, so that $x \wedge y = \bigvee_{k \in I} x_k \wedge y_k = \bigvee_{k \in I} d_k$. Thus S_{XY} is an H-closed algebraic subset of $\mathbf{S}$. □

Arbitrary joins in $\mathrm{S_p}(\mathbf{S}, H)$ can be treated as a special case of a more general problem: *Given a subset $X \subseteq S$, there is a least H-closed algebraic subset* $\mathrm{Ag}(X)$ *containing X. Describe it.* For then the join is determined as $\bigvee_i X_i = \mathrm{Ag}(\bigcup_i X_i)$. The description of $\mathrm{Ag}(X)$ is based on arguments in Gorbunov [52].

Let us introduce some operator notation. For a subset $X \subseteq S$,

- $\mathbb{D}(X)$ is the set of all joins of nonempty directed subsets of X,
- $\mathbb{M}(X)$ is the set of all meets of subsets of X, including $\bigwedge \varnothing = 1_{\mathbf{S}}$,
- $\mathbb{O}(X)$ is the set of all elements $h(x)$ with $h \in H$, $x \in X$.

Theorem 1.16 *Let $\mathbf{S}$ be an algebraic lattice with a monoid H of operators, and let $X \subseteq S$. In $\mathrm{S_p}(\mathbf{S}, H)$,*

$$\mathrm{Ag}(X) = \mathbb{DMO}(X).$$

That is, the least H-closed algebraic subset of $\mathbf{S}$ containing X is obtained by applying to X first the operators of H, then arbitrary meets, then nonempty directed joins.

Again, this is crucial for our work, so we include the proof in a series of lemmas, following closely the proof of Theorem 4-1.22 in [22].

Now, if $\mathbf{S}$ is not algebraic, it may happen that $\mathbb{D}^2(X) \supset \mathbb{D}(X)$ properly, and $\mathbb{MD}(X) \not\subseteq \mathbb{DM}(X)$, so that finding the H-closed algebraic subset $\mathrm{Ag}(X)$ generated by X requires a transfinite recursion. We want to show that, when $\mathbf{S}$ is algebraic, the process terminates with $\mathrm{Ag}(X) = \mathbb{DMO}(X)$.

Since the operators in H preserve arbitrary meets and directed joins, the first observation is obvious.

Lemma 1.17 *$\mathbb{OM}(X) \subseteq \mathbb{MO}(X)$ and $\mathbb{OD}(X) \subseteq \mathbb{DO}(X)$.*

The next inclusion requires more work.

Lemma 1.18 [12] *If* **S** *is algebraic, then $\mathbb{MD}(X) \subseteq \mathbb{DM}(X)$.*

Proof Let $m \in \mathbb{MD}(X)$, so that $m = \bigwedge_{i \in I} s_i$ where each $s_i = \bigvee_{j \in J_i} d_{ij}$ is a directed join and each $d_{ij} \in X$.

For each choice function $\varphi \in \prod_i J_i$, let $t_\varphi = \bigwedge_i d_{i\varphi(i)}$. The claim is that $\bigwedge_i s_i = \bigvee_\varphi t_\varphi$. Since the t_φ's form a directed set, this will show that $m \in \mathbb{DM}(X)$.

Clearly $\bigwedge_i s_i \geq \bigvee_\varphi t_\varphi$. For the reverse inclusion, let $c \leq \bigwedge_i s_i$ be compact. Then for each $i \in I$ we have $c \leq s_i$, and hence for each i there exists $j \in J_i$ such that $c \leq d_{ij}$ by compactness. Let $\psi(i)$ be chosen as one such j. Then $c \leq d_{i\psi(i)}$ for all i, so that $c \leq t_\psi$. The reverse inclusion follows. □

Lemma 1.19 *If* **S** *is algebraic and the subset $X \subseteq S$ satisfies $\mathbb{M}(X) = X$, then $\mathbb{D}^2(X) = \mathbb{D}(X)$.*

Proof For each $s \in S$, define $\chi(s) = \bigwedge\{x \in X : x \geq s\}$. As usual, this is an order-preserving map with the properties $s \leq \chi(s) \in X$, and for $x \in X$ we have $s \leq x$ iff $\chi(s) \leq x$.

Now consider an element $m \in \mathbb{D}^2(X)$. Then we can write $m = \bigvee_i t_i$, with each $t_i = \bigvee_j x_{ij}$, where both are up-directed joins and each $x_{ij} \in X$. Let **K** denote the semilattice of compact elements of **S**, and consider any $c \in K$ with $c \leq m$. By compactness, we have $c \leq t_{i_0}$ for some i_0, and again $c \leq x_{i_0 j_0}$ for some j_0, whence $\chi(c) \leq x_{i_0 j_0} \leq t_{i_0} \leq m$.

Let $K(m) = \{c \in K : c \leq m\}$. Then we have $m = \bigvee K(m) \leq \bigvee_{c \in K(m)} \chi(c) \leq m$, and $\{\chi(c) : c \in K(m)\}$ is an up-directed subset of X. Thus $m \in \mathbb{D}(X)$, as desired. □

Combining the three previous lemmas, we see that $\mathbb{DMO}(X)$ is closed under the operators of H, arbitrary meets, and nonempty directed joins. It follows that $\mathrm{Ag}(X) = \mathbb{DMO}(X)$, thus proving Theorem 1.16.

We will return to discussing the closure operator $\mathbb{DMO}$ in Sect. 3.2. In the meantime, let us make some useful observations.

- Every H-closed algebraic subset A of **S** has a least element. We will typically denote the least element of A by a_0, using the lower case of the same letter.
- If a_0 is the least element of A and $h \in H$, then $h(a_0) \geq a_0$.
- The least H-closed algebraic subset containing a_0 is $\mathbb{DMO}(a_0)$.

One more observation is applied often.

Corollary 1.20 *If A_i $(i \in I)$ are H-closed algebraic subsets, and a_{i0} is the least element of each A_i, then the least element of $\bigvee_i A_i$ is $\bigwedge_i a_{i0}$.*

Next we record the analogue of Theorem 1.11 when operators are included.

Theorem 1.21 *Let* **S** *be an algebraic lattice and H a monoid of operators on* **S**. *The lattice $\mathrm{S_p}(\mathbf{S}, H)$ is*

(1) *dually algebraic,*
(2) *join semidistributive,*
(3) *and it supports a weak equaclosure operator.*

Indeed, let **K** denote the join semilattice of compact elements of **S**, and let H' be the monoid of adjoints of maps in H. By Theorem 1.14(2), $\mathrm{S_p}(\mathbf{S}, H)$ is dually isomorphic to the congruence lattice $\mathrm{Con}(\mathbf{K}, \vee, 0, H')$. Since congruence lattices of semilattices are meet semidistributive (Papert [98]), and all congruence lattices are algebraic, the first two statements follow.

However, it is also enlightening to prove (1) and (2) directly. In Sect. 2.6, we derive the join semidistributivity of $\mathrm{S_p}(\mathbf{S}, H)$ as a consequence of the Jónsson-Kiefer Property, and in Sect. 2.8 we prove its dual algebraicity by identifying the dually compact elements (Theorem 2.71).

There is a natural weak equaclosure operator on $\mathrm{S_p}(\mathbf{S}, H)$, and that is the subject of Chap. 3. The closure operator involves fully invariant elements, so let us discuss those next.

1.7 Fully Invariant Elements

Let **S** be an algebraic lattice with a monoid of operators H. We say that an element $x \in S$ is *fully invariant* if $h(x) \geq x$ for all $h \in H$. Let $\mathrm{Fi}(\mathbf{S})$ denote the set of all fully invariant elements of **S**.

Theorem 1.22 *If* **S** *is an algebraic lattice with operators, then* $\mathrm{Fi}(\mathbf{S})$ *is a complete sublattice of* **S**. *Hence* $\mathrm{Fi}(\mathbf{S})$ *is itself an algebraic lattice.*

Proof First, note that $0_{\mathbf{S}}$ and $1_{\mathbf{S}}$ are fully invariant.

Consider a subset $\{x_i : i \in I\} \subseteq \mathrm{Fi}(\mathbf{S})$. For any operator $h \in H$, we have $h(\bigvee x_i) \geq h(x_i) \geq x_i$ for all $i \in I$, whence $h(\bigvee_i x_i) \geq \bigvee_i x_i$ and $\bigvee_i x_i$ is fully invariant. On the other hand, $h(\bigwedge_i x_i) = \bigwedge_i h(x_i) \geq \bigwedge_i x_i$ since h preserves arbitrary meets, so $\bigwedge_i x_i$ is also fully invariant. □

In general, $\mathrm{Fi}(\mathbf{S})$ need not be H-closed. Rather, the relevant fact is the following observation.

Lemma 1.23 *The filter* $\uparrow x$ *of* **S** *is* H*-closed if and only if* $x \in \mathrm{Fi}(\mathbf{S})$.

For each $s \in S$, let $\phi(s) = \bigwedge(\uparrow s \cap \mathrm{Fi}(\mathbf{S}))$, so that $\phi(s)$ is the least fully invariant element above s. As usual,

(1) $s \leq \phi(s)$;
(2) if $x \in \mathrm{Fi}(\mathbf{S})$, then $x \geq s$ iff $x \geq \phi(s)$;
(3) $\phi(\bigvee_j s_j) = \bigvee_j \phi(s_j)$.

These notions will play a recurring role in our study.

1.8 Finite Lower Bounded Lattices

In this section, we review the basic theory of finite lower bounded lattices. Our interest stems from Theorem 1.25 below: *If $\mathcal{K}$ is a locally finite quasivariety of finite type with only finitely many subquasivarieties, then* $\mathrm{L_q}(\mathcal{K})$ *is a finite lower bounded lattice.* For a quasivariety $\mathcal{Q}$ that is not locally finite, $\mathrm{L_q}(\mathcal{Q})$ can be finite without being lower bounded; see Sect. 8.1.

Our review of lower bounded lattices is taken from Chapter II of [46], which is based on work of Day [35], Jónsson [70], and McKenzie [85]. A version of this summary also appears in Section 4.2 of Hyndman and Nation [63].

We begin by considering lower boundedness for finitely generated lattices. A lattice homomorphism $h : \mathbf{K} \to \mathbf{L}$ is a *lower bounded homomorphism* if for every element $a \in L$, the inverse image $h^{-1}(\uparrow a) = \{w \in K : h(w) \geq a\}$ is either empty or has a least element. A finitely generated lattice $\mathbf{L}$ is a *lower bounded lattice* if it is a lower bounded homomorphic image of a finitely generated free lattice, i.e., if there is a surjective lower bounded homomorphism $h : \mathbf{FL}(X) \twoheadrightarrow \mathbf{L}$ for some finite set X. Lower bounded lattices initially arose in the context of sublattices of free lattices and projective lattices.

For an element c and a finite subset A of a lattice $\mathbf{L}$, we say that A is a *join cover* of c if $c \leq \bigvee A$. The join cover is *nontrivial* if $c \nleq a$ for all $a \in A$. Recall that for subsets A, B of $\mathbf{L}$, we say that B *refines* A if for every $b \in B$ there exists $a \in A$ with $b \leq a$. This is written $B \ll A$. The join cover $c \leq \bigvee A$ is *minimal* if whenever $c \leq \bigvee B$ and $B \ll A$, then $A \subseteq B$. Thus a minimal join cover consists of an antichain of join irreducible elements, such that no element of A can be omitted or replaced by a set of smaller elements.

Every element in a finite join semidistributive lattice has a *canonical join representation*, that is, a join representation $a = \bigvee C$ which is minimal in the above sense: whenever $a = \bigvee D$, then $C \ll D$. Finite lower bounded lattices are join semidistributive, so this applies to them.

We say that a lattice $\mathbf{L}$ has the *minimal join cover refinement property* if for every $a \in L$, there are only finitely many minimal join covers of a in $\mathbf{L}$, and every nontrivial join cover of a refines to one of these minimal ones. Of course, every finite lattice has the minimal join cover refinement property.

Let $\mathrm{D}_0(\mathbf{L})$ be the set of all join-prime elements of $\mathbf{L}$, i.e., the set of all elements that have no nontrivial join cover. Given $\mathrm{D}_k(\mathbf{L})$, define $\mathrm{D}_{k+1}(\mathbf{L})$ to be the set of all $p \in L$ such that every nontrivial join cover of p refines to a join cover contained in $\mathrm{D}_k(\mathbf{L})$, i.e., $p \leq \bigvee C$ nontrivially implies there exists $B \ll C$ with $p \leq \bigvee B$ and $B \subseteq \mathrm{D}_k(\mathbf{L})$. Note that if $p \in A$ where $A \subseteq L$ has the minimal join cover refinement property, then $p \in \mathrm{D}_{k+1}(\mathbf{L})$ if and only if every minimal nontrivial join cover of p is contained in $\mathrm{D}_k(\mathbf{L})$.

Clearly $\mathrm{D}_k(\mathbf{L}) \subseteq \mathrm{D}_{k+1}(\mathbf{L})$ for all $k \in \omega$. Let $\mathrm{D}(\mathbf{L}) = \bigcup_{k \in \omega} \mathrm{D}_k(\mathbf{L})$.

The basic theorem on lower bounded lattices can be stated thusly.

Theorem 1.24 *For a finitely generated lattice* **L**, *the following are equivalent.*

(1) **L** *is a lower bounded lattice, i.e., there is a surjective lower bounded homomorphism* $h : \mathbf{FL}(X) \twoheadrightarrow \mathbf{L}$ *for some finite set* X.
(2) *For every finitely generated lattice* **K**, *every homomorphism* $h : \mathbf{K} \to \mathbf{L}$ *is lower bounded.*
(3) $\mathrm{D}(\mathbf{L}) = \mathbf{L}$.

Moreover, if **L** *satisfies these properties, then it is join semidistributive and* **L** *has the minimal join cover refinement property.*

Examples of lower bounded lattices include any finitely generated sublattice of a free lattice [70], and the subalgebra lattice of any finite semilattice [9, 47]. For an extension of lower boundedness to lattices that may not be finitely generated, see Adaricheva and Gorbunov [17].

With these tools in hand, we return to lattices of subquasivarieties. The restriction to locally finite quasivarieties of finite type brings a strong restriction, due to Adaricheva, Dziobiak, and Gorbunov [14], based on Adaricheva [9].

Theorem 1.25 *Let* $\mathcal{K}$ *be a locally finite quasivariety of finite type. If* $\mathrm{L_q}(\mathcal{K})$ *is finite, then it is a lower bounded lattice.*

This theorem is generalized in Theorem 4.7 of [63]: *If* $\mathcal{K}$ *is a locally finite quasivariety of finite type, then* $\mathrm{L_q}(\mathcal{K})$ *is a fermentable lattice* (see below). On the other hand, there are finite lower bounded lattices that do not support any equaclosure operator, e.g., those in Fig. 4.4. A lattice that does not support an equaclosure operator cannot be represented as $\mathrm{L_q}(\mathcal{Q})$ for any quasivariety $\mathcal{Q}$.

A lattice is *fermentable* if it is $\bigvee$-generated by some set A of join irreducible elements, that is, every element is a (possibly infinite) join of members of A, such that

(a) A has the minimal join cover refinement property,
(b) $A \subseteq \mathrm{D}(\mathbf{L})$.

These lattices were introduced in Wehrung [114], generalizing Pudlák and Tůma [100]. For finite lattices, *fermentable* and *lower bounded* coincide.

Chapter 2
Varieties and Quasivarieties in General Languages

Ecrire, c'est une façon de parler sans être interrompu. – Jules Renard

Кто знает, куда смотреть, рано или поздно увидит то, что хочет увидеть. – Борис Акунин, Кладбищенские истории

In this chapter we will develop the fundamental notions of varieties and quasivarieties, starting near the beginning (just past set theory). The key idea is to link logical theories (equational theories or implicational theories) with models (varieties or quasivarieties). The tools are well-established: homomorphisms, subalgebras, direct products, ultraproducts, Galois connections, as discussed in the first chapter. It is not surprising that these tools are robust and work in a more general setting than the classical twentieth century algebra in which they arose. If at times the details seem unfamiliar, remember Bjarni Jónsson's dictum, that you see most clearly from a general perspective.

2.1 Basic Universal Algebra

Let us review the basic concepts of universal algebra in a general setting. We consider structures in a language (type) that contains a set $\mathcal{F}$ of function symbols, constants (regarded as nullary functions), and a set $\mathcal{R}$ of relation symbols. The type also contains a function that assigns an arity to each function or relation symbol. Both $\mathcal{F}$ and $\mathcal{R}$ are allowed to be empty. *The type may or may not contain an equality relation* $\approx$.

A particular structure $\mathbf{A} = \langle A, \mathcal{F}^{\mathbf{A}}, \mathcal{R}^{\mathbf{A}} \rangle$ has a carrier set A, functions $f^{\mathbf{A}}$ for $f \in \mathcal{F}$, and relations $R^{\mathbf{A}}$ for $R \in \mathcal{R}$.

In the background is a universal quasivariety $\mathcal{U}$ within which we work. Examples include the following.

- $\mathcal{U}$ is all structures of the given type, with no laws.

K. Adaricheva et al., *A Primer of Subquasivariety Lattices*, CMS/CAIMS Books in Mathematics 3, https://doi.org/10.1007/978-3-030-98088-7_2

- $\mathcal{U}$ is sets with a binary relation $\approx$, and laws stating that $\approx$ is reflexive, symmetric, and transitive.
- $\mathcal{U}$ is all groups, with the group laws and laws stating that $\approx$ is a congruence relation.
- $\mathcal{U}$ could be sets with binary relations $\approx_\varepsilon$ for all real $\varepsilon > 0$ that model the relations in a metric space where $x \approx_\varepsilon y$ means $d(x, y) < \varepsilon$. Thus each $\approx_\varepsilon$ would be reflexive and symmetric, but transitivity would be replaced by laws of the form

$$x \approx_{\varepsilon_1} y \ \ \& \ \ y \approx_{\varepsilon_2} z \longrightarrow x \approx_{\varepsilon_1+\varepsilon_2} z\,.$$

The language and the underlying quasivariety $\mathcal{U}$ are part of the specification of any quasivariety $\mathcal{Q}$. Sometimes these may be implicitly known, for example, as when discussing quasivarieties of groups or lattices, but as a general rule they should be included in the description of $\mathcal{Q}$.

Important Notice When we speak of a *language with equality*, we mean that there is a binary relation symbol that, per the laws of the base quasivariety $\mathcal{U}$, acts as an equivalence relation and preserves operations and relations. The mere presence of the symbol $\approx$, or which symbol is used to denote equality, is of course irrelevant.

We maintain the traditional distinction between *functions* and *relations* on a structure. Functions are operations on a structure, defined on some A^k and taking values in A. Relations are predicates, expressing what you can say in the language about a structure.

Let $f \in \mathcal{F}$ be a function symbol of arity $k \geq 0$, and let $\mathbf{A}$ be a structure of the type. The function $f^{\mathbf{A}}$ realizing f in $\mathbf{A}$ is a subset $f^{\mathbf{A}} \subseteq A^{k+1}$ such that

(1) for every $\mathbf{a} \in A^k$ there exists $b \in A$ such that $(\mathbf{a}, b) \in f^{\mathbf{A}}$,
(2) if $(\mathbf{a}, b) \in f^{\mathbf{A}}$ and $(\mathbf{a}, c) \in f^{\mathbf{A}}$, then $b = c$.

Note that the "=" in (2) is in the meta-language, and so it is allowed. The definition means that functions are everywhere defined on A^k and have a unique last entry. (Let us save the option to include partial function for another day.)

Let $R \in \mathcal{R}$ be a relation symbol of arity $k \geq 0$. The relation $R^{\mathbf{A}}$ realizing R in $\mathbf{A}$ is a subset $R^{\mathbf{A}} \subseteq A^k$. Of course, if $R^{\mathbf{A}}$ is just an arbitrary subset of A^k, then $\mathbf{A}$ need not satisfy the laws of our base quasivariety $\mathcal{U}$. For example, if $\mathcal{U}$ has an equality relation $\approx$, then $\approx^{\mathbf{A}}$ should be reflexive, symmetric, and transitive. When $\mathbf{A} \in \mathcal{U}$, that is, $\mathbf{A}$ satisfies the laws of $\mathcal{U}$, we call $\mathbf{A}$ a *$\mathcal{U}$-structure.*

The *terms* in a set X of variables are defined recursively:

(1) each $x \in X$ is a term,
(2) if $f \in \mathcal{F}$ is a k-ary function symbol and $t_1, \ldots, t_k$ are terms, then $f(t_1, \ldots, t_k)$ is a term,
(3) only symbol strings obtained by (1) and (2) are terms.

Only variables and function symbols are used to generate terms. The constants of the language are terms, even if X is empty. Two terms are equal if and only if they are identical.

An *atomic formula* is a symbol string $R(t_1, \ldots, t_k)$ where $R \in \mathcal{R}$ is a k-ary relation symbol and $t_1, \ldots, t_k$ are terms. Note that only relation symbols applied to terms are used to form atomic formulas, and these are the basic statements we can make about structures of the type. The case $k = 0$ is allowed. As usual, atomic formulas can be combined using quantifiers and logical connectives to make sentences.

For emphasis, *if equality is not in the language, then $s \approx t$ is not an atomic formula.*

The *term structure* $\mathbf{W}(X)$ generated by a set X is the set of all terms in X, with the obvious operations and all relations empty. The term structure is also known as the *word structure* or *absolutely free structure*. Again, the term structure need not be in our base quasivariety $\mathcal{U}$. For example, if $\mathcal{U}$ has an equality relation, then to obtain the free $\mathcal{U}$-structure we would need to add at least all relations $t \approx t$ with t a term. (More on free structures in Sect. 2.2.)

2.1.1 Substructures and Direct Products

In this general setting, nothing changes about substructures. Thus a *substructure* of $\mathbf{A} = \langle A, \mathcal{F}^{\mathbf{A}}, \mathcal{R}^{\mathbf{A}} \rangle$ is a structure $\mathbf{S} = \langle S, \mathcal{F}^{\mathbf{S}}, \mathcal{R}^{\mathbf{S}} \rangle$ where

- $S \subseteq A$,
- S is a subuniverse with respect to the operations, i.e., $\mathbf{s} \in S^k$ implies $f^{\mathbf{A}}(\mathbf{s}) \in S$ for each operation symbol, whence we can define $f^{\mathbf{S}} = f^{\mathbf{A}} \cap S^{k+1}$,
- $\mathbf{S}$ has the induced relations, i.e., for each relation symbol, $R^{\mathbf{S}} = R^{\mathbf{A}} \cap S^m$ where m is the arity of R.

If there are no constants (nullary functions) in the language, we allow $S = \varnothing$, though strictly speaking the empty set is only a *subuniverse*. For a class $\mathcal{X}$ of similar structures, $\mathbb{S}(\mathcal{X})$ denotes the class of all substructures of structures in $\mathcal{X}$.

It is convenient to have the notion of an embedding, which again is unchanged. An *embedding* of a structure $\mathbf{A}$ into $\mathbf{B}$ is a map $i : \mathbf{A} \to \mathbf{B}$ such that

- i is one-to-one,
- $if^{\mathbf{A}}(a_1, \dots, a_k) = f^{\mathbf{B}}(ia_1, \dots, ia_k)$ for each function symbol,
- $R^{\mathbf{B}}(ia_1, \dots, ia_m)$ holds iff $R^{\mathbf{A}}(a_1, \dots, a_m)$ holds, for each relation symbol.

Of course, inclusion is an embedding, and an embedding just says that $\mathbf{A}$ is isomorphic to the substructure $i(\mathbf{A}) \subseteq \mathbf{B}$. We write $\mathbf{A} \leq \mathbf{B}$ to indicate that $\mathbf{A}$ is a substructure of $\mathbf{B}$, or more generally, that there is an embedding of $\mathbf{A}$ into $\mathbf{B}$.

Likewise, direct products are unchanged. For a set of similar structures $\mathbf{A}_i$ $(i \in I)$, the *direct product* structure $\mathbf{P} = \prod_{i \in I} \mathbf{A}_i$ has the carrier set $\prod_{i \in I} A_i$, its operations are componentwise, and a relation $R^{\mathbf{P}}(\mathbf{a})$ holds in $\mathbf{P}$ if and only if $R^{\mathbf{A}_i}(\mathbf{a})_i$ holds for every $i \in I$.

Note that the empty product $\prod \varnothing$ is a 1-element structure with all possible relations of the type holding. The empty product is in every quasivariety. The class of all direct products of structures in $\mathcal{X}$ is denoted $\mathbb{P}(\mathcal{X})$.

2.1.2 Congruence Lattices from the Beginning

Our thinking on homomorphisms and congruences must be generalized. As this is somewhat unfamiliar territory, we go slowly.

Let $\mathbf{A}$ and $\mathbf{B}$ be structures of the same type. A *homomorphism* $h : \mathbf{A} \to \mathbf{B}$ is a map such that $h(f^{\mathbf{A}}) \subseteq f^{\mathbf{B}}$ for every function f and $h(R^{\mathbf{A}}) \subseteq R^{\mathbf{B}}$ for every relation R in the type. The *kernel* of a homomorphism, $\ker h$, is a function on the relation symbols of the type to subsets of A^k for the appropriate k (the arity of R) given by

$$\begin{aligned}\ker h(R) &= h^{-1}(R^{\mathbf{B}}) \\ &= \{\mathbf{a} \in A^k : h(\mathbf{a}) \in R^{\mathbf{B}}\}.\end{aligned}$$

The *homomorphism* requirement is that $f^{\mathbf{A}} \subseteq h^{-1}(f^{\mathbf{B}})$ and $R^{\mathbf{A}} \subseteq \ker h(R)$ for functions f and relations R.

If $h : \mathbf{A} \to \mathbf{B}$ is a homomorphism, then the *image* $h(\mathbf{A})$ is the set $h(A) = \{h(a) : a \in A\}$ endowed with the functions and relations of $\mathbf{B}$, restricted to $h(A)$. Clearly the image $h(\mathbf{A})$ is a substructure of $\mathbf{B}$. There is an analogue of the First Isomorphism Theorem (Theorem 2.18 below), but we do not yet have the terminology to express it properly.

Example 2.1 Suppose $\mathcal{U}$ consists of sets with a single equivalence relation E which is regarded as equality. Consider the structure $\mathbf{M} = \langle M, E^{\mathbf{M}} \rangle$ where $M = \{m, n, l\}$ and $E^{\mathbf{M}} = \{mm, nn, ll\}$, and the structure $\mathbf{T}$ with $T = \{x, y\}$ and $E^{\mathbf{T}} = \{xx, yy\}$. Let $g : \mathbf{M} \twoheadrightarrow \mathbf{T}$ be such that $g(m) = x$, $g(n) = x$, and $g(l) = y$. Then $\ker g(E) = \{mm, mn, nm, nn, ll\}$, and since $E^{\mathbf{M}} \subseteq \ker g(E)$, the map g is a homomorphism. (Here, xy denotes an ordered pair.)

Let A be any set, and let $\mathcal{R}$ be the set of all function symbols in a language. A *precongruence* on A is a map φ of $\mathcal{R}$ into $\bigcup_{k \geq 0} A^k$ such that $\varphi(R) \subseteq A^k$, where k is the arity of R, for each relation symbol R. Clearly precongruences just depend on the set, the relation symbols of the type, and their arities: they tell you only *what sort of things congruences will be.* But they are ordered pointwise by inclusion: $\varphi \leq \psi$ when $\varphi(R) \subseteq \psi(R)$ for every R. It is easy to see that with this order, the set of all precongruences on A forms a (distributive) algebraic lattice Precon A, and this gives us a starting point.

A *congruence* on a structure $\mathbf{A} = \langle A, \mathcal{F}, \mathcal{R} \rangle$ is a precongruence θ on A such that $R^{\mathbf{A}} \subseteq \theta(R)$ for every symbol R. Thus the kernel of a homomorphism is a congruence, and the intersection of any set of congruences on $\mathbf{A}$ is a congruence. There is a least congruence Δ which has $\Delta(R) = R^{\mathbf{A}}$ for all R, while the greatest congruence ∇ has $\nabla(R) = A^k$ (with the appropriate k) for all R. The set of all congruences on $\mathbf{A}$, ordered by inclusion, is just the filter $\uparrow \Delta$ in Precon A. Thus it is an algebraic (but still distributive) lattice, which we denote by Con $\mathbf{A}$.

Given a congruence θ on a structure $\mathbf{A}$, we can form the *factor structure* $\mathbf{A}/\theta = \langle A, F^{\mathbf{A}}, \theta \rangle$ by extending each $R^{\mathbf{A}}$ to $\theta(R)$. The identity map is a homomorphism from $\mathbf{A}$ to $\mathbf{A}/\theta$.

Example 2.2 Let us illustrate these notions with the structure $\mathbf{M}$ from Example 2.1. Recall that the language has a single binary relation E, and $\mathbf{M} = \langle M, E^{\mathbf{M}} \rangle$ with $M = \{m, n, l\}$ and $E^{\mathbf{M}} = \{mm, nn, ll\}$. There are 2^9 precongruences, as for a precongruence $\varphi(E)$ can be any subset of $M \times M$. Of those, 2^6 are congruences, for a congruence $\varphi(E)$ must contain $\{mm, nn, ll\}$. Of course, $\ker g$ from Example 2.1 is one of those congruences. The factor structure $\mathbf{M}/\ker g$ is the set M with the relations of $\ker g$, explicitly,

$$\mathbf{M}/\ker g = \langle \{m, n, l\}, \kappa \rangle \text{ where } \kappa(E) = \{mm, mn, nm, nn, ll\}.$$

But remember we are working in an underlying quasivariety $\mathcal{U}$, and not every factor structure will satisfy the quasi-identities of $\mathcal{U}$. For a quasivariety $\mathcal{K}$, we say that θ is a *$\mathcal{K}$-congruence* if $\mathbf{A}/\theta$ satisfies the laws of $\mathcal{K}$.

Example 2.3 In Example 2.1, the underlying quasivariety $\mathcal{U}$ was equivalence relations. There are only 5 equivalence relations on a 3-element set, so of the 64 congruence relations on $\mathbf{M}$, only 5 are $\mathcal{U}$-congruences. These are of course the ones we are interested in. You can form the factor structure $\mathbf{M}/\theta$ for the remaining 59 congruences (you cannot even do that for most precongruences), and $\theta(E)$ will be reflexive, but either not symmetric or not transitive, so not an equivalence relation.

Example 2.4 Suppose that for some reason we are working in the quasivariety $\mathcal{K}_2$ of groups satisfying $x^2 \approx 1 \rightarrow x \approx 1$. Any free group $\mathbf{F} = \mathrm{FG}(X)$ is in this quasivariety. Restricting our attention to normal group congruences, we see that a factor group $\mathbf{F}/\mathbf{N}$ is in $\mathcal{K}_2$ iff it has no element of order 2. So some group congruences are $\mathcal{K}_2$-congruences, and some are not.

This example can be extended to give the quasivariety of all aperiodic groups.

As the preceding examples illustrate, $\mathcal{U}$-congruences are our real concern. For a structure $\mathbf{A}$ in a quasivariety $\mathcal{U}$, let $\mathrm{Con}_{\mathcal{U}}\,\mathbf{A}$ denote the lattice of all $\mathcal{U}$-congruences on $\mathbf{A}$, with the inherited order for congruences: $\varphi \le \psi$ when $\varphi(R) \subseteq \psi(R)$ for every R. We want to show that $\mathrm{Con}_{\mathcal{U}}\,\mathbf{A}$ is an algebraic lattice by proving that is it an algebraic subset of $\mathrm{Con}\,\mathbf{A}$, invoking Theorem 1.10. Thus we need to show that the $\mathcal{U}$-congruences on $\mathbf{A}$ are closed under arbitrary intersections and nonempty directed joins.

Theorem 2.5 *For any $\mathcal{U}$-structure* $\mathbf{A}$, $\mathrm{Con}_{\mathcal{U}}\,\mathbf{A}$ *is an algebraic lattice.*

This will be a consequence of the next two main theorems, which interpret subdirect products and reduced products in $\mathrm{Con}_{\mathcal{U}}\,\mathbf{A}$ (Theorems 2.7 and 2.9, respectively).

Note that, as long as $\mathbf{A} \in \mathcal{U}$, we have that Δ and ∇ are $\mathcal{U}$-congruences. Also, if $h : \mathbf{A} \rightarrow \mathbf{B}$ is a homomorphism and $\mathbf{A}, \mathbf{B} \in \mathcal{U}$, then $\ker h$ is a $\mathcal{U}$-congruence. If $\mathcal{K} \le \mathcal{U}$, where $\mathcal{U}$ is our base quasivariety, then every $\mathcal{K}$-congruence is a $\mathcal{U}$-congruence.

An atomic formula can be written in the form

$$\alpha = R(t_1(x_1, \ldots, x_n), \ldots, t_k(x_1, \ldots, x_n)).$$

In the ensuing analysis, it will be useful to have a compact notation that emphasizes this form. Let us write $\alpha = R(t(\mathbf{x}))$, or when there may be more than one, $\alpha_j = R_j(t_j(\mathbf{x}))$.

Implicit in this discussion is the notion of when a structure satisfies a quasi-equation. Let $\beta_1, \ldots, \beta_n, \alpha$ be atomic formulas in a language $\mathcal{L}$ with variables from a set X, say $\beta_j = R_j(t_j(\mathbf{x}))$ and $\alpha = R_{n+1}(t_{n+1}(\mathbf{x}))$. We allow $n = 0$. Consider the quasi-equation

$$\beta_1 \,\&\, \ldots \,\&\, \beta_n \to \alpha, \tag{$\Diamond$}$$

i.e.,

$$R_1(t_1(\mathbf{x})) \,\&\, \ldots \,\&\, R_n(t_n(\mathbf{x})) \to R_{n+1}(t_{n+1}(\mathbf{x})) \,.$$

We say that a structure $\mathbf{A}$ in the type of $\mathcal{L}$ *satisfies* ($\Diamond$) if and only if the quasi-equation holds for every substitution $\sigma : X \to A$, i.e., either $R_j(t_j(\sigma(\mathbf{x})))$ is false for some $j \le n$ or $R_{n+1}(t_{n+1}(\sigma(\mathbf{x})))$ is true.

Lemma 2.6 *If φ_i $(i \in I)$ are $\mathcal{U}$-congruences on a structure $\mathbf{A}$, then $\bigcap_i \varphi_i$ is a $\mathcal{U}$-congruence.*

Proof Let $\Phi = \bigcap_i \varphi_i$, that is, $\Phi(R) = \bigcap_i \varphi_i(R)$ for each relation symbol R. Consider a quasi-equation ($\Diamond$) that holds in $\mathcal{U}$. For the hypotheses to hold in $\mathbf{A}/\Phi$ for a given substitution $\mathbf{x} \mapsto \mathbf{a}$ means that $t_j(\mathbf{a}_j) \in \Phi(R_j)$ for $1 \le j \le n$. (As usual, we allow $n = 0$.) By the definition of Φ, for every $i \in I$ we have $t_j(\mathbf{a}_j) \in \varphi_j(R_j)$. Since each φ_i is a $\mathcal{U}$-congruence that implies $t_{n+1}(\mathbf{a}) \in \varphi_i(R_{n+1})$ for every i, whence $t_{n+1}(\mathbf{a}) \in \Phi(R_{n+1})$. Thus Φ is a $\mathcal{U}$-congruence. □

Theorem 2.7 *Let $\mathbf{A}$ be a $\mathcal{U}$-structure and let φ_i $(i \in I)$ be in $\mathrm{Con}_{\mathcal{U}}\, \mathbf{A}$. Then the factor structure $\mathbf{A}/(\bigcap_i \varphi_i)$ is isomorphic to a substructure of the direct product $\prod_i \mathbf{A}/\varphi_i$.*

Although equimorphism (defined below) would suffice for our purposes, we actually get isomorphism, which is stronger, in Theorem 2.7.

Proof As observed earlier, $(\bigcap_i \varphi_i)(R) = \bigcap_i(\varphi_i(R))$ for any relation symbol R. We want to show that there is a one-to-one map g from $\mathbf{A}$ to a substructure $\widehat{\mathbf{A}}$ of the product $\prod_i \mathbf{A}/\varphi_i$ such that for any $a_1, \ldots, a_m \in A$ and m-ary relation symbol R, the tuple $(a_1, \ldots, a_m)$ is in $(\bigcap_i \varphi_i)(R)$ if and only if $(g(a_1), \ldots, g(a_m))$ is in $R^{\widehat{\mathbf{A}}}$.

Now in a direct product $\mathbf{B} = \prod_i \mathbf{B}_i$, for vectors $\mathbf{b}_1, \ldots, \mathbf{b}_m \in B$, the relation $R^{\mathbf{B}}(\mathbf{b}_1, \ldots, \mathbf{b}_m)$ holds, that is, $(\mathbf{b}_1, \ldots, \mathbf{b}_m) \in R^{\mathbf{B}}$ if and only if $(\mathbf{b}_{1i}, \ldots, \mathbf{b}_{mi}) \in R^{\mathbf{B}_i}$ for all i. For a substructure $\widehat{\mathbf{B}} \le \mathbf{B}$, the relation $R^{\widehat{\mathbf{B}}}$ is $R^{\mathbf{B}} \cap \widehat{B}^m$. In our case, $\mathbf{B}_i = \mathbf{A}/\varphi_i$, which has the carrier set A, the operations of $\mathbf{A}$, and relations $\varphi_i(R)$. Thus $\mathbf{B} = \prod_i \mathbf{A}/\varphi_i$ has the carrier set A^I, its operations componentwise, and $R^{\mathbf{B}}(\mathbf{a}_1, \ldots, \mathbf{a}_m)$ holding iff $(\mathbf{a}_{1i}, \ldots, \mathbf{a}_{mi}) \in \varphi_i(R)$ for all i.

Let $\widehat{\mathbf{A}}$ be the substructure of $\prod_i \mathbf{A}/\varphi_i$ consisting of all constant vectors $\mathbf{a}$ for $a \in A$. Naturally, the map $g : \mathbf{A} \to \widehat{\mathbf{A}}$ *via* $(g(a))_i = a$ for all i is a bijection. Moreover, $(g(a_1), \ldots, g(a_m)) \in R^{\widehat{\mathbf{A}}}$ holds iff $(a_1, \ldots, a_m) \in \varphi_i(R)$ for all i, or equivalently, $(a_1, \ldots, a_m) \in \bigcap_i(\varphi_i(R))$, as desired. □

Theorem 2.7 justifies the natural definition of subdirectly irreducible: a structure $\mathbf{A}$ in a quasivariety $\mathcal{K}$ is *$\mathcal{K}$-subdirectly irreducible* if the least congruence $\Delta^{\mathbf{A}}$ is completely meet irreducible in $\mathrm{Con}_{\mathcal{K}}\,\mathbf{A}$. This of course implies that there is a $\mathcal{K}$-congruence μ that covers $\Delta^{\mathbf{A}}$, with the property that $\varphi > \Delta^{\mathbf{A}}$ iff $\varphi \geq \mu$.

To show that $\mathrm{Con}_{\mathcal{K}}\,\mathbf{A}$ is closed under directed joins, we will use reduced products. These can be naturally translated into the general setting of a language that may not include equality.

Let $\mathbf{A}_i$ $(i \in I)$ be similar structures, and let F be a nonempty, proper filter on I. That is, F is a collection of subsets $X \subseteq I$ such that

- $I \in F$ and $\varnothing \notin F$,
- $X, Y \in F$ implies $X \cap Y \in F$,
- if $X \in F$ and $X \subseteq Z$, then $Z \in F$.

The congruence $\approx_F$ on $\prod_{i \in I} \mathbf{A}_i$ is defined so that, for each relation symbol R, we have $(\mathbf{a}_1, \ldots, \mathbf{a}_k) \in {\approx_F}(R)$ if and only if $\{i \in I : (a_{1i}, \ldots, a_{ki}) \in R^{\mathbf{A}_i}\} \in F$. The factor algebra $\prod \mathbf{A}_i/{\approx_F}$ is then the *reduced product* of the $\mathbf{A}_i$'s modulo F. When $F = \{I\}$ this becomes just the direct product $\prod_{i \in I} \mathbf{A}_i$. (If $\varnothing \in F$ were allowed, we would get the trivial structure, but that is already in every quasivariety as $\prod \varnothing$.)

The class of all reduced products of structures in $\mathcal{X}$ is denoted $\mathbb{R}(\mathcal{X})$. In the special case when F is an ultrafilter (maximal proper filter), the reduced product is known as an *ultraproduct*. The class operator for ultraproducts will be denoted $\mathbb{U}(\mathcal{X})$. (In the literature, $\mathbb{P}_u(\mathcal{X})$ is often used for ultraproducts.)

The next grand lemma records four fundamental facts about reduced products. It is based on Frayne, Morel, and Scott [43] and Łoś (see [29]). The history is included in [43, 44], and again we refer the reader to Chapter V of Burris and Sankappanavar [28].

Lemma 2.8 *Let $\mathbf{A}_i$ $(i \in I)$ be a set of similar structures in a quasivariety $\mathcal{U}$.*

(1) *Every nonempty filter on I is an intersection of ultrafilters.*
(2) *If $F = \bigcap_{j \in J} U_j$ in the lattice of filters on I, then in $\mathrm{Con}_{\mathcal{U}} \prod_i \mathbf{A}_i$ we have ${\approx_F} = \bigcap_{j \in J} {\approx_{U_j}}$ for the corresponding congruences.*
(3) *Hence if $F = \bigcap_{j \in J} U_j$ in the lattice of filters on I, then $\prod_i \mathbf{A}_i/{\approx_F}$ is a subdirect product of the structures $\prod_i \mathbf{A}_i/{\approx_{U_j}}$ over $j \in J$.*
(4) *If F is a filter, any quasi-equation satisfied by all $\mathbf{A}_i$ also holds in the reduced product $\prod_i \mathbf{A}_i/{\approx_F}$.*
(5) *If U is an ultrafilter, any first order sentence that holds in every $\mathbf{A}_i$ also holds in the ultraproduct $\prod_i \mathbf{A}_i/{\approx_U}$.*

Proof (1) Let F be a filter in any Boolean algebra $\mathbf{B}$ (in our case, the Boolean algebra $\mathbf{2}^I$ of subsets of I). Consider any $b \notin F$. By Zorn's Lemma, we can find a filter $M \supseteq F$ that is maximal with respect to the property $b \notin M$. We claim that for any $c \in B$, either c or its complement $\overline{c}$ is in M. For suppose neither c nor $\overline{c}$ is in M. By the maximality of M, b is in the filter generated by c and M, so that $b \geq c \wedge m_1$ for some $m_1 \in M$. Likewise $b \geq \overline{c} \wedge m_2$ for some $m_2 \in M$. Let $m = m_1 \wedge m_2$, which is again in M. Then $b \geq (c \wedge m) \vee (\overline{c} \wedge m) = (c \vee \overline{c}) \wedge m = 1 \wedge m = m$, whence

$b \in M$, a contradiction. We conclude that for every $c \in B$, either c or $\overline{c}$ is in M, which makes M an ultrafilter.

If we denote the ultrafilter thus obtained as M_b, then $M_b \supseteq F$ for every $b \notin F$, and thus $F = \bigcap_{b \notin F} M_b$, as desired.

(2) This is immediate from the definitions and of course applies to the intersections from (1).

(3) Apply Theorem 2.7 to (2).

(4) Consider a quasi-equation

$$R_1(t_1(\mathbf{x})) \,\&\, \ldots \,\&\, R_n(t_n(\mathbf{x})) \to R_{n+1}(t_{n+1}(\mathbf{x}))$$

that holds in every $\mathbf{A}_i$. For the hypotheses to hold in $\prod \mathbf{A}_i/{\approx_F}$ for a given substitution $\sigma : X \longrightarrow \prod A_i$ means that for each $j \leq n$, the set of indices $B_j = \{i \in I : t_j(\sigma(\mathbf{x})) \in R_j^{\mathbf{A}_i}\}$ is in the filter F. But then $B = B_1 \cap \cdots \cap B_n$ is in F. For an index $i \in B$, all the hypotheses $R^{\mathbf{A}_i}(t_j(\sigma(\mathbf{x})))$ hold, and since each $\mathbf{A}_i$ satisfies our quasi-equation, $R_{n+1}^{\mathbf{A}_i}(t_{n+1}(\sigma(\mathbf{x})))$ holds. Thus $B \subseteq \{i \in I : t_{n+1}(\sigma(\mathbf{x})) \in R_{n+1}^{\mathbf{A}_i}\}$, whence the latter set is in F. Since this is true for every substitution, our quasi-equation holds in the reduced product $\prod \mathbf{A}_i/{\approx_F}$.

(5) is the statement of the Łoś Theorem. Its proof is similar to that of (4), but requires a good deal more care with the details, and will be omitted. The crucial difference is that for a sentence $\varphi(\mathbf{x})$ and an instance $\mathbf{a} \in (\prod A_i)^k$, exactly one of $B = \{i \in I : \varphi(\mathbf{a}) \text{ holds}\}$ or its complement $\overline{B} = \{i \in I : \neg\varphi(\mathbf{a}) \text{ holds}\}$ is in the ultrafilter U. □

Now let us explain the connection between directed joins and reduced products, thereby completing the proof of Theorem 2.5 that $\mathrm{Con}_{\mathcal{U}}\,\mathbf{A}$ is an algebraic lattice.

Theorem 2.9 *Let* $\mathbf{A}$ *be a* $\mathcal{U}$*-structure and let* φ_i $(i \in I)$ *be an up-directed set of* $\mathcal{U}$*-congruences. Then there exist a filter* F *on* I *and a substructure* $\mathbf{A}^* \leq \mathbf{A}^I$ *such that* $\mathbf{A}/\bigvee_i \varphi_i$ *is isomorphic to* $\mathbf{A}^*/{\approx_F}$.

Proof Let $\mathbf{Q} = \mathbf{A}^I$, and let $\mathbf{A}^*$ be the substructure of $\mathbf{Q}$ consisting of all constant vectors $\mathbf{a}$ for $a \in A$. Again, $h : \mathbf{A} \longrightarrow \mathbf{A}^*$ *via* $(h(a))_i = a$ for all i is a bijection. To simplify notation, order the index set I by $i \leq j$ iff $\varphi_i \leq \varphi_j$.

Define a filter on the subsets of I by $F = \{X \subseteq I : \uparrow i_0 \subseteq X \text{ for some} i_0 \in I\}$. Routine arguments using directedness confirm that F is a filter. Thus we can define the reduced product: for $\mathbf{a}_1, \ldots, \mathbf{a}_m \in \mathbf{Q}$, set $(\mathbf{a}_1, \ldots, \mathbf{a}_m) \in {\approx_F}(R)$ if and only if $\{i \in I : (\mathbf{a}_{1i}, \ldots, \mathbf{a}_{mi}) \in \varphi_i(R)\}$ is in F.

Next, we claim that because the set of congruences φ_i $(i \in I)$ is up-directed, $\bigvee_i \varphi_i = \bigcup_i \varphi_i$ in $\mathrm{Con}_{\mathcal{U}}\,\mathbf{A}$. That is, if we put Φ to be the congruence such that $(a_1, \ldots, a_m) \in \Phi(R)$ if and only if $(a_1, \ldots, a_m) \in \varphi_{i_0}(R)$ for some i_0, then Φ is the least $\mathcal{U}$-congruence on $\mathbf{A}$ that contains every φ_i. Clearly Φ is the least congruence that contains every φ_i; it remains to show that Φ is a $\mathcal{U}$-congruence.

But it is apparent from the definitions that $(a_1, \ldots, a_m) \in \Phi(R)$ if and only if $(h(a_1), \ldots, h(a_m)) \in {\approx_F}(R)$. Thus $\mathbf{A}/\Phi$ is isomorphic to $\mathbf{A}^*/{\approx_F}$. Since every φ_i is a $\mathcal{U}$-congruence, each $\mathbf{A}/\varphi_i$ is in $\mathcal{U}$, so that $\prod_i \mathbf{A}/\varphi_i$ is in $\mathcal{U}$. Applying Lemma 2.8(3), we know that $\prod(\mathbf{A}/\varphi_i)/\approx_F$ is in $\mathcal{U}$. Then $\mathbf{A}^*/\approx_F$ is a substructure of that, so it is also in $\mathcal{U}$. Finally, $\mathbf{A}/\Phi$ is isomorphic to $\mathbf{A}^*/\approx_F$, so $\Phi = \bigvee_i \varphi_i$ is a $\mathcal{U}$-congruence. (Compare the proof of Theorem 2.47). □

In contrast, Viktor Gorbunov's treatment of directed joins in his book [52] used the *direct limit* of a system of structures and homomorphisms. See Section 1.2.5 of [52].

Now that we have the desired lattice $\mathrm{Con}_{\mathcal{U}}\,\mathbf{A}$, the Second Isomorphism Theorem generalizes straightforwardly.

Theorem 2.10 *If* $\mathbf{A} \in \mathcal{U}$ *and* $\theta \in \mathrm{Con}_{\mathcal{U}}\,\mathbf{A}$, *then* $\mathrm{Con}_{\mathcal{U}}(\mathbf{A}/\theta)$ *is isomorphic to the filter* $\uparrow\theta$ *in* $\mathrm{Con}_{\mathcal{U}}\,\mathbf{A}$.

In the rest of this section we use free structures $\mathbf{F}_{\mathcal{U}}(X)$. Section 2.2 gives a more thorough development of free structures, culminating in Theorem 2.27. For present purposes, we need only know that our base quasivariety $\mathcal{U}$ has free structures, and that they have the universal mapping property: for any structure $\mathbf{B} \in \mathcal{U}$, any map $h_0 : X \to \mathbf{B}$ extends to a homomorphism $h : \mathbf{F}_{\mathcal{U}}(X) \to \mathbf{B}$.

2.1.3 Equimorphism

We will use a slightly weaker notion to replace isomorphism. Structures $\mathbf{S}$ and $\mathbf{T}$ are said to be *equimorphic*, written $\mathbf{S} \equiv \mathbf{T}$, if there exist a nonempty set X and surjective homomorphisms $f : \mathbf{F}_{\mathcal{U}}(X) \twoheadrightarrow \mathbf{S}$ and $g : \mathbf{F}_{\mathcal{U}}(X) \twoheadrightarrow \mathbf{T}$ such that $\ker f = \ker g$. In this case, we say that $\mathbf{S} \equiv \mathbf{T}$ *using* X.

Lemma 2.11 *If* $\mathbf{S} \equiv \mathbf{T}$ *using* X, *and* $Y \supseteq X$, *then* $\mathbf{S} \equiv \mathbf{T}$ *using* Y.

Proof We are given $f : \mathbf{F}_{\mathcal{U}}(X) \twoheadrightarrow \mathbf{S}$ and $g : \mathbf{F}_{\mathcal{U}}(X) \twoheadrightarrow \mathbf{T}$ with $\ker f = \ker g$. Choose $x_0 \in X$, which by assumption is nonempty. Let $\hat{f} : \mathbf{F}_{\mathcal{U}}(Y) \twoheadrightarrow \mathbf{S}$ be the homomorphism such that $\hat{f}(x) = f(x)$ for $x \in X$, and $\hat{f}(y) = f(x_0)$ for $y \in Y \setminus X$. Similarly define $\hat{g} : \mathbf{F}_{\mathcal{U}}(Y) \twoheadrightarrow \mathbf{T}$ extending g, and check that $\ker \hat{f} = \ker \hat{g}$. □

Since every structure $\mathbf{M} \in \mathcal{U}$ is a homomorphic image of $\mathbf{F}_{\mathcal{U}}(X)$ for a sufficiently large set X, we have the following useful observation.

Lemma 2.12 *Structures* $\mathbf{S}$ *and* $\mathbf{T}$ *in* $\mathcal{U}$ *are equimorphic if and only if there exists a structure* $\mathbf{M} \in \mathcal{U}$ *and surjective homomorphisms* $f : \mathbf{M} \twoheadrightarrow \mathbf{S}$ *and* $g : \mathbf{M} \twoheadrightarrow \mathbf{T}$ *with* $\ker f = \ker g$.

Example 2.13 Return to the setting of Example 2.1, where $\mathcal{U}$ comprises sets with a single equivalence relation E. Consider $\mathbf{S} = \{a, b, c\}$ with the relation $E^{\mathbf{S}} = \{aa, ab, ba, bb, cc\}$ and $\mathbf{T} = \{x, y\}$ with $E^{\mathbf{T}} = \{xx, yy\}$. From above, Zeus can see that the former consists of 3 elements with a and b treated as equal, but from a perspective within the language we see only a 2-element set.

Formally, we can consider the structure $(\mathbf{M}, E^{\mathbf{M}})$ of Example 2.1 with the homomorphism $g : \mathbf{M} \twoheadrightarrow \mathbf{T}$ given there. Similarly, the map $f : \mathbf{M} \twoheadrightarrow \mathbf{S}$ such that $f(m) = a$, $f(n) = b$, $f(l) = c$ is a homomorphism with $\ker f(E) = \{mm, mn, nm, nn, ll\} = \ker g(E)$. Thus $\mathbf{S} \equiv \mathbf{T}$ by Lemma 2.12.

Example 2.14 As an extreme situation, suppose $\mathcal{U}$ has *no* relations, but possibly has functions. Then any two $\mathcal{U}$-algebras are equimorphic, since the kernel of any homomorphism is empty. But of course, in this language there are also no atomic formulas, and hence only one quasivariety/variety consisting of all the $\mathcal{U}$-structures.

Example 2.15 Looking ahead to Sect. 2.5.2, consider the variety $\mathcal{W}$ of structures in a language with one unary function f, one unary predicate P, and satisfying the law $P(f(x)) \leftrightarrow P(f^2(x))$. Figure 2.1 shows the $\mathcal{W}$-free structure on 1 generator $\mathbf{F}$, and a 2-element structure $\mathbf{G}$. These two are equimorphic, since the identity map $i : \mathbf{F} \to \mathbf{F}$ and the natural map $g : \mathbf{F} \to \mathbf{G}$ have the same kernel, with the predicate P assigned to every element $f^k(x)$ with $k > 0$, and not $P(x)$ for the generator.

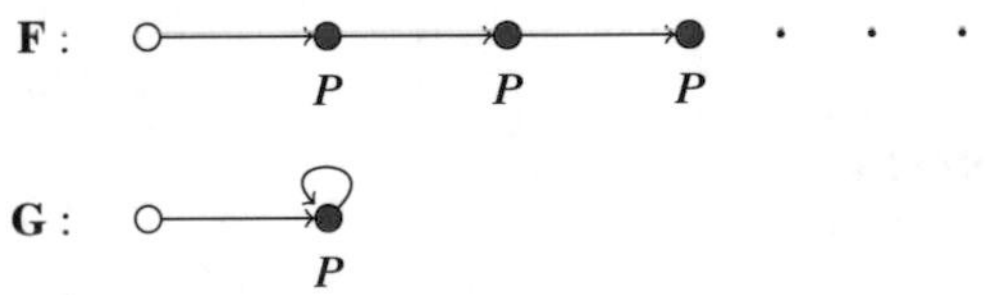

Fig. 2.1 The 1-generated $\mathcal{W}$-free structure $\mathbf{F}$ is equimorphic to the 2-element structure $\mathbf{G}$

Also note that any two $\mathcal{W}$-structures satisfying $\forall x\, P(x)$ are equimorphic. Indeed, this observation applies to the trivial quasivariety for any type of structure.

As usual, two structures are *isomorphic*, denoted $\mathbf{S} \cong \mathbf{T}$, if there exist mutually inverse homomorphisms $h : \mathbf{S} \to \mathbf{T}$ and $k : \mathbf{T} \to \mathbf{S}$. First we note that equimorphism really does generalize isomorphism.

Lemma 2.16 (1) *If* $\mathbf{S} \cong \mathbf{T}$, *then* $\mathbf{S} \equiv \mathbf{T}$.

(2) *If the language contains equality as a congruence relation, then equimorphism and isomorphism coincide, i.e.,* $\mathbf{S} \equiv \mathbf{T}$ *if and only if* $\mathbf{S} \cong \mathbf{T}$.

The structures in Example 2.13 show that, without equality, we may have $\mathbf{S} \equiv \mathbf{T}$ while $\mathbf{S} \not\cong \mathbf{T}$.

Theorem 2.17 *Equimorphism is an equivalence relation.*

Proof We must show transitivity. Assume $\mathbf{S} \equiv \mathbf{T} \equiv \mathbf{U}$ so that by Lemma 2.11, for sufficiently large X, there are homomorphisms

$$\begin{aligned} f &: \mathbf{F}_{\mathcal{U}}(X) \twoheadrightarrow \mathbf{S} \\ g_1 &: \mathbf{F}_{\mathcal{U}}(X) \twoheadrightarrow \mathbf{T} \\ g_2 &: \mathbf{F}_{\mathcal{U}}(X) \twoheadrightarrow \mathbf{T} \\ h &: \mathbf{F}_{\mathcal{U}}(X) \twoheadrightarrow \mathbf{U} \end{aligned}$$

with $\ker f = \ker g_1$ and $\ker g_2 = \ker h$.

Define an endomorphism $\varepsilon : \mathbf{F}_{\mathcal{U}}(X) \longrightarrow \mathbf{F}_{\mathcal{U}}(X)$ thusly: for each $x \in X$, choose $\varepsilon(x) \in g_2^{-1}(g_1(x))$, and extend the map to an endomorphism. This is possible because g_2 is surjective and $\mathbf{F}$ is free. This makes $g_2\varepsilon = g_1$. Similarly define η so that $g_1\eta = g_2$. Combining, $g_1 = g_2\varepsilon = g_1\eta\varepsilon$. Likewise $g_2 = g_2\varepsilon\eta$.

Let Y be the union of two disjoint copies of X, say $Y = X \dot{\cup} q(X)$ for a bijection q. Define homomorphisms $f' : \mathbf{F}_{\mathcal{U}}(Y) \twoheadrightarrow \mathbf{S}$ and $h' : \mathbf{F}_{\mathcal{U}}(Y) \twoheadrightarrow \mathbf{U}$ by extending the following maps for $y \in Y$:

$$f'(y) = \begin{cases} f(y) & \text{if } y \in X \\ f\eta(x) & \text{if } y = qx \in q(X) \end{cases}$$

$$h'(y) = \begin{cases} h\varepsilon(y) & \text{if } y \in X \\ h(x) & \text{if } y = qx \in q(X) \end{cases} .$$

We want to show that $\ker f' = \ker h'$.

Accordingly, let R be an m-ary relation symbol of the type, and let $w_1, \ldots, w_m \in F_{\mathcal{U}}(Y)$. Without loss of generality, we may assume that the variables of $w_1, \ldots, w_m$ are $x_1, \ldots, x_n, qx_1, \ldots, qx_n$, so that each $w_j = [w_j(x_1, \ldots, x_n, qx_1, \ldots, qx_n)]$. Note that

$$\begin{aligned} &f'[w_j(x_1, \ldots, x_n, qx_1, \ldots, qx_n)] \\ &\qquad = w_j(f'(x_1), \ldots, f'(x_n), f'(qx_1), \ldots, f'(qx_n)) \\ &\qquad = w_j(f(x_1), \ldots, f(x_n), f\eta(x_1), \ldots, f\eta(x_n)) \end{aligned}$$

and

$$\begin{aligned} &h'[w_j(x_1, \ldots, x_n, qx_1, \ldots, qx_n)] \\ &\qquad = w_j(h'(x_1), \ldots, h'(x_n), h'(qx_1), \ldots, h'(qx_n)) \\ &\qquad = w_j(h\varepsilon(x_1), \ldots, h\varepsilon(x_n), h(x_1), \ldots, h(x_n)) . \end{aligned}$$

Now we calculate:

$$(w_1 \ldots, w_m) \in \ker f'(R), \text{ i.e., } (f'w_1, \ldots, f'w_m) \in R^{\mathbf{S}},$$

iff

$$(w_1(f(x_1), \ldots, f(x_n), f\eta(x_1), \ldots, f\eta(x_n)), \ldots) \in R^{\mathbf{S}}$$

iff

$$\begin{aligned}
&(w_1(x_1,\ldots,x_n,\eta x_1,\ldots,\eta x_n),\ldots) \in \ker f(R)\\
&\qquad = \ker g_1(R) = \ker g_2\varepsilon(R)\\
&\qquad \text{iff}\\
&(w_1(\varepsilon x_1,\ldots,\varepsilon x_n,\varepsilon\eta x_1,\ldots,\varepsilon\eta x_n),\ldots) \in \ker g_2(R)\\
&\qquad \text{iff}\\
&(w_1(\varepsilon x_1,\ldots,\varepsilon x_n,x_1,\ldots,x_n),\ldots) \in \ker g_2(R)\\
&\qquad \text{since } g_2\varepsilon\eta = g_2,\\
&\qquad \text{iff}\\
&(w_1(\varepsilon x_1,\ldots,\varepsilon x_n,x_1,\ldots,x_n),\ldots) \in \ker h(R)\\
&\qquad \text{iff}\\
&(w_1(h\varepsilon x_1,\ldots,h\varepsilon x_n,hx_1,\ldots,hx_n),\ldots) \in R^{\mathbf{U}}\\
&\qquad \text{iff}\\
&(h'w_1,\ldots,h'w_m) \in R^{\mathbf{U}}, \text{ i.e., } (w_1,\ldots,w_m) \in \ker h'.
\end{aligned}$$

We conclude that $\ker f' = \ker h'$, whence $\mathbf{S} \equiv \mathbf{U}$, as desired. □

Let $\mathbb{E}\mathrm{q}$ denote the equimorphism class operator, i.e., $\mathbb{E}\mathrm{q}(\mathcal{X})$ is the collection of all structures $\mathbf{S}$ such that $\mathbf{S} \equiv \mathbf{T}$ for some $\mathbf{T} \in \mathcal{X}$.

Finally, we get to the analogue of the First Isomorphism Theorem.

Theorem 2.18 *If $h : \mathbf{S} \to \mathbf{T}$ is a homomorphism, then $h(\mathbf{S}) \equiv \mathbf{S}/\ker h$.*

This is an immediate consequence of Lemma 2.12, with $\mathbf{M} = \mathbf{S}$.

Example 2.19 For a simple example, consider structures in a language with one unary predicate P only. Let $\mathbf{S} = \{x, y\}$ with $P^{\mathbf{S}} = \varnothing$, and let $\mathbf{T} = \{z\}$ with $P^{\mathbf{T}} = \{z\}$, so that $\mathbf{T}$ is a singleton with Pz holding. The unique homomorphism $h : \mathbf{S} \to \mathbf{T}$ has $\ker h(P) = \{x, y\}$, so that $\mathbf{S}/\ker h$ is a 2-element set with Px and Py holding. The structures $h(\mathbf{S}) = \mathbf{T}$ and $\mathbf{S}/\ker h$ are equimorphic, but certainly not isomorphic.

Theorem 2.20 *If $\mathbf{S}$ and $\mathbf{T}$ are $\mathcal{U}$-structures with $\mathbf{S} \equiv \mathbf{T}$, then $\mathrm{Con}_{\mathcal{U}}\,\mathbf{S} \cong \mathrm{Con}_{\mathcal{U}}\,\mathbf{T}$.*

Proof Since $\mathbf{S} \equiv \mathbf{T}$, there exists a structure $\mathbf{M} \in \mathcal{U}$ and surjective homomorphisms $f : \mathbf{M} \twoheadrightarrow \mathbf{S}$ and $g : \mathbf{M} \twoheadrightarrow \mathbf{T}$ with $\ker f = \ker g$. By Theorem 2.10, using the principal filters in $\mathrm{Con}_{\mathcal{U}}\,\mathbf{M}$,

$$\mathrm{Con}_{\mathcal{U}}\,\mathbf{S} \cong {\uparrow}\ker f = {\uparrow}\ker g \cong \mathrm{Con}_{\mathcal{U}}\,\mathbf{T}.$$

Now, using equimorphism, we can formulate a proper version of subdirect products. Let us assemble all the ingredients that need to be combined.

- If $\mathbf{S} \equiv \mathbf{T}$, then $\mathrm{Con}_{\mathcal{U}}\,\mathbf{S} \cong \mathrm{Con}_{\mathcal{U}}\,\mathbf{T}$. (Theorem 2.20)
- If $\mathbf{S} \le \prod_i \mathbf{T}_i$, then the projection maps π_i are homomorphisms and $\bigcap_i \ker \pi_i = \Delta$.
- If $h : \mathbf{S} \to \mathbf{T}$, then $h(\mathbf{S}) \equiv \mathbf{S}/\ker h$. (Theorem 2.18)

- If $\Delta = \bigcap_i \varphi_i$ in $\mathrm{Con}_{\mathcal{U}}\,\mathbf{S}$, then $\mathbf{S}$ is isomorphic to a subalgebra of $\prod_i \mathbf{S}/\varphi_i$ with the projections surjective. (Theorem 2.7)

Thus we get:

Theorem 2.21 *A structure* $\mathbf{S}$ *is equimorphic to a subdirect product of structures* $\mathbf{T}_i$ $(i \in I)$ *if and only if there exist congruences* κ_i $(i \in I)$ *in* $\mathrm{Con}_{\mathcal{U}}\,\mathbf{S}$ *such that* $\mathbf{S}/\kappa_i \equiv \mathbf{T}_i$ *and* $\bigcap_i \kappa_i = \Delta$.

For the proof, just follow the bullets in order.

2.2 Freedom's Just Another Word

Since it will come up in the following discussion, we take this opportunity to remind the reader that when we speak of a *language with equality*, we mean that there is a binary relation symbol that, per the laws of the base quasivariety $\mathcal{U}$, acts as an equivalence relation and preserves operations and relations. The mere presence of the symbol $\approx$, or which symbol is used to denote equality, is irrelevant.

Let $\mathcal{K}$ be any class of structures of the same type, and let X be a set. A structure $\mathbf{S}$ is said to be *$\mathcal{K}$-freely generated by X* if

(1) $\mathbf{S} \in \mathcal{K}$,
(2) $\mathbf{S}$ is generated by X, and
(3) for any structure $\mathbf{B} \in \mathcal{K}$, any map $h_0 : X \to B$ extends to a homomorphism $h : \mathbf{S} \to \mathbf{B}$.

Of course, since X generates $\mathbf{S}$, there is at most one way to extend h_0 from X to S, *viz.*, recursively using the functions in the type. That is, $h : \mathbf{S} \to \mathbf{B}$ can only be an extension of $h_0 : X \to B$ if

$$h(x) = h_0(x) \quad \text{for } x \in X,$$
$$h(f^{\mathbf{S}}(u_1, \ldots, u_k)) = f^{\mathbf{B}}(h(u_1), \ldots, h(u_k))$$

for a basic function f and $u_1, \ldots, u_k \in S$, and this only works when the second clause is well-defined, that is, when $f^{\mathbf{S}}(u_1, \ldots, u_k) = f^{\mathbf{S}}(v_1, \ldots, v_k)$ implies $f^{\mathbf{B}}(h(u_1), \ldots, h(u_k)) = f^{\mathbf{B}}(h(v_1), \ldots, h(v_k))$.

First, let us show that a structure $\mathbf{S}$ satisfying (2) and (3) exists; for varieties and quasivarieties, $\mathbf{S}$ will also be in $\mathcal{K}$.

Theorem 2.22 *For any class* $\mathcal{K}$ *of structures and set* X, *there is a structure* $\mathbf{S}$ *satisfying properties* (2) *and* (3).

Proof Let $\mathbf{W}(X)$ be the term structure (absolutely free structure) on X in the type. Any map $h_0 : X \to B$ extends to a homomorphism $\hat{h} : \mathbf{W}(X) \to \mathbf{B}$ as usual:

$$\hat{h}(x) = h_0(x) \quad \text{for } x \in X,$$
$$\hat{h}(f(u_1, \ldots, u_k)) = f^{\mathbf{B}}(\hat{h}(u_1), \ldots, \hat{h}(u_k))$$

for a basic function f and terms $u_1, \ldots, u_k$. This is well-defined because terms are equal only if they are identical.

In Con $\mathbf{W}(X)$, let H be the set of all congruences θ such that $\theta = \ker h$ for some homomorphism $h : \mathbf{W}(X) \to \mathbf{B}$ and some $\mathbf{B} \in \mathcal{K}$. Let $\eta = \bigcap H$ and let $\mathbf{S} = \mathbf{W}(X)/\eta$.

Now consider any structure $\mathbf{B} \in \mathcal{K}$ and map $h_0 : X \to B$. As $\mathbf{W}(X)$ is absolutely free and generated by X, there is an extension $h_1 : \mathbf{W}(X) \to \mathbf{B}$. We want to factor h_1, as illustrated in Fig. 2.2.

There is the natural map $k : \mathbf{W}(X) \to \mathbf{W}(X)/\eta$, which is the identity on $W(X)$ and adds the relations of η.

By construction $\eta \le \ker h_1$, so there is a homomorphism $\ell : \mathbf{S} = \mathbf{W}(X)/\eta \to \mathbf{W}(X)/\ker h_1$, which again is the identity on $W(X)$ and just adds the relations of $\ker h_1(R)$ to $\eta(R)$, for every relation symbol R.

Then there is the natural map $m : \mathbf{W}(X)/\ker h_1 \to \mathbf{B}$ with $m(x) = h_1(x) = h_0(x)$ for all $x \in X$.

Now $m\ell k(x) = h_1(x)$ for every $x \in X$, and X generates $\mathbf{W}(X)$, so for every term t we have $m\ell k(t) = h_1(t)$. Moreover, by construction $\ker(m\ell k)(R) = \ker h_1(R)$ for every R, whence $\ker(m\ell k) = \ker h_1$. The desired map for property (3) is then $m\ell : \mathbf{S} \to \mathbf{B}$. □

Of course, Theorem 2.22 is just Birkhoff's standard construction of a $\mathcal{K}$-free algebra [24], adapted to a slightly more general setting. Lemma 2.6 shows that if $\mathcal{K}$ is a quasivariety, then $\mathbf{S} \in \mathcal{K}$. In that case, we denote the structure $\mathcal{K}$-freely generated by X as $\mathbf{F}_{\mathcal{K}}(X)$.

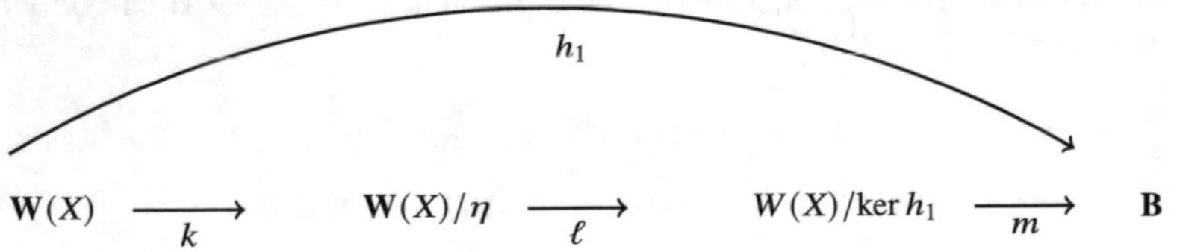

Fig. 2.2 Maps in the proof of Theorem 2.22. Here $h_1 = m\ell k$ with $k : x \mapsto x$, $\ell : x \mapsto x$, and $m : x \mapsto h_1(x) = h_0(x)$ for each $x \in X$. Thus $m\ell : \mathbf{S} = \mathbf{W}(X)/\eta \to \mathbf{B}$ with $m\ell(x) = h_0(x)$ for all $x \in X$

Aside The case $X = \varnothing$ can be handled thusly. If the language contains constants (nullary functions), then define $\mathbf{F}_{\mathcal{U}}(\varnothing)$ to be a structure generated by the constants, subject only to the laws of $\mathcal{U}$. If the language has no constants, the simplest convention is to leave $\mathbf{F}_{\mathcal{U}}(\varnothing)$ undefined.

At the other extreme, if $\mathcal{U}$ satisfies $x \approx y$, then the free structure $\mathbf{F}_{\mathcal{U}}(X)$ still contains at least $|X|$ elements, but the relation $s \approx t$ holds for all pairs of terms. Thus $\mathbf{F}_{\mathcal{U}}(X)$ and a 1-element structure are equimorphic, though not usually isomorphic.

Perhaps more relevant is Example 2.15, where we have a structure $\mathbf{G}$ that is equimorphic to a free structure $\mathbf{F}_{\mathcal{W}}(1)$, but $\mathbf{G}$ is not free; see Fig. 2.1.

So now we appear to have a problem. The characterizations of quasivarieties and varieties (Theorems 2.47 and 2.53 below) include closure under equimorphism, and $\mathcal{K}$-free structures should depend only on the variety generated by a class $\mathcal{K}$. But the universal mapping property (3) depends on equality "=" in the meta-language. We will spend the rest of this section unraveling the conundrum.

Suppose we are given

- a set X,
- structures $\mathbf{S}$ and $\mathbf{B}$ of the same type,
- an inclusion $X \subseteq S$ such that X generates $\mathbf{S}$,
- a map $h_0 : X \to \mathbf{B}$.

There is a natural (surjective) homomorphism $s : \mathbf{W}(X) \to \mathbf{S}$ from the term structure to $\mathbf{S}$, so that $\mathbf{S} \cong \mathbf{W}(X)/\ker s$ by Theorem 2.18.

Now h_0 extends to a homomorphism $\hat{h} : \mathbf{W}(X) \to \mathbf{B}$ as usual. In order for $\hat{h}$ to induce a well-defined map $h : S \to B$, it is required that for all pairs of terms u, v we have

$$s(u) = s(v) \implies \hat{h}(u) = \hat{h}(v). \qquad (\mu_1)$$

If (μ_1) holds, then the extended map h will be a homomorphism if

$$s(\mathbf{u}) \in R^{\mathbf{S}} \implies \hat{h}(\mathbf{u}) \in R^{\mathbf{B}} \qquad (\mu_2)$$

for each relation symbol in the type and vector of terms $\mathbf{u} = (u_1, \dots, u_k)$. If equality $\approx$ is in the type, then (μ_1) is a special case of (μ_2), and the condition can be written as

$$\ker s \leq \ker \hat{h}. \qquad (\mu_3)$$

If equality is not in the type, then (μ_1) must be regarded as a statement in the meta-language, and we need both (μ_1) and (μ_2) in order for h_0 to be extended to a homomorphism h from $\mathbf{S}$ to $\mathbf{B}$. However, note that (μ_1) holds trivially when $\mathbf{S}$ is the absolutely free structure $\mathbf{W}(X)$, since terms are equal only when they are identical. For all practical purposes, (μ_1) is trivial when $\mathbf{S}$ is the $\mathcal{U}$-free structure $\mathbf{F}_{\mathcal{U}}(X)$, since any two terms that evaluate the same in $\mathbf{F}_{\mathcal{U}}(X)$ will evaluate the same in any $\mathcal{U}$-structure.

Let us write $\mu(X, \mathbf{S}, \mathbf{B}, h_0)$ if X generates $\mathbf{S}$ and $h_0 : X \to \mathbf{B}$ is extendable to a homomorphism. The discussion above can be summarized thusly.

Theorem 2.23 *Let* $\mathbf{S}$ *and* $\mathbf{B}$ *be structures of the same type, with* $\mathbf{S}$ *generated by a set* X. *Let* $h_0 : X \to \mathbf{B}$. *Define the natural map extending* h_0 *to the term structure,* $\hat{h} : \mathbf{W}(X) \to \mathbf{B}$. *Then* $\mu(X, \mathbf{S}, \mathbf{B}, h_0)$ *holds if and only if conditions* (μ_1) *and* (μ_2) *are satisfied. If the language contains an equality relation, this is equivalent to* (μ_3).

As an example of the above considerations, the reason you cannot extend most maps from a free abelian group $\mathbf{F}_{\mathcal{A}}(X)$ to a nonabelian group $\mathbf{B}$ is that there will be terms u, v in the word algebra $\mathbf{W}(X)$ such that $(uv, vu) \in \approx^{\mathbf{F}_{\mathcal{A}}(X)}$ while $(uv, vu) \notin \approx^{\mathbf{B}}$.

Having dealt with specific cases of extension, let us return to classes and free structures. Without fear of confusion, for a class $\mathcal{K}$ let us write $\mu(X, \mathbf{S}, \mathcal{K})$ if $\mu(X, \mathbf{S}, \mathbf{B}, h_0)$ holds for all $\mathbf{B} \in \mathcal{K}$ and all maps $h_0 : X \to \mathbf{B}$.

Lemma 2.24 *If $\mathcal{K}$ is a class of structures for which $\mu(X, \mathbf{S}, \mathcal{K})$ holds, then $\mu(X, \mathbf{S}, \mathbb{HSP}(\mathcal{K}))$ holds.*

Indeed, it is easy to check that if $\mu(X, \mathbf{S}, \mathcal{K})$ holds and $\mathbf{C}$ is in $\mathbb{H}(\mathcal{K})$ or $\mathbb{S}(\mathcal{K})$ or $\mathbb{P}(\mathcal{K})$, then $\mu(X, \mathbf{S}, \mathbf{C}, h_0)$ holds for all h_0. Then combine the operators in the usual order.

Unfortunately, Lemma 2.24 does not extend to equimorphism. In languages without equality, this is a problem. So some adjustments will be required.

Example 2.25 Let $\mathcal{K}$ consist of 1-unary algebras in a language with no relations. As in Example 2.14, any two algebras in $\mathcal{K}$ are equimorphic. Take $\mathbf{S} = \mathbf{B}$ to be a 1-element algebra with $fx = x$, and take $\mathbf{C}$ to be the countable chain $x, fx, f^2x, \cdots$. Then $\mathbf{B} \equiv \mathbf{C}$ and $\mu(X, \mathbf{S}, \mathbf{B})$ holds with $X = \{x\}$. But there is no homomorphism $h : \mathbf{S} \to \mathbf{C}$, so $\mu(X, \mathbf{S}, \mathbf{C})$ fails.

One way to look at the example is this: $\mathbf{S}$ is $\mathcal{K}$-free for the class $\mathcal{K} = \mathbb{I}(\mathbf{B})$ of one-element structures (where $\mathbb{I}$ denotes isomorphism), but not free for the "variety" $\mathbb{E}\mathrm{q}\mathbb{HSP}(\mathbf{B})$ of structures that satisfy the same atomic formulas as $\mathbf{B}$, which in this case is all 1-unary structures (since there are no predicates, and hence no atomic formulas, in the language!).

Indeed, the construction from the proof of Theorem 2.22 does respect equimorphism. The key lemma is the following.

Lemma 2.26 *Let φ be a congruence on the word algebra $\mathbf{W}(X)$, and let $\mathbf{B}$, $\mathbf{C}$ be structures in the type. If $\mu(X, \mathbf{W}(X)/\varphi, \mathbf{B})$ holds and $\mathbf{B} \equiv \mathbf{C}$, then $\mu(X, \mathbf{W}(X)/\varphi, \mathbf{C})$ holds.*

Proof Let us dissect the hypotheses of the lemma. The assumption $\mu(X, \mathbf{W}(X)/\varphi, \mathbf{B})$ means that any map $k_0 : X \to B$ can be extended to a homomorphism $\hat{k} : \mathbf{W}(X) \to \mathbf{B}$ with $\ker \hat{k} \geq \varphi$, that is, $\mathbf{t} \in \varphi(R)$ implies $R^{\mathbf{B}}(\mathbf{t})$ for any relation symbol R. The condition $\mathbf{B} \equiv \mathbf{C}$ means that for some set Y there are surjective homomorphisms $f : \mathbf{W}(Y) \twoheadrightarrow \mathbf{B}$ and $g : \mathbf{W}(Y) \twoheadrightarrow \mathbf{C}$ with $\ker f = \ker g$, that is, $R^{\mathbf{B}}(f(\mathbf{t}))$ holds iff $R^{\mathbf{C}}(g(\mathbf{t}))$ holds.

Now let $h_0 : X \to C$. This map extends to a homomorphism $\hat{h} : \mathbf{W}(X) \to \mathbf{C}$. We want to show that $\varphi \leq \ker \hat{h}$. Since g is surjective, it has a right inverse map $r_g : C \to W(Y)$, which need not be a homomorphism. Let $k_0 : X \to B$ be defined by $k_0 = f r_g h_0$. By the first assumption, k_0 extends to a homomorphism $\hat{k} : \mathbf{W}(X) \to \mathbf{B}$. Moreover, $\hat{k}$ factors as $\hat{k} = f\hat{\ell}$, where $\hat{\ell}$ extends $r_g h_0 : W(X) \to W(Y)$. See Fig. 2.3.

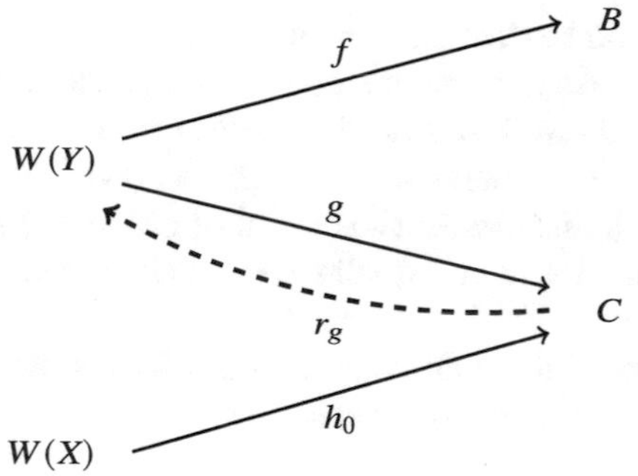

Fig. 2.3 Maps in the proof of Lemma 2.26

Consider a vector $\mathbf{t} \in \varphi(R)$. Since $\varphi \leq \ker \hat{k}$, we have $\hat{k}(\mathbf{t}) \in R^{\mathbf{B}}$, which can be written as $f\hat{\ell}(\mathbf{t}) \in R^{\mathbf{B}}$. Since $\ker f = \ker g$, this implies $g\hat{\ell}(\mathbf{t}) \in R^{\mathbf{C}}$. But on the set X of generators, $g\hat{\ell}(x) = gr_g h_0(x) = h_0(x)$, so that $g\hat{\ell} = \hat{h}$, and we conclude that $\hat{h}(\mathbf{t}) \in R^{\mathbf{C}}$. Thus $\varphi \leq \ker \hat{h}$, as desired. □

This shows that the standard construction of free structures, starting with the term structure and factoring by the intersection of all congruences whose factor is in $\mathbb{S}(\mathcal{K})$, works and produces a structure that is free for the variety generated by $\mathcal{K}$ (in view of Theorems 2.7 and 2.53); the $\mathbb{S}$ is required because homomorphisms into a structure in $\mathcal{K}$ need not be surjective.

Theorem 2.27 *Let $\mathcal{K}$ be a class of structures of the same type, and let X be a set. In the congruence lattice* $\operatorname{Con} \mathbf{W}(X)$*, let*

$$\Phi = \bigcap \{\varphi : \mathbf{W}(X)/\varphi \in \mathbb{S}(\mathcal{K})\}.$$

Then $\mathbf{W}(X)/\Phi$ is $\mathcal{V}$-freely generated by X, where $\mathcal{V} = \mathbb{E}q\mathbb{HSP}(\mathcal{K})$.

When $\mathcal{K}$ is a subquasivariety of our base quasivariety $\mathcal{U}$, then $\mathbf{W}(X)/\varphi \in \mathcal{K}$ implies $\mathbf{W}(X)/\varphi \in \mathcal{U}$, so that

$$\bigcap \{\varphi : \mathbf{W}(X)/\varphi \in \mathcal{U}\} \leq \bigcap \{\varphi : \mathbf{W}(X)/\varphi \in \mathcal{K}\}.$$

Thus the term structure $\mathbf{W}(X)$ in the construction could be replaced by the $\mathcal{U}$-free structure $\mathbf{F}_{\mathcal{U}}(X)$, and in practice this is what we often do.

2.3 Theories

Fix a type of structure and a countable set of variables X. Fix the base quasivariety $\mathcal{U}$. Using the function symbols, form the term structure $\mathbf{W}(X)$ and the free structure

$\mathbf{F}_{\mathcal{U}}(X)$. Theories should be defined using terms from $\mathbf{W}(X)$, but again it sometimes simplifies notation to use the free structure $\mathbf{F}_{\mathcal{U}}(X)$. Throughout this section, we denote the universe of $\mathbf{F}_{\mathcal{U}}(X)$ by F, and the universe of $\mathbf{W}(X)$ by W.

An *atomic formula* is a statement of the form $R(\mathbf{s})$ with R a relation symbol and $\mathbf{s} \in F^k$ (or W^k). An *implication* is a statement of the form $\beta_1 \ \& \cdots \& \ \beta_{n-1} \rightarrow \beta_n$ with each β_j an atomic formula. We allow $n = 1$, in which case the implication is just an atomic formula.

At this point, we need to recall what it means for a structure to satisfy a quasi-equation. Let the quasi-equation in question be

$$R_1(\mathbf{s}_1) \ \& \ \ldots \ \& \ R_{n-1}(\mathbf{s}_{n-1}) \rightarrow R_n(\mathbf{s}_n). \qquad (\beta)$$

A structure $\mathbf{T}$ *satisfies* (β) if for every homomorphism $h : \mathbf{W}(X) \rightarrow \mathbf{T}$ we have that $R_i^{\mathbf{T}}(h(\mathbf{s}_i))$ for all $i < n$ implies $R_n^{\mathbf{T}}(h(\mathbf{s}_n))$. That is, $\mathbf{T}$ satisfies (β) if $\mathbf{s}_n \in \ker h(R_n)$ whenever $\mathbf{s}_i \in \ker h(R_i)$ for all $i < n$. In terms of factor structures, this means that for φ a congruence, the structure $\mathbf{T}/\varphi$ satisfies (β) when for every homomorphism $h : \mathbf{W}(X) \rightarrow \mathbf{T}$, if $h(\mathbf{s}_i) \in \varphi(R_i)$ for all $i < n$, then $h(\mathbf{s}_n) \in \varphi(R_n)$.

An *endomorphism* of a structure $\mathbf{A}$ is a homomorphism $h : \mathbf{A} \rightarrow \mathbf{A}$. The endomorphisms of a structure form a monoid under composition, which we usually denote by $\mathcal{E}_{\mathbf{A}}$, or when the context is clear, just $\mathcal{E}$.

Note that the endomorphisms of a free structure $\mathbf{F} = \mathbf{F}_{\mathcal{U}}(X)$ are determined by their values on the generators. Thus an endomorphism of a free structure is just a homomorphism $\varepsilon : \mathbf{F} \rightarrow \mathbf{F}$ that extends a substitution $\varepsilon_0 : X \rightarrow F$ of terms for variables.

The endomorphism monoid $\mathcal{E}$ of a structure $\mathbf{A}$ acts on $\mathrm{Con}_{\mathcal{U}}\,\mathbf{A}$ in the following way. For $\varepsilon \in \mathcal{E}$, $\varphi \in \mathrm{Con}_{\mathcal{U}}\,\mathbf{A}$, and a relation symbol R,

$$\begin{aligned}(\varepsilon^*(\varphi))(R) &= \varepsilon^{-1}(\varphi(R)) \\ &= \{\mathbf{t} \in A^k : \varepsilon(\mathbf{t}) \in \varphi(R)\}.\end{aligned}$$

Thus $\mathbf{t} \in (\varepsilon^*(\varphi))(R)$ if and only if $\varepsilon(\mathbf{t}) \in \varphi(R)$.

Lemma 2.28 *Let* $\mathbf{A}$ *be a structure in* $\mathcal{U}$ *and* ε, η, κ *endomorphisms of* $\mathbf{A}$.

(1) *If* φ *is a* $\mathcal{U}$*-congruence, then* $\varepsilon^*\varphi$ *is a* $\mathcal{U}$*-congruence.*
(2) $(\eta\kappa)^* = \kappa^*\eta^*$.
(3) *The operator* ε^* *on* $\mathrm{Con}_{\mathcal{U}}\,\mathbf{A}$ *preserves the largest congruence* ∇, *arbitrary meets, and nonempty directed joins.*

Proof Parts (2) and (3) are straightforward calculations, but (1) requires a little care.

Consider a quasi-equation (β) : $R_1(\mathbf{s}_1) \ \& \ \ldots \ \& \ R_{n-1}(\mathbf{s}_{n-1}) \rightarrow R_n(\mathbf{s}_n)$ that holds in $\mathcal{U}$. Let $h : \mathbf{W}(X) \rightarrow \mathbf{A}$ where X contains all the variables occurring in (β). Given that $\mathbf{A}/\varphi$ satisfies (β), we want to show that $\mathbf{A}/\varepsilon^*\varphi$ satisfies (β).

So assume $h(\mathbf{s}_i) \in (\varepsilon^*\varphi)(R_i)$ for all $i < n$. Then $\varepsilon h(\mathbf{s}_i) \in \varphi(R_i)$ for $i < n$. Since $\varepsilon h : \mathbf{W}(X) \rightarrow \mathbf{A}$ is also a homomorphism and $\mathbf{A}/\varphi$ satisfies (β), we conclude that $\varepsilon h(\mathbf{s}_n) \in \varphi(R_n)$. Thence $h(\mathbf{s}_n) \in (\varepsilon^*\varphi)(R_n)$, as desired. □

Let $\mathcal{E}^* = \{\varepsilon^* : \varepsilon \in \mathcal{E}\}$. Given a groupoid $\mathbf{T} = \langle T, \cdot\rangle$, the *opposite* groupoid is $\mathbf{T}^{\mathrm{opp}} = \langle T, \star\rangle$ where $s \star t = t \cdot s$. Part (2) of Lemma 2.28 says that $\mathcal{E}^*$ is a homomorphic image of $\mathcal{E}^{\mathrm{opp}}$.

2.3.1 Atomic Theories

An *atomic theory*, relative to a quasivariety $\mathcal{U}$, is a $\mathcal{U}$-congruence θ on the free structure $\mathbf{F}_{\mathcal{U}}(X)$ such that $\mathbf{s} \in \theta(R)$ implies $\varepsilon(\mathbf{s}) \in \theta(R)$ for every relation symbol R, endomorphism ε, and $\mathbf{s} \in F^k$.

An atomic theory is the natural generalization of an *equational theory*. As a congruence, θ is a map on relation symbols, but we can also think of the corresponding set of atomic formulas:

$$\hat{\theta} = \{R(\hat{\mathbf{s}}) : \mathbf{s} \in \theta(R)\},$$

where $\hat{\mathbf{s}}$ denotes a tuple of terms that evaluate to $\mathbf{s}$ under the natural map $\mathbf{W}(X) \to \mathbf{F}_{\mathcal{U}}(X)$. Then we say that a structure $\mathbf{T}$ *satisfies* θ if $\mathbf{T}$ satisfies $R(\hat{\mathbf{s}})$ for every atomic formula in $\hat{\theta}$, i.e., for every homomorphism $h : \mathbf{W}(X) \to \mathbf{T}$ we have $h(\hat{\mathbf{s}}_n) \in R^{\mathbf{T}}$.

The next lemma characterizes atomic theories.

Lemma 2.29 *Let θ be a $\mathcal{U}$-congruence on $\mathbf{F}_{\mathcal{U}}(X)$. Let ε an endomorphism and $\mathbf{s} \in F^k$. The following are equivalent.*

(1) $\mathbf{s} \in \theta(R)$ *implies* $\varepsilon(\mathbf{s}) \in \theta(R)$ *for all relation symbols* R.
(2) $\varepsilon^*(\theta) \geq \theta$.

The proof just applies the definitions.

A $\mathcal{U}$-congruence θ on a structure $\mathbf{A}$ is *fully invariant* if $\varepsilon^*(\theta) \geq \theta$ for all endomorphisms ε of $\mathbf{A}$. As a consequence of Lemma 2.29, a fully invariant $\mathcal{U}$-congruence θ on a free structure $\mathbf{F}_{\mathcal{U}}(X)$ defines an atomic theory relative to $\mathcal{U}$.

Theorem 2.30 *Let* $\mathbf{A}$ *be a structure in* $\mathcal{U}$. *The set of fully invariant* $\mathcal{U}$-*congruences is a complete sublattice of* $\mathrm{Con}_{\mathcal{U}}\,\mathbf{A}$.

Proof Let θ_i $(i \in I)$ be fully invariant congruences, and let ε be an endomorphism. Then $\varepsilon^*(\bigwedge \theta_i) = \bigwedge \varepsilon^*(\theta_i) \geq \bigwedge \theta_i$. Also $\varepsilon^*(\bigvee \theta_i) \geq \varepsilon^*(\theta_i) \geq \theta_i$ for all i, whence $\varepsilon^*(\bigvee \theta_i) \geq \bigvee \theta_i$, as desired. □

Corollary 2.31 *Let θ be a fully invariant $\mathcal{U}$-congruence on $\mathbf{F} = \mathbf{F}_{\mathcal{U}}(X)$. If φ is a $\mathcal{U}$-congruence, then $\mathbf{F}/\varphi$ satisfies θ if and only if $\varphi \geq \theta$.*

Proof This is a situation where it is convenient to suppress the natural homomorphism $f : \mathbf{W}(X) \to \mathbf{F}_{\mathcal{U}}(X)$ and think of terms as elements of F. Certainly if $\varphi \ngeq \theta$, i.e., there exist $\mathbf{s} \in F^k$ and a relation R such that $\mathbf{s} \in \theta(R)$ but $\mathbf{s} \notin \varphi(R)$, so that $\mathbf{F}/\varphi$ does not satisfy θ as witnessed by the atomic formula $R(\mathbf{s})$.

Conversely, assume that $\varphi \geq \theta$, and let ε be an endomorphism of $\mathbf{F}$. For any pair with $\mathbf{s} \in \theta(R)$, we also have $\varepsilon(\mathbf{s}) \in \theta(R)$, since θ is fully invariant. Thence we get $\varepsilon(s) \in \varphi(R)$, so that $\mathbf{F}/\varphi$ satisfies $R(\mathbf{s})$ whenever $\mathbf{s} \in \theta(R)$, i.e., $\mathbf{F}/\varphi$ satisfies θ. □

A structure **S** is said to be *relatively freely generated by* X if it is $\mathcal{K}$-freely generated by X for *some* class $\mathcal{K}$. The next result shows in detail that a $\mathcal{U}$-congruence θ on **F** is fully invariant if and only if $\mathbf{F}/\theta$ is relatively free.

Theorem 2.32 *The following are equivalent for a* $\mathcal{U}$*-congruence* θ *on a free structure* $\mathbf{F} = \mathbf{F}_{\mathcal{U}}(X)$.

(1) $\mathbf{F}/\theta$ *is* $\mathcal{K}_1$*-freely generated by* X *for some class* $\mathcal{K}_1$.
(2) $\mathbf{F}/\theta$ *is* $\mathcal{K}_2$*-freely generated by* X *for* $\mathcal{K}_2 = \{\mathbf{F}/\theta\}$.
(3) $\mathbf{F}/\theta$ *is* $\mathcal{K}_3$*-freely generated by* X *for* $\mathcal{K}_3 = \mathbb{E}q\mathbb{HSP}(\mathbf{F}/\theta)$.
(4) $\mathbf{F}/\theta$ *is* $\mathcal{K}_4$*-freely generated by* X *for* $\mathcal{K}_4 = \{\mathbf{F}/\varphi : \varphi \geq \theta\}$.
(5) θ *is a fully invariant* $\mathcal{U}$*-congruence.*

Proof Lemmas 2.24 and 2.26 combine to show that (2) and (3) are equivalent, while (4) is just a lattice interpretation of (3) using Theorem 2.27. So it suffices to prove the equivalence of (1), (2), and (5). Of course, (2) implies (1).

Assuming (1), let us show that (5) holds. Since $\mathbf{F}/\theta$ is $\mathcal{K}_1$-freely generated by X, in particular, $\mathbf{F}/\theta \in \mathcal{K}_1$. An endomorphism ε of **F** is determined by its values on the generators, i.e., $\{\varepsilon x : x \in X\}$. The corresponding map on $\mathbf{F}/\theta$, *viz.*, $\varepsilon_0(x) = \varepsilon x/\theta$, can be extended to a homomorphism $\hat{\varepsilon} : \mathbf{F}/\theta \to \mathbf{F}/\theta$. It is of course the same map ε on the elements, but with the relations required to contain $\theta(R)$ for each R. Thus we have $\ker \hat{\varepsilon} \geq \theta$. In other words, $\mathbf{t} \in \theta(R)$ implies $\varepsilon(\mathbf{t}) \in \theta(R)$, whence $\theta \leq \varepsilon^*(\theta)$, making θ fully invariant.

This same calculation in reverse shows that (5) implies (2). □

Corollary 2.33 *The lattice* $\mathrm{L_v}(\mathcal{U})$ *of relative varieties of* $\mathcal{U}$ *is isomorphic to the lattice of* $\mathcal{E}^*$*-closed principal filters of* $\mathrm{Con}_{\mathcal{U}}\,\mathbf{F}_{\mathcal{U}}(\omega)$*, ordered by set containment. These are each of the form* $\uparrow\theta$ *for a fully invariant congruence.*

Larger congruences determine smaller filters, yielding the standard statement.

Corollary 2.34 *For a variety* $\mathcal{V}$*, the lattice* $\mathrm{L_v}(\mathcal{V})$ *is dually isomorphic to the lattice of fully invariant congruences of* $\mathbf{F}_{\mathcal{V}}(\omega)$*. Thus* $\mathrm{L_v}(\mathcal{V})$ *is a dually algebraic lattice.*

2.3.2 Implicational Theories

If $\alpha_1, \ldots, \alpha_{n-1}$ and β are atomic formulas and $A = \{\alpha_1, \ldots, \alpha_{n-1}\}$, then $A \to \beta$ denotes the implication $\alpha_1 \,\&\, \ldots \,\&\, \alpha_{n-1} \to \beta$. As usual, we allow A to be empty.

A collection $\mathcal{T}$ of implications is an *implicational theory* if it has the following properties.

(1) For each finite set B of atomic formulas, $B \to \beta$ is in $\mathcal{T}$ for every $\beta \in B$.
(2) If $A \to \beta$ is in $\mathcal{T}$ for all $\beta \in B$, and $B \to \gamma$ is in $\mathcal{T}$, then $A \to \gamma$ is in $\mathcal{T}$.
(3) If $A \to \beta$ is in $\mathcal{T}$ and ε is an endomorphism of $\mathbf{F}_{\mathcal{U}}(\omega)$, then $\varepsilon A \to \varepsilon\beta$ is in $\mathcal{T}$.

The set of all implicational theories containing a given theory $\mathcal{U}$ forms a complete lattice $\mathrm{ITh}(\mathcal{U})$. Theorem 2.36 below shows that $\mathrm{ITh}(\mathcal{U})$ is an algebraic lattice.

It is useful to introduce some shorthand notation. For a set of atomic formulas $R_1(\mathbf{s}_1), \ldots, R_n(\mathbf{s}_n)$, an endomorphism ε, and a congruence θ:

- β_i stands for $R_i(\mathbf{s}_i)$, that is, $\mathbf{s}_i \in R_i$.
- $\varepsilon\beta_i$ stands for $R_i(\varepsilon\mathbf{s}_i)$, that is, $\varepsilon\mathbf{s}_i \in R_i$.
- $\beta_i(\theta)$ means $\mathbf{s}_i \in \theta(R_i)$.
- $\varepsilon\beta_i(\theta)$ means $\varepsilon\mathbf{s}_i \in \theta(R_i)$.
- $\langle\beta_1, \ldots, \beta_n\rangle$ is short for $\beta_1 \,\&\, \cdots \,\&\, \beta_{n-1} \to \beta_n$.
- $\langle\beta_1, \ldots, \beta_n\rangle(\theta)$ is short for $\beta_1(\theta) \,\&\, \cdots \,\&\, \beta_{n-1}(\theta) \to \beta_n(\theta)$.

In this terminology, $\mathbf{F}_{\mathcal{U}}(X)/\theta$ satisfies $\langle\beta_1, \ldots, \beta_n\rangle$ if $\langle\varepsilon\beta_1, \ldots, \varepsilon\beta_n\rangle(\theta)$ holds for all endomorphisms ε. In an abuse of notation, we write $\theta \models \langle\beta_1, \ldots, \beta_n\rangle$ to denote this.

Let $\mathbf{A}$ be a $\mathcal{U}$-structure, R an n-ary relation symbol in the language of $\mathcal{U}$, and $\mathbf{a} \in A^n$. The *principal congruence* generated by $R(\mathbf{a})$ is the least $\mathcal{U}$-congruence containing $R(\mathbf{a})$, i.e.,

$$\mathrm{Cg}(R(\mathbf{a})) = \bigcap\{\theta \in \mathrm{Con}_{\mathcal{U}}\,\mathbf{A} : \mathbf{a} \in \theta(R)\}.$$

Clearly this generalizes the standard notion for pairs and equality. In particular, every $\mathcal{U}$-congruence is the join of the principal congruences it contains, and a $\mathcal{U}$-congruence is compact if and only if it is a join of finitely many principal congruences.

With the notation in place, we can state some elementary equivalences.

Corollary 2.35 *Let $\beta_1, \ldots, \beta_n$ be atomic formulas in the language of $\mathcal{U}$. Let θ be in $\mathrm{Con}_{\mathcal{U}}\,\mathbf{F}_{\mathcal{U}}(X)$, and let ε be an endomorphism of $\mathbf{F}_{\mathcal{U}}(X)$. The following are equivalent (to $\theta \models \langle\beta_1, \ldots, \beta_n\rangle$).*

(1) $\langle\varepsilon\beta_1, \ldots, \varepsilon\beta_n\rangle(\theta)$ *holds.*
(2) $\langle\beta_1, \ldots, \beta_n\rangle(\varepsilon^*\theta)$ *holds.*
(3) $\theta \geq \mathrm{Cg}(\varepsilon\beta_1) \vee \cdots \vee \mathrm{Cg}(\varepsilon\beta_{n-1})$ *implies* $\theta \geq \mathrm{Cg}(\varepsilon\beta_n)$.
(4) $\varepsilon^*\theta \geq \mathrm{Cg}(\beta_1) \vee \cdots \vee \mathrm{Cg}(\beta_{n-1})$ *implies* $\varepsilon^*\theta \geq \mathrm{Cg}(\beta_n)$.

Let $\mathcal{B}$ denote the set of all formal implications $\langle\beta_1, \ldots, \beta_n\rangle$. Define a relation $G \subseteq \mathrm{Con}_{\mathcal{U}}\,\mathbf{F}_{\mathcal{U}}(\omega) \times \mathcal{B}$ thusly:

$$(\theta, \langle\beta_1, \ldots, \beta_n\rangle) \in G \text{ if } \mathbf{F}/\theta \text{ satisfies } \langle\beta_1, \ldots, \beta_n\rangle$$
$$\text{i.e., } \langle\beta_1, \ldots, \beta_n\rangle(\varepsilon^*\theta) \text{ holds for all endomorphisms } \varepsilon.$$

In shorthand, $(\theta, \beta) \in G$ iff $\theta \models \beta$. The relation G determines a Galois correspondence, and the basic result just characterizes the closed sets.

Theorem 2.36 *The closed sets for the Galois correspondence determined by G are described as follows.*

(1) *The closed sets on the left are the $\mathcal{E}^*$-closed algebraic subsets of* $\mathrm{Con}_{\mathcal{U}}\,\mathbf{F}_{\mathcal{U}}(\omega)$.

(2) *The closed sets on the right are implicational theories containing the theory of* $\mathcal{U}$.

Hence $\mathrm{L_q}(\mathcal{U}) \cong \mathrm{S_p}(\mathrm{Con}_{\mathcal{U}}\,\mathbf{F}_{\mathcal{U}}(\omega), \mathcal{E}^*)$, *since both are dually isomorphic to the lattice of implicational theories* $\mathrm{ITh}(\mathcal{U})$.

Example 2.37 An example before the proof might help. Let the type of $\mathcal{U}$ be 1-unary algebras with a binary predicate $\approx$ for equality. Take the base quasivariety $\mathcal{U}$ to be the one stating that $\approx$ is a congruence relation, i.e., reflexive, symmetric, transitive, compatible with the operation. For convenience let X be a countably infinite set and $\mathbf{F} = \mathbf{F}_{\mathcal{U}}(X)$. Let us start with the identity $\beta_0 : fx \approx x$, a member the right-hand side $\mathcal{B}$.

A congruence $\theta \in \mathrm{Con}\,\mathbf{F}$ satisfies $fx \approx x$ if and only if $(u, fu) \in \theta$ for all $u \in F$. The congruence

$$\varphi = \{(u, f^k u) : u \in F,\ k \geq 0\}$$

is the least such congruence; it is $\mathcal{E}^*$-closed, and indeed fully invariant. The $\mathcal{E}^*$-closed algebraic subset referred to in part (1) of Theorem 2.36 is $\uparrow\varphi$.

The implicational theory determined by β_0 contains not only all equations $u \approx fu$ consisting of pairs from φ, but other implications that are consequences of these, such as $fx \approx fy \rightarrow x \approx y$. The collection of all these is the theory referred to in part (2).

Proof As always for Galois connections, to describe closed sets it suffices to consider the closures of singletons and take intersections.

Fix an implication $\beta = \langle \beta_1, \ldots, \beta_n \rangle$ and let

$$\lambda(\beta) = \{\theta \in \mathrm{Con}_{\mathcal{U}}\,\mathbf{F} : \theta \models \beta\}.$$

The claim is that $\lambda(\beta)$ is an $\mathcal{E}^*$-closed algebraic subset. Closure of $\lambda(\beta)$ under arbitrary intersections is clear, and closure under $\mathcal{E}^*$ is built into the definition.

For directed joins, we argue as in the proof of Theorem 2.9. Let θ_j $(j \in J)$ be a nonempty directed subset of $\lambda(\beta)$. Then $\Theta = \bigcup_{j \in J} \theta_j$ satisfies $\beta_1 \,\&\, \cdots \,\&\, \beta_{n-1} \rightarrow \beta_n$ (using the directedness) and is a $\mathcal{U}$-congruence (since $\mathcal{U}$ is a quasivariety, by the same argument). Hence $\Theta = \bigvee_{j \in J} \theta_j$ and $\Theta \in \lambda(\beta)$. Hence $\lambda(\beta)$ is closed under directed joins.

For the second part, fix $\theta \in \mathrm{Con}_{\mathcal{U}}\,\mathbf{F}$ and let

$$\rho(\theta) = \{\beta \in \mathcal{B} : \theta \models \beta\}.$$

The claim is that $\rho(\theta)$ is an implicational theory. Checking the definition, items (1) and (2) are straightforward, though the notation is unwieldy for (2). Item (3) uses Corollary 2.35.

Initially the maps ρ and λ are defined for elements. Extend the domains to subsets, so that for $A \subseteq \mathrm{Con}_{\mathcal{U}}\,\mathbf{F}$ and $B \subseteq \mathcal{B}$,

$$\rho(A) = \bigcap_{\theta \in A} \rho(\theta)$$
$$\lambda(B) = \bigwedge_{\beta \in B} \lambda(\beta).$$

First, let us prove that if A is an $\mathcal{E}^*$-closed algebraic subset of $\mathrm{Con}_{\mathcal{U}}\,\mathbf{F}$, then $\lambda\rho(A) = A$ (i.e., $\lambda\rho(A) \subseteq A$), and if $\mathcal{T}$ is an implicational theory, then $\rho\lambda(\mathcal{T}) = \mathcal{T}$ (i.e., $\rho\lambda(\mathcal{T}) \subseteq \mathcal{T}$).

Let A be an $\mathcal{E}^*$-closed algebraic subset of $\mathrm{Con}_{\mathcal{U}}\,\mathbf{F}$. Then

$$\rho(A) = \{\beta \in \mathcal{B} : \theta \models \beta \text{ for all } \theta \in A\}$$
$$\lambda\rho(A) = \{\varphi \in \mathrm{Con}_{\mathcal{U}}\,\mathbf{F} : \varphi \models \beta \text{ for all } \beta \in \rho(A)\}.$$

We want to show that $\lambda\rho(A) \subseteq A$, i.e., if $\varphi \models \beta$ whenever $\theta \models \beta$ for all $\theta \in A$, then $\varphi \in A$.

For any $\zeta \in \mathrm{Con}_{\mathcal{U}}\,\mathbf{F}$, define $\alpha(\zeta) = \bigwedge\{\theta \in A : \theta \geq \zeta\}$ which is the least member of A above ζ. Assume $\varphi \notin A$, so that $\alpha_0 = \alpha(\varphi) > \varphi$. As a matter of notation, for an atomic formula γ, let $\hat{\gamma}$ denote the principal congruence $\mathrm{Cg}(\gamma)$ in $\mathrm{Con}_{\mathcal{U}}\,\mathbf{F}$. Since $\alpha_0 > \varphi$, we can choose a principal $\mathcal{U}$-congruence $\hat{\gamma}_0$ that is below α_0 but not below φ.

Let K denote the set of compact $\mathcal{U}$-congruences below φ. For each $\kappa \in K$ we have $\kappa \leq \alpha(\kappa) \leq \alpha_0$. Moreover, $\alpha(K)$ is a directed subset of A. Thus $\varphi \leq \bigvee \alpha(K) \in A$, whence $\bigvee \alpha(K) = \alpha_0$. In particular, $\bigvee \alpha(K) \geq \hat{\gamma}_0$ which is compact, so $\alpha(\kappa_0) \geq \hat{\gamma}_0$ for some $\kappa_0 \in K$.

Write κ_0 as a join of principal $\mathcal{U}$-congruences $\kappa_0 = \hat{\gamma}_1 \vee \cdots \vee \hat{\gamma}_m$, and consider $\langle \gamma_1, \ldots, \gamma_m, \gamma_0 \rangle$. (Note: $\kappa_0 = \Delta$ with $m = 0$ is allowed.) We claim that every $\theta \in A$ satisfies $\langle \gamma_1, \ldots, \gamma_m, \gamma_0 \rangle$. For if $\theta \in A$, then for each endomorphism $\varepsilon^*\theta \in A$. So if $\varepsilon^*\theta \geq \hat{\gamma}_1 \vee \cdots \vee \hat{\gamma}_m$, then $\varepsilon^*\theta \geq \alpha(\kappa_0) \geq \hat{\gamma}_0$.

On the other hand, $\varphi \geq \hat{\gamma}_1 \vee \cdots \vee \hat{\gamma}_m$ but $\varphi \not\geq \hat{\gamma}_0$, so φ fails the implication. Thus $\varphi \notin \lambda\rho(A)$, as desired.

It remains to show that if $\mathcal{T}$ is an implicational theory, then $\rho\lambda(\mathcal{T}) \subseteq \mathcal{T}$. The next lemma is useful.

Lemma 2.38 *A set $\mathcal{T}$ of formal implications is an implicational theory with respect to $\mathcal{U}$ if and only if*

(0) the theory of $\mathcal{U}$ is contained in $\mathcal{T}$,
(1) $\langle \beta, \gamma \rangle \in \mathcal{T}$ whenever $\hat{\beta} \geq \hat{\gamma}$,
(2) if $\langle \beta_1, \ldots, \beta_m \rangle \in \mathcal{T}$ and $\hat{\gamma}_1 \vee \cdots \vee \hat{\gamma}_{n-1} \geq \hat{\beta}_1 \vee \cdots \vee \hat{\beta}_{m-1}$ and $\hat{\beta}_n \geq \hat{\gamma}_n$, then $\langle \gamma_1, \ldots \gamma_n \rangle \in \mathcal{T}$,
(3) if $\langle \beta_1, \ldots, \beta_m \rangle \in \mathcal{T}$ then $\langle \varepsilon\beta_1, \ldots, \varepsilon\beta_m \rangle \in \mathcal{T}$ for every endomorphism ε. □

Proof This is the $\mathcal{U}$-congruence interpretation of the definition. But see the discussion after the end of the proof of Theorem 2.36. □

Now, to complete the proof of Theorem 2.36, let $\mathcal{T}$ be an implicational theory. We want to show that $\rho\lambda(\mathcal{T}) \subseteq \mathcal{T}$, i.e., that $\gamma \notin \mathcal{T}$ implies $\gamma \notin \rho\lambda(\mathcal{T})$. Consider an implication $\gamma = \langle \gamma_1, \ldots, \gamma_n \rangle$ that is not in $\mathcal{T}$. We want to find $\theta \in \mathrm{Con}_{\mathcal{U}}\,\mathbf{F}$ such that $\theta \models \beta$ for all $\beta \in \mathcal{T}$ but $\gamma(\theta)$ does not hold, i.e., $\theta \geq \hat{\gamma}_1 \vee \cdots \vee \hat{\gamma}_{n-1}$ while $\theta \not\geq \hat{\gamma}_n$.

So we can ask: what *does* $\theta \geq \hat{\gamma}_1 \vee \cdots \vee \hat{\gamma}_{n-1}$ imply under $\mathcal{T}$?

Let $\overline{\delta} = \bigvee\{\hat{\delta} : \langle \gamma_1, \ldots, \gamma_{n-1}, \delta \rangle \in \mathcal{T}\}$. Note that $\overline{\delta} \geq \hat{\gamma}_1 \vee \cdots \vee \hat{\gamma}_{n-1}$ by Lemma 2.38(1). If perchance $\overline{\delta} \geq \hat{\gamma}_n$, then by the compactness of $\hat{\gamma}_n$ we have $\hat{\delta}_1 \vee \cdots \vee \hat{\delta}_k \geq \hat{\gamma}_n$ for some finite subset. But then $\langle \gamma_1, \ldots, \gamma_{n-1}, \delta_j \rangle \in \mathcal{T}$ for all j, and $\langle \delta_1, \ldots, \delta_k, \gamma_n \rangle \in \mathcal{T}$ (in fact it is in the theory of $\mathcal{U}$), whence $\langle \gamma_1, \ldots, \gamma_{n-1}, \gamma_n \rangle \in \mathcal{T}$ by Lemma 2.38(2), a contradiction. So $\overline{\delta} \not\geq \hat{\gamma}_n$, and $\theta = \overline{\delta}$ witnesses that $\gamma \notin \rho\lambda(\mathcal{T})$, since $\overline{\delta} \geq \hat{\gamma}_1 \vee \cdots \vee \hat{\gamma}_{n-1}$ while $\overline{\delta} \not\geq \hat{\gamma}_n$. □

Discussion The description of implicational theories in Lemma 2.38 is a variation on the Don relation from [20]. Let $\mathbf{K} = \langle K, \vee, \Delta, \widehat{\mathcal{E}} \rangle$ be the semilattice with operators of compact $\mathcal{U}$-congruences of the free structure $\mathbf{F}$. Because [20] uses the *join* semilattice with operators $\mathbf{K}$, the monoid of operators $\widehat{\mathcal{E}}$ on $\mathbf{K}$ induced by endomorphisms is not $\mathcal{E}^*$; the connection is explained in [64]. A binary relation τ on $\mathbf{K}$ is a *dongruence* if

(1) $\gamma \geq \delta$ implies $\tau(\gamma, \delta)$,
(2) $\tau(\gamma, \delta)$ implies $\tau(\gamma \vee \beta, \delta \vee \beta)$,
(3) $\tau(\gamma, \delta)$ implies $\tau(\hat{\varepsilon}\gamma, \hat{\varepsilon}\delta)$ for all ε.

Think $\tau(\gamma, \delta)$ means $\gamma \geq \delta \bmod \tau$. Congruences correspond to implicational theories. In [20] it is shown that $\mathrm{Don}_{\mathcal{U}}(\mathbf{F}, \widehat{\mathcal{E}}) \cong \mathrm{Con}_{\mathcal{U}}(\mathbf{F}, \widehat{\mathcal{E}})$, and in [64] it is shown that this is dually isomorphic to $\mathrm{S_p}(\mathcal{I}(\mathbf{K}), \mathcal{E}^*)$. So the preceding proof is a reprise.

We can now describe $\mathrm{L_q}(\mathcal{K})$ via implicational theories for several $\mathcal{K}$.

Example 2.39 For the quasivariety $\mathcal{K}_1$ with one unary relation A and no equality, the lattice of subquasivarieties is a 3-element chain: $\langle Ax \rangle$, $\langle Ax \to Ay \rangle$, and $\mathcal{K}_1$.

Example 2.40 For the quasivariety $\mathcal{R}_1$ with one unary relation A and $\approx$, the lattice of subquasivarieties is given in Fig. 2.4, from [63].

Example 2.41 For the quasivariety $\mathcal{K}_2$ with one unary relation A, a constant e such that Ae holds (as a law of $\mathcal{K}_2$), and no equality, the lattice of subquasivarieties is a 2-element chain: $\langle Ax \rangle$ and $\mathcal{K}_2$.

Example 2.42 Now add $\approx$ to the relations of $\mathcal{K}_2$ to form $\mathcal{E}_1$. There are four subquasivarieties, ordered as $\mathbf{2} \times \mathbf{2}$, which are $\langle x \approx e \rangle$, $\langle Ax \rangle$, $\langle Ax \to x \approx e \rangle$, and $\mathcal{E}_1$.

2.3.3 The Converse

Theorem 2.36 says that a lattice of implicational theories $\mathrm{ITh}(\mathcal{K})$ is dually isomorphic to a lattice $\mathrm{S_p}(\mathbf{L}, H)$ of H-closed algebraic subsets of an algebraic lattice with operators. A model-theoretic form of the same result is Theorem 2.54 below. It is reasonable then to ask about the converse: *Is every lattice* $\mathrm{S_p}(\mathbf{L}, H)$ *isomorphic to a lattice of implicational theories?* The answer is YES, by means of a construction in [91].

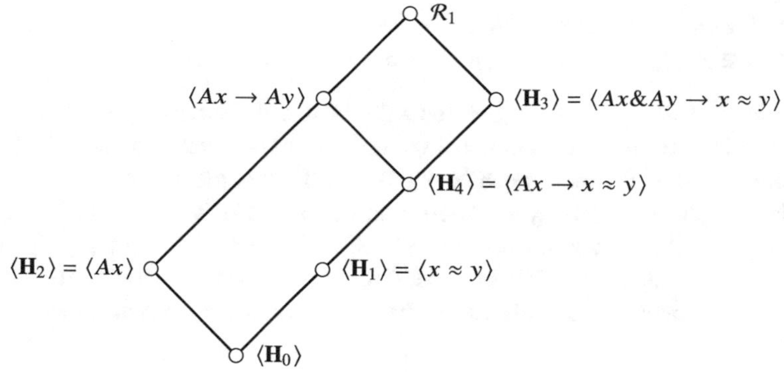

Fig. 2.4 Quasivariety lattice $\mathrm{L_q}(\mathcal{R}_1)$

Theorem 2.43 *Let* $\mathbf{L}$ *be an algebraic lattice with a monoid of operators* H*. Then there exists an implicational theory* $\mathcal{K}$ *(in a language without equality) such that* $\mathrm{S_p}(\mathbf{L}, H)$ *is dually isomorphic to* $\mathrm{ITh}(\mathcal{K})$*. Thus* $\mathrm{S_p}(\mathbf{L}, H) \cong \mathrm{L_q}(\mathcal{K})$.

While we will not repeat the proof here, we will use the construction from [91] in Chap. 7, which is perhaps the best way to understand it anyway. Theorem 2.60 and Example 2.61 show that the converse is not true if we insist upon equality as a relation.

2.4 Models

In order to talk about varieties and quasivarieties as classes of structures, we need to adapt the terminology to a general setting, so as to include languages that may not have equality.

Recall that for an implication $\beta = \langle \beta_1, \ldots, \beta_m\rangle$, we say that a structure $\mathbf{S}$ *satisfies* β, and write $\mathbf{S} \models \beta$, if for every homomorphism $h : \mathbf{F}_{\mathcal{U}}(\omega) \to \mathbf{S}$ we have $\ker h \in \lambda(\beta)$, where $\lambda(\beta)$ is as in the proof of Theorem 2.36:

$$\lambda(\beta) = \{\theta \in \mathrm{Con}_{\mathcal{U}}\, \mathbf{F} : \theta \models \beta\}.$$

That is, in the notation of page 47, $\mathbf{S} \models \beta$ if and only if $\langle \varepsilon\beta_1, \ldots, \varepsilon\beta_m\rangle(\ker h)$ holds for every endomorphism ε of $\mathbf{F}_{\mathcal{U}}(\omega)$ and homomorphism $h : \mathbf{F}_{\mathcal{U}}(\omega) \to \mathbf{S}$.

If $T \subseteq \mathcal{B}$ is a set of implications, we say that a structure $\mathbf{S}$ is a *model* of T if $\mathbf{S} \models \beta$ for every $\beta \in T$, i.e., $\ker h \in \bigcap_{\beta\in T} \lambda(\beta)$.

Lemma 2.44 *The following are equivalent.*

(1) $\ker h \in \lambda(T)$ *for every* $h : \mathbf{F}_{\mathcal{U}}(n) \to \mathbf{S}$ *and all finite* n.

(2) $\ker h \in \lambda(T)$ *for every* $h : \mathbf{F}_{\mathcal{U}}(\omega) \to \mathbf{S}$.
(3) $\ker h \in \lambda(T)$ *for every* $h : \mathbf{F}_{\mathcal{U}}(\kappa) \to \mathbf{S}$ *for* κ *arbitrarily large.*

Our task in this section is to extend the standard description of varieties and quasivarieties. Recall that $\mathbb{E}\mathrm{q}$ denotes the equimorphism class operator, i.e., $\mathbb{E}\mathrm{q}(\mathcal{X})$ is the collection of all structures $\mathbf{S}$ such that $\mathbf{S} \equiv \mathbf{T}$ for some $\mathbf{T} \in \mathcal{X}$.

In languages with equality we often omit the isomorphism operator $\mathbb{I}$, implicitly assuming that model classes are closed under isomorphism, writing, for example, $\mathbb{SPU}(\mathcal{X})$ rather than $\mathbb{ISPU}(\mathcal{X})$. Likewise, later on we may sometimes omit the equimorphism operator $\mathbb{E}\mathrm{q}$ and assume that model classes are closed under equimorphism.

Theorem 2.45 *If* $\mathbf{S} \equiv \mathbf{T}$ *and* $\beta \in \mathcal{B}$*, then* $\mathbf{S} \models \beta$ *if and only if* $\mathbf{T} \models \beta$.

Proof Assume that $\mathbf{S} \equiv \mathbf{T}$ with surjective homomorphisms $f : \mathbf{F}_{\mathcal{U}}(X) \twoheadrightarrow \mathbf{S}$ and $g : \mathbf{F}_{\mathcal{U}}(X) \twoheadrightarrow \mathbf{T}$ and $\ker f = \ker g$. Consider an implication $\beta = \langle \beta_1, \ldots, \beta_n \rangle$. By Lemma 2.11 we may assume that X contains all the variables in β. Now $\mathbf{S}$ satisfies β if and only if for every endomorphism ε of $\mathbf{F}_{\mathcal{U}}(X)$ and $\mathbf{s} \in F^k$, $\beta_i(\varepsilon\mathbf{s}) \in \ker f$ for all $i < n$ implies $\beta_n(\varepsilon\mathbf{s}) \in \ker f$ (cf. Corollary 2.35). Since $\ker f = \ker g$, $\mathbf{S}$ satisfies β if and only if $\mathbf{T}$ satisfies β. □

Recall that $\mathbb{H}$ denotes homomorphic images, $\mathbb{S}$ denotes substructures, $\mathbb{R}$ denotes reduced products (which includes direct products $\mathbb{P}$).

Lemma 2.46 *The class operators satisfy the following inclusions.*

(1) $\mathbb{H}\mathbb{E}\mathrm{q}(\mathcal{X}) \subseteq \mathbb{E}\mathrm{q}\mathbb{H}(\mathcal{X})$
(2) $\mathbb{S}\mathbb{E}\mathrm{q}(\mathcal{X}) \subseteq \mathbb{E}\mathrm{q}\mathbb{S}(\mathcal{X})$
(3) $\mathbb{R}\mathbb{E}\mathrm{q}(\mathcal{X}) \subseteq \mathbb{E}\mathrm{q}\mathbb{R}(\mathcal{X})$

The reverse inclusions may also be true, but these are the directions we want.

Proof In each case below we assume that $\mathbf{S} \equiv \mathbf{T}$ is witnessed by a structure $\mathbf{M}$ and surjective homomorphisms $f : \mathbf{M} \twoheadrightarrow \mathbf{S}$ and $g : \mathbf{M} \twoheadrightarrow \mathbf{T}$ with $\ker f = \ker g$.

(1) Assume that $\mathbf{A} \in \mathbb{H}\mathbb{E}\mathrm{q}(\mathcal{X})$, which means that there exist $\mathbf{S} \equiv \mathbf{T} \in \mathcal{X}$ and a surjective homomorphism $h : \mathbf{S} \twoheadrightarrow \mathbf{A}$. Then $hf : \mathbf{M} \twoheadrightarrow \mathbf{A}$, and since $\ker g \leq \ker hf$, there exists $k : \mathbf{T} \twoheadrightarrow \mathbf{A}$ such that $kg = hf$. So $\mathbf{A} \in \mathbb{H}(\mathcal{X}) \subseteq \mathbb{E}\mathrm{q}\mathbb{H}(\mathcal{X})$.

(2) Assume that $\mathbf{A} \in \mathbb{S}\mathbb{E}\mathrm{q}(\mathcal{X})$, which means that there exist $\mathbf{S} \equiv \mathbf{T} \in \mathcal{X}$ and $\mathbf{A} \leq \mathbf{S}$. Take $\mathbf{N} = f^{-1}(\mathbf{A})$, so that $\mathbf{N} \leq \mathbf{M}$, and let $\mathbf{B} = g(\mathbf{N})$. Then we have $\mathbf{B} \leq \mathbf{T}$, and $f|_{\mathbf{N}} : \mathbf{N} \twoheadrightarrow \mathbf{A}$ and $g|_{\mathbf{N}} : \mathbf{N} \twoheadrightarrow \mathbf{B}$ and $\ker f|_{\mathbf{N}} = \ker g|_{\mathbf{N}}$. Hence $\mathbf{A} \equiv \mathbf{B} \leq \mathbf{T}$, so that $\mathbf{A} \in \mathbb{E}\mathrm{q}\mathbb{S}(\mathcal{X})$.

(3) Assume that $\mathbf{A} \in \mathbb{R}\mathbb{E}\mathrm{q}(\mathcal{X})$, which means that there exist $\mathbf{S}_i \equiv \mathbf{T}_i \in \mathcal{X}$ for $i \in I$, and a filter G on I, such that $\mathbf{A} = \prod \mathbf{S}_i/{\approx_G}$. Let

$$\mathbf{B} = \prod \mathbf{T}_i/{\approx_G} \qquad \mathbf{M} = \prod \mathbf{M}_i$$
$$f = \prod f_i \qquad g = \prod g_i.$$

Let $\gamma : \prod \mathbf{S}_i \to \prod \mathbf{S}_i/{\approx_G}$ and $\gamma' : \prod \mathbf{T}_i \to \prod \mathbf{T}_i/{\approx_G}$ be the canonical maps. Then

- $\mathbf{B} \in \mathbb{R}(\mathcal{X})$,
- $\gamma f : \mathbf{M} \twoheadrightarrow \mathbf{A}$,
- $\gamma' g : \mathbf{M} \twoheadrightarrow \mathbf{B}$,
- $\ker \gamma f = \ker \gamma' g$,

whence $\mathbf{A} \equiv \mathbf{B} \in \mathbb{R}(\mathcal{X})$. We conclude that $\mathbf{A} \in \mathbb{E}\mathrm{q}\mathbb{R}(\mathcal{X})$, as desired. □

This brings us to the basic results on quasivarieties and varieties. We begin with quasivarieties. Recall from the proof of Theorem 2.36 that for a subset $A \subseteq \mathrm{Con}_{\mathcal{U}}\, \mathbf{F}$,

$$\rho(A) = \{\beta : \theta \models \beta \text{ for all } \theta \in A\}.$$

Also remember from Lemma 2.8(3) that every reduced product is a subdirect product of ultraproducts, while direct products are type of reduced product, which allows us to state the next theorem using either $\mathbb{E}\mathrm{q}\mathbb{S}\mathbb{R}$ or $\mathbb{E}\mathrm{q}\mathbb{S}\mathbb{P}\mathbb{U}$, equivalently.

Theorem 2.47 *Let $T \subseteq \mathcal{B}$ be a set of implications, and let $\mathcal{K}$ denote the class of all models of T. Then $\mathcal{K}$ is closed under equimorphism, substructures and reduced products, so that $\mathbb{E}\mathrm{q}\mathbb{S}\mathbb{R}(\mathcal{K}) = \mathbb{E}\mathrm{q}\mathbb{S}\mathbb{P}\mathbb{U}(\mathcal{K}) = \mathcal{K}$.*

Conversely, assume that $\mathcal{M}$ is a class of $\mathcal{U}$-structures which is closed under equimorphism, substructures, and reduced products (or equivalently, equimorphism, substructures, direct products, and ultraproducts). Let

$$A = \{\theta \in \mathrm{Con}_{\mathcal{U}}\, \mathbf{F}(\omega) : \mathbf{F}/\theta \in \mathcal{M}\}.$$

Then A is an $\mathcal{E}^*$-closed algebraic subset of $\mathrm{Con}_{\mathcal{U}}\, \mathbf{F}$*, and hence $\mathcal{M}$ is the set of all models of $\rho(A)$.*

Proof For the first part, let us check reduced products. Assume that $\mathbf{S}_i$ $(i \in I)$ are in $\mathcal{K}$, let G be a filter on I, and let $\beta \in T$. To say that $\beta_1(\varepsilon\mathbf{s})\ \&\ \cdots\ \&\ \beta_{n-1}(\varepsilon\mathbf{s})$ holds in $\mathbf{F}/{\approx_G}$ means that for all $j < n$ the set $H_j = \{i \in I : (\varepsilon\mathbf{s})_i \in R_j^{\mathbf{S}_i}\}$ is in G. Then $H = \bigcap_{j<n} H_j$ is in G and $(\varepsilon\mathbf{s})_i \in R_j^{\mathbf{S}_i}$ for all $j < n$ and $i \in H$. Since each $\mathbf{S}_i$ satisfies β, we conclude that $(\varepsilon\mathbf{s})_i \in R_n^{\mathbf{S}_i}$ for all $i \in H$, wherefore $\beta_n(\varepsilon\mathbf{s})$ holds in $\mathbf{F}/{\approx_G}$.

Closure under equimorphism is the second statement of Theorem 2.17. Substructure closure is clear, and that the trivial structure is in $\mathcal{K}$.

For the converse, the closure of A under intersections follows from the closure of $\mathcal{M}$ under subdirect products. (Note that the empty direct product is by definition in $\mathcal{M}$.)

Given a directed set $\{\varphi_i : i \in I\}$ of $\mathcal{U}$-congruences on A, let $\Phi = \bigcup_{i\in I} \varphi_i$. Thus for any relation symbol R in the type of $\mathcal{U}$ (including possibly $\approx$) and any $\mathbf{t} \in F^k$ of the appropriate arity, $\mathbf{t} \in \Phi(R)$ if and only if there exists $i_0 \in I$ such that $\mathbf{t} \in \varphi_{i_0}(R)$. Then Φ is a congruence (contains the relations $R^{\mathbf{F}}$); we must show that it is in A.

Consider $\prod_{i\in I} \mathbf{F}/\varphi_i$, recalling that $\mathbf{F}/\varphi_i$ is in $\mathcal{M}$ for all i. For a subset $X \subseteq I$, let X be in the set G if there exists i_0 such that $X \supseteq \{i \in I : \varphi_i \geq \varphi_{i_0}\}$. By the directedness of the set of congruences, the set G is a filter, so we can form the reduced product $\prod_{i\in I}(\mathbf{F}/\varphi_i)/{\approx_G}$. There is a natural embedding $h : \mathbf{F} \to \prod_{i\in I}(\mathbf{F}/\varphi_i)/{\approx_G}$, and $\ker h = \Phi$ by the remarks in the preceding paragraph. Thus F/Φ is a substructure of a reduced product of structures in $\mathcal{M}$. This implies $\mathbf{F}/\Phi$ is in $\mathcal{M}$, whence $\Phi \in A$, as desired. It remains to show that A is $\mathcal{E}^*$-closed. We are given that $\mathcal{M}$ is closed under $\mathbb{E}\mathrm{q}$, $\mathbb{S}$ and $\mathbb{R}$. We want to show that if $\theta \in A$, i.e., $\mathbf{F}/\theta \in \mathcal{M}$, and $\varepsilon \in \operatorname{End}\mathbf{F}$, then $\mathbf{F}/\varepsilon^*\theta \in \mathcal{M}$, so that $\varepsilon^*\theta \in A$. Recall that, for a relation symbol R, $\varepsilon^*\theta(R) = \{\mathbf{t} \in F^k : \varepsilon\mathbf{t} \in \theta(R)\}$, whence $\mathbf{t} \in \varepsilon^*\theta(R)$ iff $\varepsilon\mathbf{t} \in \theta(R)$. Claim:

(i) $\varepsilon(\mathbf{F})/\theta|_{\varepsilon(\mathbf{F})} \leq \mathbf{F}/\theta$,
(ii) $\mathbf{F}/\varepsilon^*\theta \equiv \varepsilon(\mathbf{F})/\theta|_{\varepsilon(\mathbf{F})}$.

The identity mapping provides the embedding for (i), along with the definition of the induced relations on a substructure.

For (ii), let $f : \mathbf{F} \twoheadrightarrow \mathbf{F}/\varepsilon^*\theta$ be the natural homomorphism (which is the identity on the elements but expands the relations). For any relation symbol, $\ker f(R) = \{\mathbf{t} \in F^k : \varepsilon\mathbf{t} \in \theta(R)\}$, by definition. On the other hand, we have the natural map $f : \mathbf{F} \twoheadrightarrow \mathbf{F}/\theta$ and $\varepsilon : \mathbf{F} \to \mathbf{F}$. Let $g = h\varepsilon$, so that $g : \mathbf{F} \twoheadrightarrow \varepsilon(\mathbf{F})/\theta|_{\varepsilon(\mathbf{F})}$ with exactly the same kernel as f. The fact that $\ker f = \ker g$ witnesses (ii). □

Corollary 2.48 *Let $\mathcal{X}_i$ $(i \in I)$ be collections of structures, all of the same similarity type. Then the quasivariety generated by $\bigcup_{i\in I} \mathcal{X}_i$ is $\mathbb{E}q\mathbb{S}\mathbb{R}(\bigcup_i \mathcal{X}_i) = \mathbb{E}q\mathbb{S}\mathbb{P}\mathbb{U}(\bigcup_i \mathcal{X}_i)$.*

In particular, if Q_i $(i \in I)$ are subquasivarieties of a quasivariety $\mathcal{K}$, then in $\mathrm{L_q}(\mathcal{K})$ *we have $\bigvee_{i\in I} Q_i = \mathbb{E}q\mathbb{S}\mathbb{R}(\bigcup_{i\in I} Q_i) = \mathbb{E}q\mathbb{S}\mathbb{P}\mathbb{U}(\bigcup_{i\in I} Q_i)$.*

Since a structure which is an ultraproduct of structures from the union of finitely many sets, $\mathcal{X}_1 \cup \cdots \cup \mathcal{X}_n$, is in fact in $\mathbb{U}(\mathcal{X}_j)$ for some j, the situation simplifies when there are only finitely many classes.

Corollary 2.49 *Let $\mathcal{X}_1, \ldots, \mathcal{X}_n$ be finitely many collections of structures, all of the same similarity type. Then the quasivariety generated by the union $\mathcal{X}_1 \cup \cdots \cup \mathcal{X}_n$ is given by $\mathbb{E}q\mathbb{S}\mathbb{P}(\mathcal{X}_1 \cup \cdots \cup \mathcal{X}_n)$.*

In particular, if $Q_1 \cup \cdots \cup Q_n$ are subquasivarieties of a quasivariety $\mathcal{K}$, then in $\mathrm{L_q}(\mathcal{K})$ *we have $Q_1 \vee \ldots \vee Q_n = \mathbb{E}q\mathbb{S}\mathbb{P}(Q_1 \cup \cdots \cup Q_n)$.*

We can describe finite joins of quasivarieties a bit more explicitly. Let $\mathcal{K}$ be a quasivariety and $\mathbf{S} \in \mathcal{K}$. For any subquasivariety $Q \leq \mathcal{K}$, there is a least congruence θ on $\mathbf{S}$ such that $\mathbf{S}/\theta \in Q$, *viz.*,

$$\rho_Q = \bigwedge\{\varphi \in \mathrm{Con}_{\mathcal{K}}\mathbf{S} : \mathbf{S}/\varphi \in Q\}.$$

In [63] this is called the *reflection congruence* of $\mathbf{S}$ into Q. Familiar examples include the commutator subgroup of a group or the least distributive congruence on a lattice, determined by its prime ideals. Note that every Q-congruence is above ρ_Q, but not every congruence above ρ_Q need be a Q-congruence unless Q is equational relative to $\mathcal{K}$.

Let us note some elementary properties of reflection congruences, as found in Lemma 2.17 of [63].

Lemma 2.50 *Let $\mathbf{S}$ be a structure in a quasivariety $\mathcal{K}$. Reflection congruences in* $\mathrm{Con}_{\mathcal{K}}\mathbf{S}$ *satisfy the following.*

(1) *If* $\mathcal{Q} \leq \mathcal{R}$, *then* $\rho_{\mathcal{Q}} \geq \rho_{\mathcal{R}}$.
(2) $\rho_{\mathcal{Q} \vee \mathcal{R}} = \rho_{\mathcal{Q}} \wedge \rho_{\mathcal{R}}$.
(3) *If* $\rho_{\mathcal{Q}_i} = \theta$ *for all* $i \in I$, *then* $\rho_{\bigwedge \mathcal{Q}_i} = \theta$.

The lemma in [63] assumes that $\mathbf{S}$ is finite and gives slightly stronger properties. Item (2) is a consequence of the next theorem.

Theorem 2.51 *The following are equivalent for a structure* $\mathbf{S}$ *in a quasivariety* $\mathcal{K}$ *and subquasivarieties* $\mathcal{Q}_1, \ldots, \mathcal{Q}_n \leq \mathcal{K}$.

(1) $\mathbf{S} \in \mathcal{Q}_1 \vee \ldots \vee \mathcal{Q}_n$.
(2) $\mathbf{S} \in \mathbb{E}q\mathbb{S}\mathbb{P}(\mathcal{Q}_1 \cup \cdots \cup \mathcal{Q}_n)$.
(3) *There exist congruences* $\kappa_1, \ldots, \kappa_n \in \mathrm{Con}_{\mathcal{K}} \mathbf{S}$ *such that* $\mathbf{S}/\kappa_i \in \mathcal{Q}_i$ *for all* i *and* $\kappa_1 \wedge \ldots \wedge \kappa_n = \Delta$.
(4) $\rho_{\mathcal{Q}_1} \wedge \ldots \wedge \rho_{\mathcal{Q}_n} = \Delta$ *in* $\mathrm{Con}_{\mathcal{K}} \mathbf{S}$.

Proof The equivalence of (1) and (2) is Corollary 2.49, while (3) implies (2) by Theorem 2.21. If (2) holds, then $\Delta = \bigwedge_{j \in J} \theta_j$ say, with each $\mathbf{S}/\theta_j$ in some $\mathcal{Q}_i$, again by Theorem 2.21. To get (3), put $\kappa_i = \bigwedge\{\theta_j : \mathbf{S}/\theta_j \in \mathcal{Q}_i\}$. Clearly (4) implies (3), and (3) implies (4) since $\rho_{\mathcal{Q}_i} \leq \kappa_i$ for all i. □

Example 2.52 To see that Theorem 2.51 does not extend to infinite joins, let p be a prime number and consider the cyclic groups $\mathbb{Z}_{p^k}$ $(k \geq 1)$ and the quasicyclic group $\mathbb{Z}_{p^\infty}$. For each $k \geq 1$, let $\mathcal{Q}_k = \mathbb{Q}(\mathbb{Z}_{p^k})$. It is not hard to see that $\mathbb{Z}_{p^\infty}$ is in $\bigvee_{k \geq 1} \mathcal{Q}_k$. But the subgroups of $\mathbb{Z}_{p^\infty}$ are all finite and cyclic, and every nontrivial factor group is isomorphic to the whole group $\mathbb{Z}_{p^\infty}$. Thus there are no proper congruences of $\mathbb{Z}_{p^\infty}$ with $\mathbb{Z}_{p^\infty}/\kappa \in \mathcal{Q}_k$ for any $k \geq 1$, so that the analogue of (3) fails.

Likewise, corresponding to Theorem 2.47, there is an extension for general structures of Birkhoff's theorem characterizing varieties.

Theorem 2.53 *Let* T *be a set of atomic formulas, and let* $\mathcal{K}$ *denote the class of all models of* T. *Then* $\mathcal{K}$ *is closed under equimorphism, homomorphic images, substructures, and direct products, so that* $\mathbb{E}q\mathbb{H}\mathbb{S}\mathbb{P}(\mathcal{K}) = \mathcal{K}$.

Conversely, assume that $\mathcal{M}$ *is a class of* $\mathcal{U}$*-structures which is closed under equimorphism, homomorphic images, substructures, and direct products. Let*

$$G = \{\theta \in \mathrm{Con}_{\mathcal{U}} \mathbf{F}(\omega) : \mathbf{F}/\theta \in \mathcal{M}\}.$$

Then G *is a principal filter* $\uparrow \varphi$ *of* $\mathrm{Con}_{\mathcal{U}} \mathbf{F}$ *with* φ *fully invariant. Hence* $\mathcal{M}$ *is the set of all models of* $\rho(G)$.

Proof The first statement is clear.

For the second statement, the closure of $\mathcal{M}$ under $\mathbb{S}$ and $\mathbb{P}$ implies that G has a least element, say φ. Closure of $\mathcal{M}$ under $\mathbb{H}$ ensures that $\uparrow \varphi \subseteq G$. The argument from the last part of the proof of Theorem 2.47 shows that φ is fully invariant. □

Putting all this together yields our basic result, generalizing Hoehnke [60] and Adaricheva and Nation [20].

Theorem 2.54 *Let $\mathcal{U}$ be a quasivariety, and let* **F** *be the countably generated free $\mathcal{U}$-structure, $\mathcal{E}$ its endomorphism monoid.*

(1) $\mathrm{L_v}(\mathcal{U})$ *is isomorphic to the lattice of fully invariant ($\mathcal{E}^*$-closed) principal filters of* $\mathrm{Con}_{\mathcal{U}}\,\mathbf{F}$*, ordered by set containment.*
(2) $\mathrm{L_q}(\mathcal{U})$ *is isomorphic to* $\mathrm{S_p}(\mathrm{Con}_{\mathcal{U}}\,\mathbf{F}, \mathcal{E}^*)$.

Corollary 2.55 *For any quasivariety $\mathcal{U}$, the lattices* $\mathrm{L_v}(\mathcal{U})$ *and* $\mathrm{L_q}(\mathcal{U})$ *are dually algebraic.*

For $\mathrm{L_v}(\mathcal{U})$ we appeal to Corollary 2.30, while $S_p(\mathbf{L}, H)$ is dually isomorphic to the congruence lattice of a semilattice with operators when **L** is algebraic ([64], but related facts were known in antiquity).

Note These are not the standard versions of the model theorem for implicational classes and atomic classes in languages without equality. The following version is Theorem 9 from Dellunde and Jansana [36]. It uses the expansion and contraction class operators, $\mathbb{E}\mathrm{x}$ and $\mathbb{C}$.

A homomorphism $h : \mathbf{M} \to \mathbf{N}$ is *strict* provided $\mathbf{a} \in R^{\mathbf{M}}$ iff $h(\mathbf{a}) \in R^{\mathbf{N}}$ for every $\mathbf{a} \in M^k$ and k-ary relation symbol R. For a class $\mathcal{K}$, let $\mathbf{M} \in \mathbb{E}\mathrm{x}(\mathcal{K})$ if there is a surjective strict homomorphism $h : \mathbf{M} \to \mathbf{N}$ with $\mathbf{N} \in \mathcal{K}$, and let $\mathbf{N} \in \mathbb{C}(\mathcal{K})$ if there is a surjective strict homomorphism $h : \mathbf{M} \to \mathbf{N}$ with $\mathbf{M} \in \mathcal{K}$.

Theorem 2.56 *A class $\mathcal{Q}$ is a quasivariety, i.e., the class of all models of an implicational theory, if and only if it is closed under the class operators* $\mathbb{E}\mathrm{x}$, $\mathbb{C}$, $\mathbb{S}$, $\mathbb{P}$, $\mathbb{U}$. *The smallest quasivariety containing a class $\mathcal{K}$ is* $\mathbb{E}\mathrm{x}\mathbb{CSPU}(\mathcal{K})$.

It is not hard to see that this characterization is equivalent to our formulation using equimorphism.

2.5 Two Quasivarieties Without Equality

It is useful to pause at this point to see a couple of examples of how one works in a quasivariety without equality, using the generalized notions of congruences and equimorphism.

2.5.1 Quasivarieties of Prequivalences

Chapter 2 discusses quasivarieties in languages without equality. One option is to have nothing by way of equality; another is to have a multitude of equivalence relations $\approx_k$ with no least one. What if we have just one relation, which is reflexive and symmetric, but not necessarily transitive?

A *prequivalence* structure is a structure $\mathbf{S} = \langle S, E \rangle$ with no functions, and one binary relation that is reflexive and symmetric. The class of all prequivalence structures forms a quasivariety $\mathcal{P}$. Figure 2.5 shows the lattice of subquasivarieties of $\mathcal{P}$. In $\mathcal{T}$ the relation $x\, E\, y$ holds for all pairs; in $\mathcal{E}$, the relation is transitive and hence an equivalence relation; while $\mathcal{P}$ is the quasivariety of all prequivalence structures. We will prove this in Theorem 2.57.

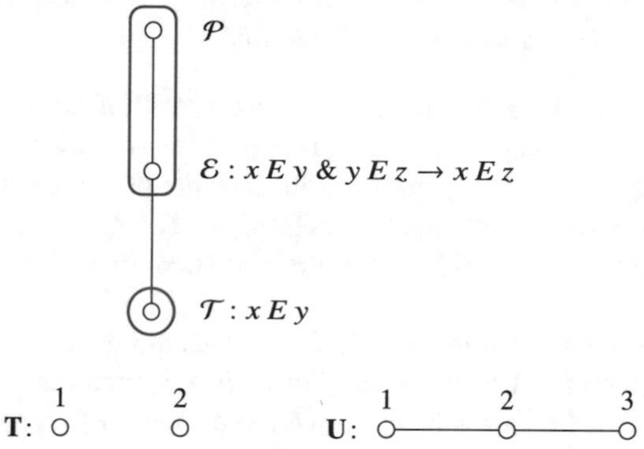

Fig. 2.5 The lattice $\mathrm{L_q}(\mathcal{P})$ of subquasivarieties of prequivalences (structures with a reflexive, symmetric, binary relation). The structures **T** and **U** are quasicritical, with $\mathcal{E} = \mathbb{Q}(\mathbf{T})$ and $\mathcal{P} = \mathbb{Q}(\mathbf{U})$. Red curves indicate the equational closure operator; see Sect. 3.1

Prequivalences provide an exercise in adapting the methods for locally finite varieties in the book [63] to the no-equality situation. One thing different here is that a structure could be equimorphic to a proper substructure.

Recall that a finite structure is *quasicritical* if it is not in the quasivariety generated by its proper substructures. More generally, a structure **T** is *quasicritical* if **T** is finitely generated and $\mathbb{Q}(\mathbf{T})$-subdirectly irreducible. The importance of quasicritical structures is easy to see. In general, a subquasivariety lattice $\mathrm{L_q}(\mathcal{U})$ is dually algebraic, which implies that every subquasivariety $\mathcal{Q} \leq \mathcal{U}$ is a join of completely join irreducible quasivarieties. When $\mathcal{U}$ is locally finite and of finite type, completely join irreducible subquasivarieties of $\mathcal{U}$ are precisely those generated by a single quasicritical structure. This is Theorem 7.4 below. For a more complete discussion of quasicriticality, see Sect. 7.4 and Appendix A.3, also Section 2.1 of [63].

Let us consider some small prequivalence structures.

- The 1-element structure **1** generates the quasivariety $x\, E\, y$.
- The 2-element structure $\{1, 2\}$ with $1\, E\, 2$ is equimorphic to **1**.
- The 2-element structure **T** on $\{1, 2\}$ with $E = \{(1, 1), (2, 2)\}$, which is $\mathbf{F}_{\mathcal{P}}(2)$, generates $\mathcal{E}$.
- The 3-element structure **U** on $\{1, 2, 3\}$ with $1\, E\, 2\, E\, 3$, but not $1\, E\, 3$, is the smallest structure that is a prequivalence but not transitive, and it generates $\mathcal{P}$.

That gives us the 1-element structure, plus the 2 quasicritical structures in Fig. 2.5. But how to see that there are no more quasicritical prequivalences?

To avoid abuse of notation, let $\Delta_{\mathbf{S}}$ denote the least congruence of a structure $\mathbf{S}$, so that $\Delta_{\mathbf{S}}(E) = E^{\mathbf{S}}$, the relation on $\mathbf{S}$. If $\mathbf{S}$ has congruences ψ_i with $\bigwedge \psi_i = \Delta_{\mathbf{S}}$ and each $\mathbf{S}/\psi_i$ equimorphic to a proper substructure of $\mathbf{S}$, then it is not quasicritical, by Theorem 2.7.

Theorem 2.57 *Every prequivalence structure is either equimorphic to* $\mathbf{1}$ *or a subdirect product of factors equimorphic to* $\mathbf{T}$ *and/or* $\mathbf{U}$.

Proof Let $\mathbf{S} = \langle S, E^{\mathbf{S}} \rangle$ be a prequivalence structure with $E^{\mathbf{S}} \neq S^2$.

For each connected component C of $\mathbf{S}$, let φ_C be the congruence such that $(x, y) \in \varphi_C(E)$ if and only if either x, y are both in C or neither is. Clearly $E^{\mathbf{S}} \subseteq \varphi_C(E)$, and if $\mathbf{S}$ has more than 1 component, then $\mathbf{S}/\varphi_C \equiv \mathbf{T}$, which is a substructure of $\mathbf{S}$. If every component of $\mathbf{S}$ has all its elements E-related, then $\bigcap \varphi_C = \Delta^{\mathbf{S}}$ and we are done.

So assume that $\mathbf{S}$ has a component C_0 that is not equimorphic to $\mathbf{1}$. Note that this makes $\mathbf{U}$ a substructure of $\mathbf{S}$. For each pair $a, b \in S$ such that a and b are in the same component but $(a, b) \notin E^{\mathbf{S}}$, let $\psi_{ab}(E) = S^2 \setminus \{(a, b), (b, a)\}$. Then $\mathbf{S}/\psi_{ab}$ is equimorphic to $\mathbf{U}$.

Finally, $\bigcap_C \varphi_C \cap \bigcap_{(a,b)\notin E^{\mathbf{S}}} \psi_{ab} = E^{\mathbf{S}}$. Thus every structure in $\mathcal{P}$ is a subdirect product of substructures that are equimorphic to $\mathbf{1}$, $\mathbf{T}$ or $\mathbf{U}$. □

As a sidelight, we can relate Theorem 2.57 to Theorem 2.54, that $\mathrm{L_q}(\mathcal{P})$ is isomorphic to the lattice of $\mathcal{E}^*$-closed subsets of Con $\mathbf{F}_{\mathcal{P}}(\omega)$, where $\mathcal{E}^*$ is the monoid of operators induced by the endomorphisms of $\mathbf{F}_{\mathcal{P}}(\omega)$. Since there are no operations or further axioms, any reflexive, symmetric relation on ω is a congruence, and any map $h : \omega \longrightarrow \omega$ is an endomorphism (since the relation is empty on the free structure). The algebraic subset corresponding to $\mathcal{E}$ consists of all equivalence relations on ω, as expected.

2.5.2 A Variety of Unary Structures

Let us consider the variety $\mathcal{W}$ of structures in a language with one unary function f, one unary predicate P, and satisfying the law $P(f(x)) \leftrightarrow P(f^2(x))$. Our goal is to find the lattice of subquasivarieties $\mathrm{L_q}(\mathcal{W})$. This exercise will illustrate how one uses congruences and equimorphism in languages without equality.

Now the laws of $\mathcal{W}$ do not - indeed cannot - say that $f^2(x) \approx f(x)$; see Example 2.15. But they *do* say that we cannot distinguish between $f(x)$ and $f^2(x)$, because the only available predicate either holds for both or neither. Thus it suffices to consider models where $f(x) = f^2(x)$, because every structure in $\mathcal{W}$ is equimorphic to one of those. Moreover, we easily see that any two structures satisfying $\forall x\, P(x)$ are equimorphic, since the kernel of the homomorphism from any sufficiently large free structure just adds the predicate P to every element.

For an example of how to compute a congruence lattices, consider the structure **R** of Fig. 2.6. The predicate $P(c)$ holds in **R**; congruences can add $P(a)$ or $P(b)$ or both.

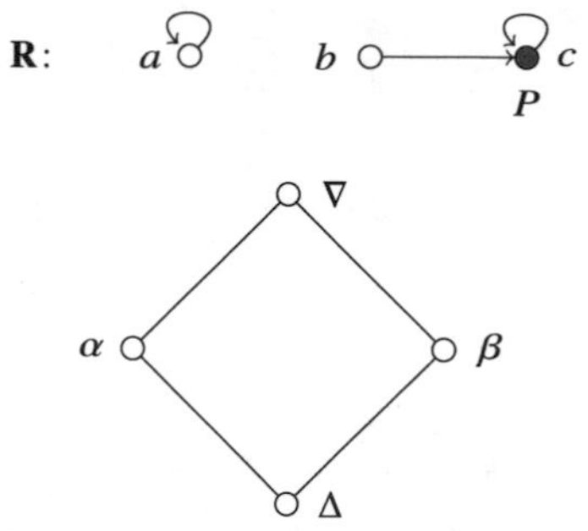

$\mathrm{Con}_{\mathcal{W}}\, \mathbf{R}$

Fig. 2.6 A small $\mathcal{W}$-structure and its congruence lattice. The congruences are given by $\Delta(P) = \{c\}, \alpha(P) = \{a, c\}, \beta(P) = \{b, c\}, \nabla(P) = \{a, b, c\}$

A second problem in determining quasicritical structures is that it is possible to have a proper substructure $\mathbf{S} < \mathbf{T}$ with $\mathbf{S}$ equimorphic to $\mathbf{T}$. In Fig. 2.7, the identity map on $\mathbf{T}$ and the map $h : \mathbf{T} \to \mathbf{S}$ with $h(a) = h(c) = c$ and $h(b) = b$ have the same kernel.

Fig. 2.7 $\mathbf{S} \equiv \mathbf{T}$ even though $\mathbf{S} < \mathbf{T}$. Red (solid) nodes indicate the predicate P

It is now a straightforward exercise to find all the quasicritical structures in $\mathcal{W}$. These are given in Fig. 2.8, along with the trivial structure **U** which is in every subquasivariety. In fact, all the quasicritical structures are simple: if we use the simplifying assumption that $f^2(x) = f(x)$, then in a quasicritical structure there is only one element without the predicate P, and the unique nontrivial congruence adds that element to P. The reader must check that for structures not in Fig. 2.8, either the structure is equimorphic to a proper substructure (like **T** in Fig. 2.7) or it is a subdirect product of images that are equimorphic to proper substructures (like **R** in Fig. 2.6).

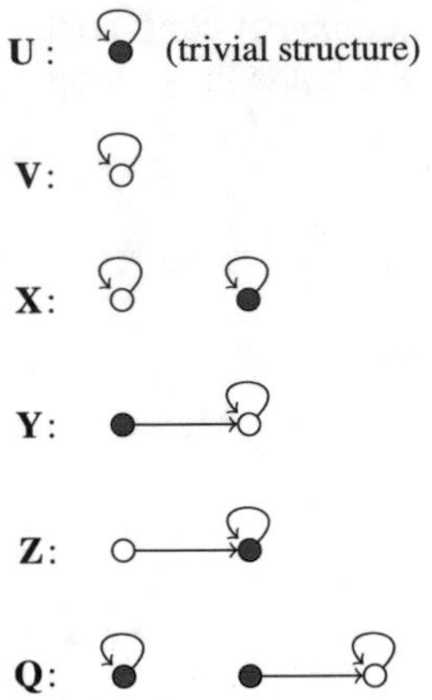

Fig. 2.8 Quasicritical $\mathcal{W}$-structures. Red (solid) nodes indicate the predicate P

Note that $\mathbf{V} \leq \mathbf{X}$ and $\mathbf{V} \leq \mathbf{Y} \leq \mathbf{Q}$. Also, the fact that the quasicritical algebras in $\mathcal{W}$ are simple makes the quasivarieties they generate join prime in $\mathrm{L_q}(\mathcal{W})$. Combining these observations allows us to draw $\mathrm{L_q}(\mathcal{W})$, as is done in Fig. 2.9.

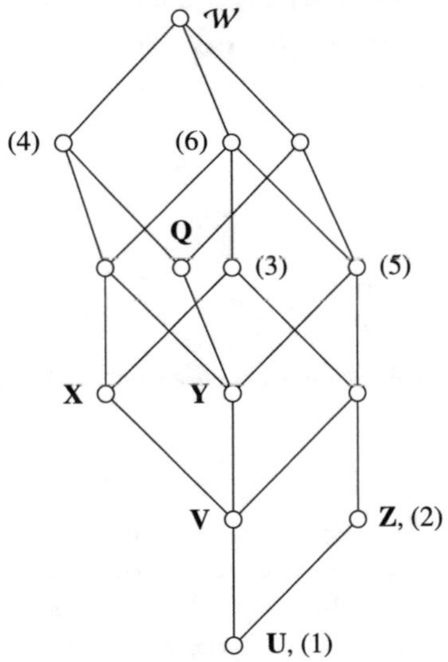

Fig. 2.9 $\mathrm{L_q}(\mathcal{W})$. Join irreducible subquasivarieties are labeled by a generating quasicritical structure, and meet irreducible ones by the numbered laws in the text

The easy way to identify the meet irreducible subquasivarieties of $\mathcal{W}$ is to write down all the quasi-equations in at most 2 variables, and then see which quasicritical structures satisfy which quasi-equations. Here are the nontrivial quasi-equations of $\mathcal{W}$:

(1) $P(x)$ (the least quasivariety, not meet irreducible)
(2) $P(f(x))$
(3) $P(x) \to P(f(x))$
(4) $P(f(x)) \to P(x)$
(5) $P(f(x)) \to P(f(y))$
(6) $P(x) \,\&\, P(f(y)) \to P(f(x))$

For example, only **U** and **Z** satisfy $P(f(x))$.

Note that the only "equational" subquasivarieties (i.e., relative subvarieties) of $\mathcal{W}$ are $\langle P(x)\rangle = \mathbb{Q}(\mathbf{U})$, $\langle P(f(x))\rangle = \mathbb{Q}(\mathbf{Z})$, and $\mathcal{W}$ itself.

2.6 Basic Properties of Subquasivariety Lattices

Now let us return to the general consequences of the analysis leading up to Theorem 2.54. This section and the next are modified from Sections A.3 and A.5 of the Appendix of [63].

The congruence lattice of a semilattice is meet semidistributive and algebraic [98]. So the fact that a subquasivariety lattice $\mathrm{L_q}(\mathcal{K})$ is dually isomorphic to the congruence lattice of a semilattice with operators immediately implies that $\mathrm{L_q}(\mathcal{K})$ is dually algebraic and join semidistributive, i.e., satisfies

$$x \vee y \approx x \vee z \to x \vee y \approx x \vee (y \wedge z). \qquad (\mathrm{SD}_\vee)$$

Indeed, since being dually algebraic implies lower continuity, $\mathrm{L_q}(\mathcal{K})$ satisfies the more general join semidistributive law

$$u \approx x \vee z_i \text{ for all } i \in I \text{ implies } u \approx x \vee \bigwedge_{i \in I} z_i .$$

A complete lattice **K** has the *Jónsson-Kiefer Property* if every element $a \in K$ is the join of elements that are (finitely) join prime in the ideal $\downarrow a$. This property holds in all finite join semidistributive lattices [69].

An easy argument shows that in any complete lattice, the Jónsson-Kiefer Property implies join semidistributivity. On the other hand, there is a dually algebraic, join semidistributive lattice that has no join-prime elements [19]. Thus the Jónsson-Kiefer Property is strictly stronger than join semidistributivity.

Gorbunov showed that the property holds in lattices of subquasivarieties [51].

Corollary 2.58 *For any quasivariety $\mathcal{Q}$, the lattice $\mathrm{L_q}(\mathcal{Q})$ has the Jónsson-Kiefer Property.*

This is a consequence of Theorem 2.54(2) and the next result.

Theorem 2.59 *If* **S** *is an algebraic lattice and H a monoid of operators on* **S***, then* $\mathrm{S_p}(\mathbf{S}, H)$ *has the Jónsson-Kiefer Property.*

The proof of Theorem 2.59 is given in the Appendix of the monograph [63]. For the sake of completeness we include it here.

Proof Consider a lattice $\mathrm{S_p}(\mathbf{S}, H)$ with **S** an algebraic lattice and H a set of operators. Recall from Theorem 1.15 that the join of finitely many algebraic subsets in the lattice $\mathrm{S_p}(\mathbf{S}, H)$ is given by

$$X_1 \vee \cdots \vee X_n = \{x_1 \wedge \cdots \wedge x_n : x_j \in X_j \text{ for } 1 \leq j \leq n\}.$$

Hence if $q \in S$ is meet irreducible, then $q \in X_1 \vee \cdots \vee X_n$ implies $q \in X_j$ for some j. Therefore, the H-closed algebraic subset generated by q is join prime in $\mathrm{S_p}(\mathbf{S}, H)$. But every element of an algebraic lattice is a meet of completely meet irreducible elements. Thus the join of all the join-prime algebraic sets in $\mathrm{S_p}(\mathbf{S}, H)$ is **S** itself. Finally, an algebraic subset of an algebraic lattice is itself an algebraic lattice. Relativizing the argument to a principal ideal $\downarrow A$ in $\mathrm{S_p}(\mathbf{S}, H)$ gives the statement in the property. □

For a more complete discussion of the Jónsson-Kiefer property, see [19]. In particular, a straightforward modification of the proof of Theorem 17 there yields that *if a dually algebraic lattice* **L** *admits a preclop with the property that* $\mathrm{CJI}(\mathbf{L}) \subseteq \tau(\mathbf{L})$*, then* **L** *has the Jónsson-Kiefer Property.* See also Sect. 5.3.

On the other hand, other notions of join semidistributivity that play a role (at 0) in lattices of subvarieties are automatic for subquasivariety lattices. Dual $*$-distributivity is a strengthening of the Zipper Condition (Erné [41]). Wehrung [114], refining arguments from [41] and Reinhold [101], showed that *a complete lattice is dually $*$-distributive if and only if it is lower continuous and join semidistributive,* which subquasivariety lattices are. The proofs are routine once you formulate the correct statement, and we refer the reader to the references for details.

Recall that a lattice is *atomic* if it has a least element 0 and for every $x > 0$ there exists an atom a such that $x \geq a \succ 0$. Gorbunov's argument that $\mathrm{L_q}(\mathcal{U})$ is atomic [52] uses the special quasivariety $x \approx y$ and need not apply to implicational theories without equality.

Theorem 2.60 *If the quasivariety $\mathcal{K}$ has an equality relation, then* $\mathrm{L_q}(\mathcal{K})$ *is atomic.*

Proof Let $\mathcal{T}$ denote the least subquasivariety of $\mathcal{K}$, and let $\mathcal{T} < \mathcal{Q} \leq \mathcal{K}$. If it happens that $\mathcal{Q} \cap \langle x \approx y \rangle = \mathcal{T}$ (including the case when $\mathcal{T} = \langle x \approx y \rangle$), then since $\langle x \approx y \rangle$ is finitely based and hence dually compact, there is a quasivariety $\mathcal{H}$ such that $\mathcal{T} \prec \mathcal{H} \leq \mathcal{Q}$. For we can use Zorn's Lemma to find a quasivariety that is minimal with respect to $\mathcal{H} \leq \mathcal{Q}$ but $\mathcal{H} \nleq \langle x \approx y \rangle$, and that will be an atom.

On the other hand, if $\mathcal{T} < \mathcal{Q} \cap \langle x \approx y \rangle$, then $\mathcal{Q}$ contains a 1-element structure **S** in which not all the relations of the language of $\mathcal{K}$ hold. Letting $\mathbf{T}_0$ denote the 1-element structure in which all the relations hold, we see that $\{\mathbf{S}, \mathbf{T}_0\}$ is a subquasivariety of $\mathcal{Q}$ (it is closed under $\mathbb{S}$, $\mathbb{P}$, $\mathbb{U}$) which is of course an atom of $\mathrm{L_q}(\mathcal{K})$. □

Example 2.61 We note that a lattice $\mathrm{S_p}(\mathbf{L}, H)$ of H-closed algebraic sets need not be atomic. Here is an example from [20]. Let $\mathbf{\Omega}$ be $\omega + 1$ with the single operator h such that $h(k) = k + 1$ for $k < \omega$ and $h(\omega) = \omega$. Then $\mathrm{S_p}(\mathbf{\Omega}, \{h\}) \cong (\omega + 1)^d$, which is not atomic. However, by Theorem 2.43, this lattice can be represented as $\mathrm{L_q}(\mathcal{U})$ for an implicational theory *without* equality.

2.7 Atomistic and Finite Distributive Subquasivariety Lattices

There are partial converses to Theorem 2.54 from [20, 91]. The first is just a restatement of Theorem 2.43.

Theorem 2.62 *For every algebraic lattice with operators* $\langle \mathbf{S}, H \rangle$, *there is a quasivariety* $\mathcal{U}$ without equality *such that* $\mathrm{S_p}(\mathbf{S}, H) \cong \mathrm{L_q}(\mathcal{U})$.

Theorem 2.63 *Let* $\mathbf{S}$ *be an algebraic lattice with* $1_{\mathbf{S}}$ *compact, and let* G *be a* group *of operators on* $\mathbf{S}$ *such that every element of* G *fixes both* $0_{\mathbf{S}}$ *and* $1_{\mathbf{S}}$. *Then there is a quasivariety* $\mathcal{Q}$ with equality *such that* $\mathrm{S_p}(\mathbf{S}, G) \cong \mathrm{L_q}(\mathcal{Q})$.

Recall that a complete lattice is *atomistic* if every element is a join of atoms. Note that any lattice $\mathrm{S_p}(\mathbf{S})$ (with no operators) is atomistic. Two basic results apply to atomistic subquasivariety lattices. The first is a classic result of Gorbunov and Tumanov [53].

Theorem 2.64 *The following are equivalent for a dually algebraic lattice* $\mathbf{L}$.

(1) $\mathbf{L} \cong \mathrm{S_p}(\mathbf{S})$ *for some algebraic lattice* $\mathbf{S}$.
(2) $\mathbf{L} \cong \mathrm{L_q}(\mathcal{K})$ *for some quasivariety* $\mathcal{K}$ *of one-element relational structures.*

Any quasivariety $\mathcal{K}$ of relational structures in a language with equality has a subquasivariety $\mathcal{N}$ consisting of its 1-element members, and Theorem 2.64 tells us about the principal ideal $\downarrow \mathcal{N}$ in $\mathrm{L_q}(\mathcal{K})$. This is reflected in property (I8) of the definition of an equaclosure operator; see Sect. 3.1. That being the case, it behooves us to sketch the proof.

Let $\mathcal{K}$ be a quasivariety of 1-element structures. This implies that equality is in the language and $\mathcal{K}$ satisfies $x \approx y$ (otherwise we could do an expansion). Moreover, we can ignore the operations of $\mathcal{K}$, and we may as well assume that the relations of $\mathcal{K}$ are unary. These relations serve as unary predicates which either hold or do not hold, in a structure in $\mathcal{K}$. Quasi-equations in these predicates determine the subquasivarieties of $\mathcal{K}$.

Both types of lattices under consideration are atomistic. In $\mathrm{S_p}(\mathbf{S})$, for any $x \neq 1$ in S, the set $\{1, x\}$ is an atom. In $\mathrm{L_q}(\mathcal{K})$ where $\mathcal{K}$ satisfies $x \approx y$, if we let $\mathbf{U}$ denote the 1-element structure in which *all* relations hold, then for any $\mathbf{A} \neq \mathbf{U}$ in $\mathcal{K}$, the set $\{\mathbf{U}, \mathbf{A}\}$ is closed under $\mathbb{SPU}$, and hence an atom. (Of course, these statements would not be true if $\mathbf{S}$ had operators, or if $\mathcal{K}$ had structures with more than 1 element.) Clearly, every element of the respective lattices is a join of the atoms just described, making them atomistic.

By Theorem 2.36, we have $\mathrm{L_q}(\mathcal{K}) \cong \mathrm{S_p}(\mathrm{Con}_{\mathcal{K}}\,\mathbf{F}_{\mathcal{K}}(\omega), \mathcal{E}^*)$. But now $\mathbf{F}_{\mathcal{K}}(\omega)$ is equimorphic to a 1-element structure, which has no proper endomorphisms! (It can, however, have lots of $\mathcal{K}$-congruences due to the relations in the language.) Thus $\mathrm{L_q}(\mathcal{K}) \cong \mathrm{S_p}(\mathrm{Con}_{\mathcal{K}}\,\mathbf{F}_{\mathcal{K}}(\omega))$, showing that (2) implies (1).

If that seems a bit sneaky, well, it is, and a direct proof indicates how we should go about proving the other direction. Starting with $\mathbf{L} = \mathrm{L_q}(\mathcal{K})$, we identify S as the set of "equational" subquasivarieties of $\mathcal{K}$, that is, the set of relative subvarieties of $\mathcal{K}$. (This is a bit of a misnomer in the context of 1-element structures. Each $\mathcal{V} \in S$ is determined by a collection of atomic formulas. To call them "atomic" subquasivarieties would be an even worse abuse of terminology.) Ordering S by reverse inclusion yields an algebraic lattice $\mathbf{S}$. There are two ways of looking at $\mathbf{S}$:

- $\mathbf{S}$ is the companion lattice $\gamma^d(\mathbf{L})$ from Sect. 3.3, and it is algebraic by Corollary 3.22
- $\mathbf{S}$ is the lattice $\mathrm{ETh}(\mathcal{K})$ of "equational theories" (really, atomic theories) contained in the theory of $\mathcal{K}$ and is algebraic for that reason.

Recall that $\mathrm{ETh}(\mathcal{K}) \cong \mathrm{FiCon}_{\mathcal{K}}\,\mathbf{F}_{\mathcal{K}}(\omega)$. Again because there are no proper endomorphisms, every $\mathcal{K}$-congruence is fully invariant, whence we have $\mathrm{FiCon}_{\mathcal{K}}\,\mathbf{F}_{\mathcal{K}}(\omega) = \mathrm{Con}_{\mathcal{K}}\,\mathbf{F}_{\mathcal{K}}(\omega)$. Therefore,

$$\mathrm{L_q}(\mathcal{K}) \cong \mathrm{S_p}(\mathrm{Con}_{\mathcal{K}}\,\mathbf{F}_{\mathcal{K}}(\omega)) \cong \mathrm{S_p}(\mathrm{FiCon}_{\mathcal{K}}\,\mathbf{F}_{\mathcal{K}}(\omega)) \cong \mathrm{S_p}(\mathrm{ETh}(\mathcal{K})) = \mathrm{S_p}(\mathbf{S}).$$

Moreover, an inclusion $\mathcal{U} \leq \bigvee \mathcal{V}_j$ in $\mathrm{L_q}(\mathcal{K})$, involving members of S, translates into $\mathcal{U} \geq \bigwedge \mathcal{V}_j$ in $\mathrm{ETh}(\mathcal{K})$, which in the simplest case says that the quasi-equation $\&_j\, V_j(x) \to U(x)$ holds in $\mathcal{K}$ for certain predicates. The work required to show directly that $\mathbf{L} \cong \mathrm{S_p}(\mathbf{S})$, without invoking Theorem 2.36, just duplicates the proof of the theorem for this special case.

But in view of the preceding analysis, starting with an algebraic lattice $\mathbf{S}$, we can see how to construct a quasivariety $\mathcal{K}$ of 1-element structures such that $\mathrm{L_q}(\mathcal{K}) \cong \mathrm{S_p}(\mathbf{S})$. Let $\mathbf{T}$ denote the semilattice of compact elements of $\mathbf{S}$. For each nonzero $a \in T$, the language of $\mathcal{K}$ should contain a unary predicate Ax. The laws of $\mathcal{K}$ should be, in addition to $x \approx y$,

(1) $Bx \to Ax$ whenever $a \leq b$ in $\mathbf{T}$,
(2) $\&_j\, B_j x \to Ax$ whenever $a \leq \bigvee_j b_j$ in $\mathcal{T}$.

These laws make $\mathrm{ETh}(\mathcal{K}) \cong \mathbf{S}$, so we can argue as above that $\mathrm{L_q}(\mathcal{K}) \cong \mathrm{S_p}(\mathbf{S})$. Thus (1) implies (2) in Theorem 2.64.

The second important result for atomistic subquasivariety lattices is a nice variation on this theme, which assumes that $\mathbf{L}$ is also algebraic, and includes finite atomistic lattices. It is due to Adaricheva, Dziobiak, and Gorbunov [14].

Theorem 2.65 *Let* $\mathbf{L}$ *be an algebraic, atomistic lattice. The following are equivalent.*

(1) $\mathbf{L} \cong \mathrm{L_q}(\mathcal{K})$ *for some quasivariety* $\mathcal{K}$ *with equality.*
(2) $\mathbf{L} \cong \mathrm{L_q}(\mathcal{U})$ *for a quasivariety* $\mathcal{U}$ *of one-element structures (with equality in the language and* $\mathcal{U}$ *satisfying* $x \approx y$*).*

(3) $\mathbf{L} \cong \mathrm{S_p}(\mathbf{S})$ *for some algebraic lattice* $\mathbf{S}$ *satisfying the descending chain condition and in which the join of every infinite ascending chain is* 1.
(4) $\mathbf{L}$ *is a dually algebraic lattice that supports an equaclosure operator.*

It is not known whether the equivalence of parts (1), (2), and (4) of Theorem 2.65 can be extended to atomistic, dually algebraic lattices. The version for finite atomistic lattices is Theorem 4.11.

By direct construction, Tumanov [112] represented finite distributive lattices as lattices of subquasivarieties; see also [64].

Theorem 2.66 *Every finite distributive lattice is isomorphic to* $\mathrm{L_q}(\mathcal{K})$ *for some quasivariety* $\mathcal{K}$.

In Chap. 9 we will show that every distributive dually algebraic lattice can be represented as $\mathrm{S_p}(\mathbf{S}, H)$ with $\mathbf{S}$ an algebraic lattice and H a monoid of operators (Theorem 9.1). Moreover, if a distributive lattice is both algebraic and dually algebraic, and its least element is dually compact, then it can be represented as $\mathrm{L_q}(\mathcal{K})$ for some quasivariety $\mathcal{K}$ (Theorem 9.18). On the other hand, we have already seen that the dually algebraic, distributive lattice $(\omega + 1)^d$ is not isomorphic to any subquasivariety lattice $\mathrm{L_q}(\mathcal{K})$ with equality, since it is not atomic (Example 2.61).

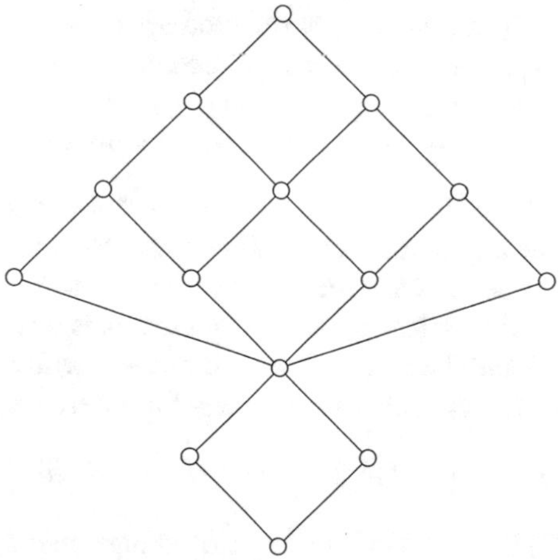

Fig. 2.10 The near-leaf lattice, which is isomorphic to $\mathrm{L_q}(\mathcal{K})$ for a quasivariety $\mathcal{K}$ with equality, but not to $\mathrm{L_q}(\mathcal{Q})$ for any locally finite quasivariety $\mathcal{Q}$ of finite type

Other constructions yield representations of some particular lattices. The *near-leaf lattice* in Fig. 2.10 is isomorphic to $\mathrm{L_q}(\mathcal{K})$ for a quasivariety $\mathcal{K}$ with equality constructed in [20], but it is not lower bounded since it contains the lattice $\mathrm{Co}(\mathbf{4})$ of convex subsets of a 4-element chain. Hence $\mathbf{L} \not\cong \mathrm{L_q}(\mathcal{Q})$ for any locally finite quasivariety $\mathcal{Q}$ of finite type. Later, in Sect. 8.1, we will show that the *leaf lattice* $\mathbf{1} + \mathrm{Co}(\mathbf{4})$ of Fig. 8.1 is also representable as $\mathrm{L_q}(\mathcal{K})$ for a quasivariety with equality, and again $\mathcal{K}$ cannot be locally finite since the leaf lattice is not lower bounded.

2.8 The Lattice $S_p(\mathbf{S}, H)$ Is Dually Algebraic

While we can show that $S_p(\mathbf{S}, H)$ is dually algebraic whenever $\mathbf{S}$ is algebraic by translating from congruence lattices of semilattices with operators, it is instructive to do so directly. This section extends arguments from Adaricheva, Gorbunov, and Tumanov for lattices of algebraic sets without operators [18].

For compact elements $c, d \in S$ let

$$[c \to d] = \{x \in S : c \leq h(x) \text{ implies } d \leq h(x) \text{ for all } h \in H\}.$$

Here we assume that the identity map is in the monoid H. Note that $[c \to d] = [c \to c \vee d]$, so without loss of generality we can take $c \leq d$.

Lemma 2.67 *If c, d are compact elements in an algebraic lattice $\mathbf{S}$ with a monoid of operators H, then $[c \to d]$ is in $S_p(\mathbf{S}, H)$, i.e., $[c \to d]$ is an H-closed algebraic subset of $\mathbf{S}$.*

Proof Let us first show that $[c \to d]$ is a dually compact H-closed subset when c, d are compact; it then follows from general principles that finite meets of these are so. It is straightforward that the set $[c \to d]$ is closed under arbitrary meets and the operators of H.

Suppose $\{e_j : j \in J\}$ is a nonempty directed set with each $e_j \in [c \to d]$, and let $h \in H$. Then $\{h(e_j) : j \in J\}$ is also directed. If $c \leq \bigvee_j h(e_j)$, then by the compactness of c we have $c \leq h(e_{j_0})$ for some j_0. As $e_{j_0} \in [c \to d]$ this implies $d \leq h(e_{j_0}) \leq \bigvee_j h(e_j)$. Thus $[c \to d]$ is closed under directed joins. □

Now for $x \in S$ and $A \in S_p(\mathbf{S}, H)$, let $\alpha(x) = \bigwedge(\uparrow x \cap A)$, which is the least element of A above x. A routine argument shows that α preserves directed joins, as follows. Let D be an up-directed set in $\mathbf{S}$ and let $x = \bigvee D$. Then $x = \bigvee D \leq \bigvee_{d \in D} \alpha(d) \in A$, whence $\alpha(x) \leq \bigvee_{d \in D} \alpha(d)$, while the reverse inclusion is obvious.

Note that if $x, y \in S$ and there is some element $a \in A$ with $a \geq x$ and $a \ngeq y$, then $\alpha(x) \ngeq y$. We use this observation several times, for different choices of A.

Lemma 2.68 *If c, d are compact in $\mathbf{S}$, then $[c \to d]$ is dually compact in $S_p(\mathbf{S}, H)$.*

Proof Let B_k $(k \in K)$ be a collection of H-closed algebraic subsets of $\mathbf{S}$. Suppose that for every finite subset $F \subseteq K$ we have $\bigcap_{k \in F} B_k \nsubseteq [c \to d]$. For each finite $F \subseteq K$, let $A_F = \bigcap_{k \in F} B_k$, which is an H-closed algebraic subset. For $x \in S$, define $\alpha_F(x) = \bigwedge(\uparrow x \cap A_F)$. By assumption, each A_F contains an element that is above c and not above d. Hence each $\alpha_F(c) \ngeq d$.

Note that $F \subseteq G$ implies $A_F \supseteq A_G$, whence $\alpha_F(c) \leq \alpha_G(c)$. Thus the set of elements $\alpha_F(c)$ form an up-directed set in $\mathbf{S}$. Let $z = \bigvee_F \alpha_F(c)$. For any fixed finite $G \subseteq K$, $\{\alpha_F(c) : F \supseteq G\}$ is a directed subset of A_G, so

$$z = \bigvee_F \alpha_F(c) = \bigvee_{F \supseteq G} \alpha_F(c)$$

is in A_G. Therefore, letting G vary over finite subsets, $z \in \bigcap_G A_G = \bigcap_{k \in K} B_k$.

But $c \leq z$, while by compactness $d \nleq z$, because $d \leq \bigvee_F \alpha_F(c)$ would imply $d \leq \alpha_G(c)$ for some G, a contradiction. Hence $z \in \bigcap_{k \in K} B_k \setminus [c \to d]$, so that $\bigcap_{k \in K} B_k \nsubseteq [c \to d]$. Thus we have shown that $[c \to d]$ satisfies the contrapositive of the definition of dual compactness. □

Lemma 2.69 *Let* $\mathbf{S}$ *be an algebraic lattice and* H *a monoid of operators on* $\mathbf{S}$. *For any* H*-closed algebraic subset* A *of* $\mathbf{S}$,

$$A = \bigcap \{\, [c \to d] : c, d \text{ are compact in } \mathbf{S} \text{ and } A \subseteq [c \to d] \,\}.$$

Proof Consider any $A \in \mathrm{S_p}(\mathbf{S}, H)$. Let us show that there are compact elements such that $A = \bigcap_i [c_i \to d_i]$. Since $S = [0 \to 0]$, we may assume $A \subset S$ properly.

Let $x \notin A$, and let $\alpha(x) = \bigwedge(\uparrow x \cap A)$. Since $x \notin A$ we have $x < \alpha(x)$. Choose a compact element d with $d \leq \alpha(x)$, $d \nleq x$.

Let $\{e_i : i \in I\}$ be the set of compact elements below x. We claim that there is an i such that $A \subseteq [e_i \to d]$, while of course $x \notin [e_i \to d]$ as $x \geq e_i$, $x \ngeq d$. Suppose to the contrary that $A \nsubseteq [e_i, d]$ for all i. For each i, let $a_i = \alpha(e_i)$. By assumption, $a_i \geq e_i$ but $a_i \ngeq d$. Since the elements e_i form a directed set and α preserves directed joins, we get $\bigvee a_i = \alpha(x)$. But then, by the compactness of d we have $d \leq a_i$ for some i, a contradiction. Thus there is an i_0 such that $A \subseteq [e_{i_0}, d]$.

Doing this for all elements $x \notin A$ yields a collection such that $A = \bigcap_i [c_i \to d_i]$, as desired. □

Corollary 2.70 *An* H*-closed algebraic set* A *is dually compact in* $\mathrm{S_p}(\mathbf{S}, H)$ *if and only if* $A = \bigcap_{i=1}^n [c_i \to d_i]$ *for finitely many pairs of compact elements* $c_i, d_i \in S$.

Combining the preceding lemmas yields our theorem.

Theorem 2.71 *If* $\mathbf{S}$ *is an algebraic lattice and* H *a monoid of operators on* $\mathbf{S}$, *then the lattice* $\mathrm{S_p}(\mathbf{S}, H)$ *is dually algebraic.*

The case when $c = 0_{\mathbf{S}}$ is of particular interest. Note that

$$[0_{\mathbf{S}} \to d] = \{x \in S : d \leq h(x) \text{ for all } h \in H\}.$$

These algebraic sets are easy to describe, giving a rather abstract precursor to Theorem 3.6.

Theorem 2.72 *Let* d *be a compact element in an algebraic lattice* $\mathbf{S}$ *with a monoid of operators* H, *and let* $\phi(d)$ *be the least fully invariant element above* d. *Then* $[0_{\mathbf{S}} \to d] = \uparrow \phi(d)$.

Proof The least element of any H-closed algebraic set is fully invariant, and every element of $[0_{\mathbf{S}} \to d]$ is above d since the identity map is in H. Therefore, the least element of $[0_{\mathbf{S}} \to d]$ is above $\phi(d)$. On the other hand, if x is fully invariant and $x \geq d$, then $h(x) \geq x \geq d$ for all $h \in H$, so that $x \in [0_{\mathbf{S}} \to d]$. In particular, $\phi(d) \in [0_{\mathbf{S}} \to d]$. The conclusion follows. □

Chapter 3
Equaclosure Operators

Prouver que j'ai raison serait accorder que je puis avoir tort. – Pierre Augustin Caron de Beaumarchais

Снявши пробу с двух океанов и континентов, я чувствую то же почти, что глобус. То есть дальше некуда. Дальше — ряд звёзд. – Иосиф Бродский, Собрание стихотворений

Wiesław Dziobiak [39] observed that there is a natural closure operator Γ on the lattice $L_q(\mathcal{K})$ of subquasivarieties of a quasivariety $\mathcal{K}$, *viz.*, for $Q \leq \mathcal{K}$ let $\Gamma(Q) = \mathcal{K} \cap \mathbb{HSP}(Q) = \mathcal{K} \cap \mathbb{H}(Q)$. Moreover, there is a least subquasivariety $\mathcal{L} = \tau(Q)$ with $\Gamma(\mathcal{L}) = \Gamma(Q)$, which is the quasivariety generated by the Q-free structure $\mathbf{F}_Q(\omega)$. The map Γ is called the *natural equaclosure operator* on $L_q(\mathcal{K})$.

Adaricheva and Gorbunov [16] defined an equaclosure operator abstractly to have those properties that are known to hold for the natural equaclosure operator on the lattice of subquasivarieties of a quasivariety. The modern version is given below. Not every lattice supports an equaclosure operator, e.g., $\mathbf{M}_3$ and the lattices $\mathbf{H}$, $\mathbf{J}$ in Fig. 4.4 do not. The existence of an equaclosure operator has structural consequences for those lattices that do support one. Lattices of subquasivarieties are dually algebraic, and since property (I7) refers to that, equaclosure operators are defined for dually algebraic lattices. The property (K9) implies a couple of the original conditions (*biatomicity* and the so-called 4-*atom condition*) [20], so those are omitted from our definition. Property (K9) strengthens a previous version called (I9) in [92], and properties (K10) and (K11) are new.

A lattice is *biatomic* if whenever a is an atom and $a \leq u \vee v$ in $\mathbf{L}$, then there exist atoms $b \leq u$ and $c \leq v$ such that $a \leq b \vee c$. Theorem 3.9 will show that (K9) implies biatomicity. The 4-*atom condition* is Lemma 4.14 for $n = 2$.

An *equaclosure operator* on a dually algebraic lattice $\mathbf{L}$ is a map $\gamma : \mathbf{L} \to \mathbf{L}$ satisfying (I1)–(I8) and (K1)–(K11).

(I1) $x \leq \gamma(x)$.
(I2) $x \leq y$ implies $\gamma(x) \leq \gamma(y)$.

K. Adaricheva et al., *A Primer of Subquasivariety Lattices*, CMS/CAIMS Books in Mathematics 3, https://doi.org/10.1007/978-3-030-98088-7_3

(I3) $\gamma^2(x) = \gamma(x)$.
(I4) $\gamma(0) = 0$.
(I5) $\gamma(x) = u$ for all $x \in X$ implies $\gamma(\bigwedge X) = u$.
(I6) $\gamma(x) \wedge (y \vee z) = (\gamma(x) \wedge y) \vee (\gamma(x) \wedge z)$.
(I7) The image $\gamma(L)$ is the complete meet subsemilattice of **L** generated by $\gamma(L) \cap K$, where **K** is the semilattice of dually compact elements of **L**.
(I8) There is a dually compact element $w \in L$ such that $\gamma(w) = w$ and the interval $[0, w]$ is isomorphic to the lattice $\mathrm{S_p}(\mathbf{S})$ of all algebraic subsets of an algebraic lattice **S**.

Property (I5) implies that for each $x \in L$, there is a least element z such that $\gamma(z) = \gamma(x)$. Denoting this element by $\tau(x)$, we have

$$\tau(x) = \bigwedge\{z \in L : \gamma(z) = \gamma(x)\}.$$

Note that the operation τ is implicitly defined by γ, *via* the above formula. The last three properties defining an *equaclosure operator* refer to $\tau(x)$.

(K9) For any index set I, $\gamma(x \wedge \bigvee_{i \in I} \tau(x \vee z_i)) \geq x \wedge \bigvee_{i \in I} \tau(z_i)$.
(K10) If $\tau a \leq \tau b$ and $\gamma s \leq \gamma b \leq \gamma(r \vee s)$, then

$$a \leq \gamma(\tau(r \vee s) \wedge (\tau r \vee (\tau s \wedge \gamma a))).$$

(K11) If x and c_i $(i \in I)$ are such that

(a) $\gamma c_i = c_i$,
(b) $\{c_i : i \in I\}$ is down-directed,
(c) $\tau c_i \leq x$ for all $i \in I$,

then $\tau(\bigwedge_i c_i) \leq x$.

The special case of (K9) when $|I| = 1$ is

$$\gamma[x \wedge \tau(x \vee z)] \geq x \wedge \tau(z). \qquad (\ddagger)$$

The condition $(\ddagger)$ is the form in which (K9) is most frequently used, and it is equivalent to (K9) for any finite index set [20, 92].

Section 3.6.1 verifies some additional properties of the natural equaclosure operator that we have found, which turned out to be consequences of those listed above.

(E1) If $\tau a < \tau b$, then for every $s \in L$ there exists $s' \leq \tau s$ such that $\tau(a \vee s') \leq \tau(b \vee s)$.
(E2) If $\tau a < \tau b$ and $\gamma b = \gamma(r \vee s)$, then $\tau a \leq \tau(\tau r \wedge \gamma a) \vee \tau(\tau s \wedge \gamma a)$.
(E3) If $\gamma a \vee \gamma b = \gamma z > \gamma c$ is a minimal join cover in $\gamma(\mathbf{L})$, then $\tau z \not\geq \tau c$.

3.1 Natural Equaclosure Operators

Let us dissect the definition of an equaclosure operator.

Properties (I1)–(I4) say that γ is a closure operator with $\gamma(0) = 0$.

Property (I5) allows us to define the operation τ so that $\tau(x)$ is the least element z with $\gamma(z) = \gamma(x)$. More generally, $\gamma(z) = \gamma(x)$ if and only if $\tau(x) \leq z \leq \gamma(x)$. The equivalence relation given by $x \equiv y$ iff $\gamma(x) = \gamma(y)$ is called the *equapartition* of $\mathbf{L}$ determined by γ.

It is important to note that the map τ need not be order preserving. Instead, we have only an inclusion:

Lemma 3.1 *If γ satisfies* (I1)–(I5)*, then $\tau(\bigvee X) \leq \bigvee_{x\in X} \tau x$ for any $X \subseteq L$.*

Figure 3.1 illustrates the lemma and shows that equality need not hold. In the figure, red curves enclose the classes of the equapartition.

Proof For any closure operator on a complete lattice $\mathbf{L}$ and elements $y, z \in L$, the equation $\gamma(y) = \gamma(z)$ is equivalent to $y \leq \gamma(z)$ and $z \leq \gamma(y)$. If γ satisfies (I5), then we also have $\gamma(\tau x) = \gamma(x)$ for any $x \in L$.

Considering a subset $X \subseteq L$, take $y = \bigvee_{x\in X} \tau x$ and $z = \bigvee X$. For each $x \in X$, we have

$$\tau(x) \leq \gamma(x) \leq \gamma(\bigvee X)$$
$$x \leq \gamma(x) = \gamma(\tau x) \leq \gamma(\bigvee_{u\in X} \tau u)$$

whence $y \leq \gamma(z)$ and $z \leq \gamma(y)$, that is, $\gamma(\bigvee_{x\in X} \tau x) = \gamma(\bigvee X)$. Thus $\bigvee_{x\in X} \tau x \geq \tau(\bigvee X)$, since the latter is the least element in that block of the equapartition. □

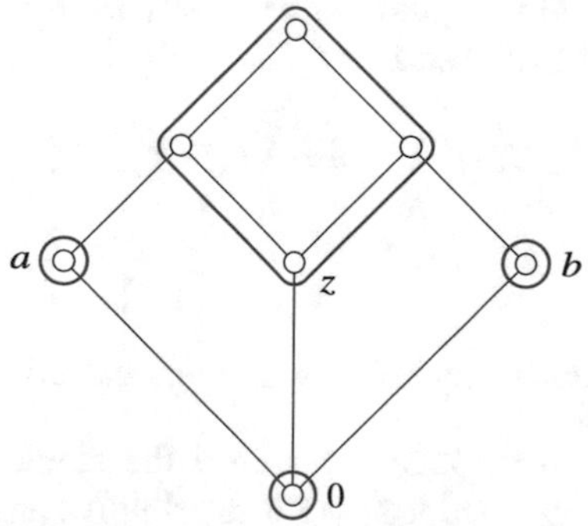

Fig. 3.1 Lattice with equaclosure operator illustrating Lemma 3.1: $\tau(a \vee b) \leq \tau(a) \vee \tau(b)$

An element $a \in L$ is *meet distributive* if $a \wedge (y \vee z) = (a \wedge y) \vee (a \wedge z)$ for all $y, z \in L$. Property (I6) says that the γ-closed elements of $\mathbf{L}$ are meet distributive. Section 4.1 will deal with the consequences of (I6) for finite lattices. For now, let us see why the natural equaclosure operators satisfy it.

First, let us consider the natural weak equaclosure operator on $S_p(\mathbf{S}, H)$. Given H-closed algebraic subsets X, Y, Z, we have $\Gamma(X) = \uparrow x_0$ where x_0 is the least element of X. Let $w \in \Gamma(X) \wedge (Y \vee Z)$. Then $w \geq x_0$, while by Gorbunov's Lemma 1.15 we have $w = y \wedge z$ for some $y \in Y$, $z \in Z$. Then $y \geq w \geq x_0$ and $z \geq w \geq x_0$, so $w \in (\Gamma(X) \wedge Y) \vee (\Gamma(X) \wedge Z)$, as desired.

But Lemma 1.15 only applies to finite joins: to find $\bigvee Y_i$ for an infinite collection, you need to worry about directed joins from $\bigcup Y_i$, and the weaving trick from Gorbunov's proof no longer works. *Property* (I6) *is only valid for finite joins.* An easy example below with quasivarieties shows that $\Gamma(X)$ needs to distribute over infinite joins.

To verify property (I6) for the natural equaclosure operator on $L_q(\mathcal{K})$, let $\mathcal{X}$, $\mathcal{Y}$, $\mathcal{Z}$ be subquasivarieties of $\mathcal{K}$, and consider a structure $\mathbf{W} \in \Gamma(\mathcal{X}) \wedge (\mathcal{Y} \vee \mathcal{Z})$. Now $\mathbf{W} \in \Gamma(\mathcal{X}) = \mathbb{H}(\mathcal{X}) \cap \mathcal{K}$ means that $\mathbf{W} \in \mathcal{K}$ and there exist $\mathbf{A} \in \mathcal{X}$ and a surjective homomorphism $h : \mathbf{A} \twoheadrightarrow \mathbf{W}$. This implies that $\mathrm{Con}_{\mathcal{K}}\,\mathbf{W} \cong \uparrow(\ker h)$ in $\mathrm{Con}_{\mathcal{K}}\,\mathbf{A}$, so that any homomorphic image of $\mathbf{W}$ is a homomorphic image of $\mathbf{A}$. Meanwhile, $\mathbf{W} \in \mathcal{Y} \vee \mathcal{Z}$ means that $\mathbf{W}$ is a subdirect product of a structure $\mathbf{U} \in \mathcal{Y}$ and a structure $\mathbf{V} \in \mathcal{Z}$ by Theorem 2.51. (That theorem is for finite joins only and ultimately goes back to the fact that in the join $\mathcal{Y} \vee \mathcal{Z} = \mathbb{Q}(\mathcal{Y} \cup \mathcal{Z}) = \mathbb{E}\mathrm{q}\mathbb{SPU}(\mathcal{Y} \cup \mathcal{Z})$, an ultraproduct of structures from $\mathcal{Y} \cup \mathcal{Z}$ is in either $\mathcal{Y}$ or $\mathcal{Z}$.) Therefore, in $\mathrm{Con}_{\mathcal{K}}\,\mathbf{W}$ we have $\Delta = \alpha \wedge \beta$ with $\mathbf{W}/\alpha \in \mathcal{Y}$ and $\mathbf{W}/\beta \in \mathcal{Z}$. Since $\mathbf{W} \in \mathbb{H}(\mathcal{X})$, indeed $\mathbf{W}/\alpha \in \Gamma(\mathcal{X}) \wedge \mathcal{Y}$ and $\mathbf{W}/\beta \in \Gamma(\mathcal{X}) \wedge \mathcal{Z}$, whence their subdirect product $\mathbf{W}$ is in $(\Gamma(\mathcal{X}) \wedge \mathcal{Y}) \vee (\Gamma(\mathcal{X}) \wedge \mathcal{Z})$, as desired.

When $\mathcal{K}$ is a locally finite quasivariety of finite type, then $L_q(\mathcal{K})$ is an algebraic lattice, and hence upper continuous. So for those quasivarieties, the infinite version of (I6) does work, which we faithfully record.

Theorem 3.2 *If $\mathcal{K}$ is a locally finite quasivariety of finite type, then the natural equaclosure operator on $L_q(\mathcal{K})$ satisfies*

$$\Gamma(\mathcal{X}) \wedge \bigvee_{i \in I} \mathcal{Y}_i = \bigvee_{i \in I} (\Gamma(\mathcal{X}) \wedge \mathcal{Y}_i) \qquad \text{(I6}')$$

for any index set I.

But otherwise, (I6) needs only hold for finite index sets.

Example 3.3 Let $\mathbf{E} = \mathbf{1} + (\omega \times \mathbf{2}) + \mathbf{1}$ with γ the identity map, as illustrated in Fig. 3.2. Now the filter $\uparrow y_0 = (\omega \times \mathbf{2}) + \mathbf{1}$ is distributive and dually algebraic, so by Theorem 9.1 it can be represented as a lattice of H-closed algebraic subsets $S_p(\mathbf{S}, H)$. The construction in that proof results in the natural weak equaclosure on $\uparrow y_0$ being the identity, as indicated by the red circles in the figure. Then by Theorem 9.12, which uses Corollary 7.10, the pair $(\mathbf{E}, \gamma)$ can be represented as $(L_q(\mathcal{K}), \Gamma)$ for a quasivariety $\mathcal{K}$. With the labeling in the figure, we see that the infinite version (I6′) fails.

Property (I7) has the following consequences.

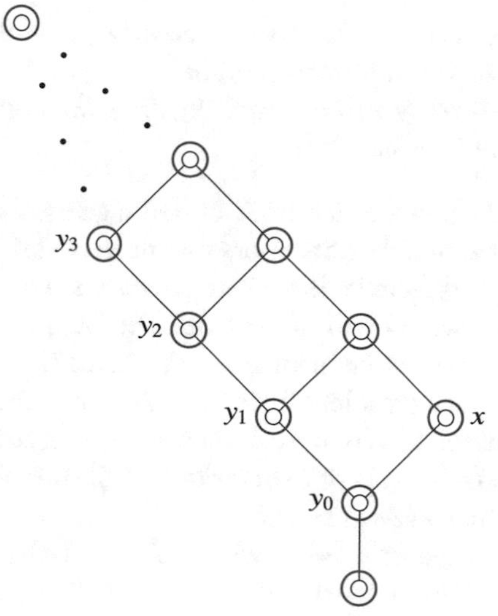

Fig. 3.2 Pair $(\mathbf{E}, \gamma)$ for Example 3.3

Lemma 3.4 *If* γ *satisfies* (I1)–(I7), *then the image* $\gamma(\mathbf{L})$ *is a dually algebraic lattice, and* x *is dually compact in* $\gamma(\mathbf{L})$ *if and only if* $x \in \gamma(\mathbf{L})$ *and* x *is dually compact in* $\mathbf{L}$.

Property (I8) holds for quasivarieties in languages with equality, and not necessarily otherwise. Let $\mathcal{K}$ be a quasivariety with equality, and let $\mathcal{W}$ be the subquasivariety of 1-element structures in $\mathcal{K}$, i.e., all structures satisfying $x \approx y$. The element w in (I8) corresponds to the subquasivariety $\mathcal{W}$, which for types that include relations need not be the least subquasivariety. Now $\mathcal{W}$ is finitely based relative to $\mathcal{K}$, so it is dually compact in $\mathrm{L_q}(\mathcal{K})$, and it is closed under homomorphic images, so $\Gamma(\mathcal{W}) = \mathcal{W}$. Moreover, Theorem 2.64 says that $\downarrow \mathcal{W}$ is isomorphic to $\mathrm{S_p}(\mathbf{S})$ for an algebraic lattice $\mathbf{S}$. Thus the natural equaclosure operator on $\mathrm{L_q}(\mathcal{K})$ satisfies (I8).

In particular, (I8) implies that the ideal $\downarrow w$ is atomistic; see Theorems 1.11 and 1.12.

Property (I8) may not hold for the natural closure operator Γ on $\mathrm{S_p}(\mathbf{S}, H)$, described below following Theorem 3.5; see Example 3.8. These lattices are isomorphic to some $\mathrm{L_q}(\mathcal{K})$ in a language that may or may not contain equality. However, the remaining properties of an equaclosure operator are satisfied by $\mathrm{S_p}(\mathbf{S}, H)$. Thus we make these distinctions.

- A *preclop* (derived from pre-equaclosure operator) satisfies (I1)–(I7).
- A *weak equaclosure operator* satisfies (I1)–(I7) and (K9)–(K11).
- An *equaclosure operator* satisfies (I1)–(I8) and (K9)–(K11).

Fortunately, for finite lattices we can always take $w = 0_{\mathbf{L}}$ in (I8), making the distinction between equaclosure operators and weak equaclosure operators moot.

To clarify the situation, here is the proper statement.

Theorem 3.5 (1) *If $\mathcal{K}$ is a quasivariety with equality, then the natural equaclosure operator on* $\mathrm{L_q}(\mathcal{K})$ *is an equaclosure operator.*

(2) *If $\mathcal{U}$ is a quasivariety without equality, then the natural weak equaclosure operator on* $\mathrm{L_q}(\mathcal{U})$ *need not satisfy* (I8).

Recall that $\mathrm{S_p}(\mathbf{S}, H)$ is the lattice of H-closed algebraic subsets of an algebraic lattice $\mathbf{S}$, where H is a monoid of operators that preserve arbitrary meets (including $1 = \bigwedge \varnothing$) and nonempty directed joins. There is a *natural weak equaclosure operator* on the lattice $\mathrm{S_p}(\mathbf{S}, H)$, which will also be denoted by a capital Γ. Each H-closed algebraic subset A has a least element $a_0 \in S$. Then $\Gamma(A) = \uparrow a_0$ is the filter in $\mathbf{S}$ generated by a_0. Not every principal filter $\uparrow x$ is H-closed, but $\uparrow a_0$ is. For if A is an H-closed algebraic subset and $a_0 = \bigwedge A$, then $h(a_0) \in A$ for every $h \in H$, whence $h(a_0) \geq a_0$. In other words, a_0 is *fully invariant*. Moreover, for any $x \in S$, if $x \geq a_0$ then $h(x) \geq h(a_0) \geq a_0$, so $\uparrow a_0$ is H-closed.

The least H-closed algebraic set T with $\Gamma(T) = \Gamma(A)$ is $T = \tau(A) = \mathrm{Ag}(a_0)$, the H-closed algebraic subset generated by a_0. (Note that we use the notation τ for the least object map for generic (weak) equaclosure operators γ and natural (weak) equaclosure operators Γ.) This in turn is given by $\tau(A) = \mathbb{DMO}(a_0)$ where $\mathbb{D}$ is joins of nonempty directed sets, $\mathbb{M}$ is meets, $\mathbb{O}$ is applying the operators of H; see Theorem 1.16. The equapartition on $\mathrm{S_p}(\mathbf{S}, H)$ has blocks consisting of all H-closed algebraic subsets with the same least element. We formalize this as follows.

Theorem 3.6 *Let* $\mathbf{S}$ *be an algebraic lattice and H a monoid of operators on S. There is a natural weak equaclosure operator on* $\mathrm{S_p}(\mathbf{S}, H)$ *given by* $\Gamma(A) = \uparrow a_0$ *where a_0 is the least element of A. For this weak equaclosure operator,* $\tau(A) = \mathrm{Ag}(a_0)$, *the H-closed algebraic subset generated by a_0.*

Proof Properties (I1)–(I5) are clear, remembering that the least element of $\mathrm{S_p}(\mathbf{S}, H)$ is the set $\{1_S\}$. The proof that (I6) holds was given in the discussion of that property above.

To prove (I7), consider an H-closed algebraic subset A with least element a_0, so that $\Gamma(A) = \uparrow a_0$. In $\mathbf{S}$, write a_0 as a join of compact elements, $a_0 = \bigvee_i c_i$. For each i, let C_i be the dually compact algebraic set $[0 \to c_i]$, as in Sect. 2.8. By Lemma 2.72, each $C_i = \uparrow \phi(c_i)$ where $\phi(c_i)$ is the least fully invariant element above c_i; thus $\Gamma(C_i) = C_i$. But a_0 is fully invariant, so $c_i \leq \phi(c_i) \leq a_0$, whence $a_0 = \bigvee_i \phi(c_i)$ and $\Gamma(A) = \bigwedge_i \Gamma(C_i)$, as the condition requires.

Example 2.61 shows that (I8) can fail in $\mathrm{S_p}(\mathbf{S}, H)$.

That property (K9) holds is the main result of [92], strengthening an earlier version in [20]. That argument is somewhat involved, but a simple proof of (‡), which is equivalent to (K9) for finite index sets, is given in Theorem 3.14 below.

Finally, the facts that properties (K10) and (K11) hold in $\mathrm{S_p}(\mathbf{S}, H)$ are our Theorems 3.26 and 3.32. □

Theorem 2.54 represents every subquasivariety lattice $\mathrm{L_q}(\mathcal{K})$ as a lattice $\mathrm{S_p}(\mathbf{S}, H)$; see also Theorem 2.36. We claim that under this representation of $\mathrm{L_q}(\mathcal{K})$ as $\mathrm{S_p}(\mathbf{S}, H)$, these natural closure operators correspond. This fact is crucial, so let us write it out carefully. For this argument only (that is, until we show they are the same), we use Γ on $\mathrm{L_q}(\mathcal{K})$ and Γ' on $\mathrm{S_p}(\mathbf{S}, H)$.

Fix the quasivariety $\mathcal{K}$, and let $\mathbf{F} = \mathbf{F}_{\mathcal{K}}(\omega)$ be the countably generated free structure. Take $\mathbf{S} = \mathrm{Con}_{\mathcal{K}} \mathbf{F}$ and $H = \mathcal{E}^*$, the monoid of operators on $\mathbf{S}$ induced by endomorphisms of $\mathcal{F}$, introduced just before Lemma 2.28. The isomorphism $\sigma : \mathrm{L_q}(\mathcal{K}) \cong \mathrm{S_p}(\mathbf{S}, H)$ from Theorem 2.54 is given by, for $Q \leq \mathcal{K}$,

$$\sigma(Q) = \{\theta \in \mathrm{Con}_{\mathcal{K}} \mathbf{F} : \mathbf{F}/\theta \in Q\}$$

which is an H-closed algebraic subset of $\mathbf{S}$. The least member of $\sigma(Q)$ is a congruence φ_Q such that $\mathbf{F}/\varphi_Q$ is isomorphic to $\mathbf{F}_Q(\omega)$.

On $\mathrm{L_q}(\mathcal{K})$ we have $\Gamma(Q) = \mathcal{K} \cap \mathbb{H}(Q)$. Meanwhile, on $\mathrm{S_p}(\mathbf{S}, H)$ the closure is $\Gamma'(Q) = \uparrow q_0$, where q_0 is the least member of Q. But the least member of $\sigma(Q)$ is φ_Q, and by the First Isomorphism Theorem, a $\mathcal{K}$-congruence θ satisfies $\mathbf{F}/\theta \in \mathcal{K} \cap \mathbb{H}(Q)$ if and only if $\theta \geq \psi$ for some $\mathcal{K}$-congruence ψ with $\mathbf{F}/\psi \in Q$, or equivalently, $\theta \geq \varphi_Q$. Thus $\sigma\Gamma(Q) = \Gamma'\sigma(Q)$ for each $Q \leq \mathcal{K}$, which was to be established. We summarize this as:

Theorem 3.7 *Let $\mathcal{K}$ be a quasivariety. Under the representation of a subquasivariety lattice $\mathrm{L_q}(\mathcal{K})$ as a lattice of H-closed algebraic subsets $\mathrm{S_p}(\mathbf{S}, H)$ given by Theorem 2.54, the equaclosure operator on $\mathrm{L_q}(\mathcal{K})$, given by $\Gamma(Q) = \mathbb{H}(Q) \cap \mathcal{K}$, corresponds to the natural weak equaclosure operator on $\mathrm{S_p}(\mathbf{S}, H)$, given by $\Gamma(A) = \uparrow a_0$ where $a_0 = \bigwedge A$.*

Thus properties of the natural equaclosure operator on $\mathrm{L_q}(\mathcal{K})$ can be derived from those of the natural weak equaclosure operator on $\mathrm{S_p}(\mathbf{S}, H)$. Moreover, in languages without equality, the natural equaclosure operator on $\mathrm{L_q}(\mathcal{K})$ need only be a *weak* equaclosure operator, as (I8) can fail.

Example 3.8 In Theorem 9.1, we prove that every distributive dually algebraic lattice can be represented as $\mathrm{S_p}(\mathbf{S}, H)$ with $\mathbf{S}$ an algebraic lattice and H a monoid of operators. However, many infinite distributive dually algebraic lattices fail to have a candidate for the element w of (I8); failing (I8), they cannot be represented as subquasivariety lattices $\mathrm{L_q}(\mathcal{K})$ with equality. In particular, this applies to

- $(\omega + 1)^d$ (Example 2.61),
- more generally, lattices $\mathcal{O}(\mathbf{P})$ where $\mathbf{P}$ has no or infinitely many minimal elements (cf. Theorem 9.18),
- $\mathbf{2}^\kappa$ with κ uncountable (Theorem 1.12).

Let us prove that (‡) implies biatomicity, a result from [20].

Theorem 3.9 *If $\mathbf{L}$ is a dually algebraic lattice that supports a weak equaclosure operator, then $\mathbf{L}$ is biatomic.*

Note that the biatomic property does not assume that $\mathbf{L}$ is atomic; it is rather a property that applies to such atoms as may exist.

Proof Let a be an atom of $\mathbf{L}$, and assume $a \leq u \vee v$. We first show that there exists an atom $b \leq u$ such that $a \leq b \vee v$. Applying the argument a second time shows that there exists an atom $c \leq v$ with $a \leq b \vee c$, as desired. Assume $a \nleq u$ and $a \nleq v$, as the conclusion is trivial otherwise. Moreover, by (I6) we have $a \leq \gamma a \wedge (u \vee v) = (\gamma a \wedge u) \vee (\gamma a \wedge v)$, so we may assume that $u, v \leq \gamma a$.

By lower continuity and Zorn's Lemma, there is an element $b \leq u$ that is minimal with respect to the property that $a \leq b \vee v$. If b is an atom, we are done. Suppose to the contrary that b is not an atom. Then there is an element $e < b$ such that $0 < e \leq \tau b$. Apply (‡) with $x = v \vee e$ and $z = b$. Since $a \leq b \vee v = x \vee z \leq u \vee v \leq \gamma a$, we have $\tau(x \vee z) = a$. Then (‡) yields

$$e \leq x \wedge \tau z \leq \gamma[(v \vee e) \wedge a] = \gamma 0 = 0,$$

a contradiction. Therefore, b is an atom, as desired. □

Before moving on, it is useful to see examples of how properties (K9) and (K10) can fail. Figure 3.3 gives 3 pairs $(\mathbf{L}, \gamma)$ consisting of a finite lattice and a preclop on it. The red curves indicate the equapartition determined by γ. Recall that (‡) is a special case of (K9). As the pairs in the figure fail (‡), they cannot be represented as $\mathrm{L_q}(\mathcal{K})$ or $\mathrm{S_p}(\mathbf{S}, H)$.

The Boolean algebra can be represented as a lattice of subquasivarieties, as can all finite distributive lattices [64, 112], but not with the preclop in Fig. 3.3. The second lattice supports four preclops. The operator in the figure fails (‡); the other three, with $\gamma(x \wedge t) = x$ or two versions with $\gamma(x \wedge t) = 1$, satisfy all the conditions and can be represented as $\mathrm{S_p}(\mathbf{S}, H)$. (The construction project of Chap. 7 easily represents two of these pairs as $\mathrm{L_q}(\mathcal{K})$, and possibly the third; see Sect. A.1.) The third lattice in the figure, from [20], supports only this preclop, which fails (‡).

In fact, the pairs in Fig. 3.3 fail both (K9) and (K10). Proposition 3.28 shows that (K10) implies a property (†) that is only slightly weaker than (‡), and (†) fails for all the pairs in the figure. However, Fig. 3.8 gives a modification of the third example that satisfies all the properties of an equaclosure operator except (K9).

3.2 The H-Closed Algebraic Subset Generated by a Set

Suppose we are given a lattice of H-closed algebraic subsets $\mathrm{S_p}(\mathbf{S}, H)$ with $\mathbf{S}$ an algebraic lattice and H a monoid of operators. For $X \subseteq S$, we know that $\mathrm{Ag}(X) = \mathbb{DMO}(X)$, the H-closed algebraic subset generated by X, consists of all directed joins of meets of elements $h(x)$ with $x \in X$. It is convenient to think of $\mathbb{DMO}$ as a derived set of maps, like polynomials. This requires a little care.

The maps in H are ordered pointwise, so we can take meets and joins of subsets of H. Moreover, it makes sense to talk about an up-directed set of maps. For convenience, we assume that the identity map id and the constant map 1 are in H.

Let $\mathbb{DMO}$ consist of all maps $k : S \to S$ such that k is the join of an up-directed set of meets of maps from H. Obviously $\mathbb{DMO}$ depends on H, but since we deal

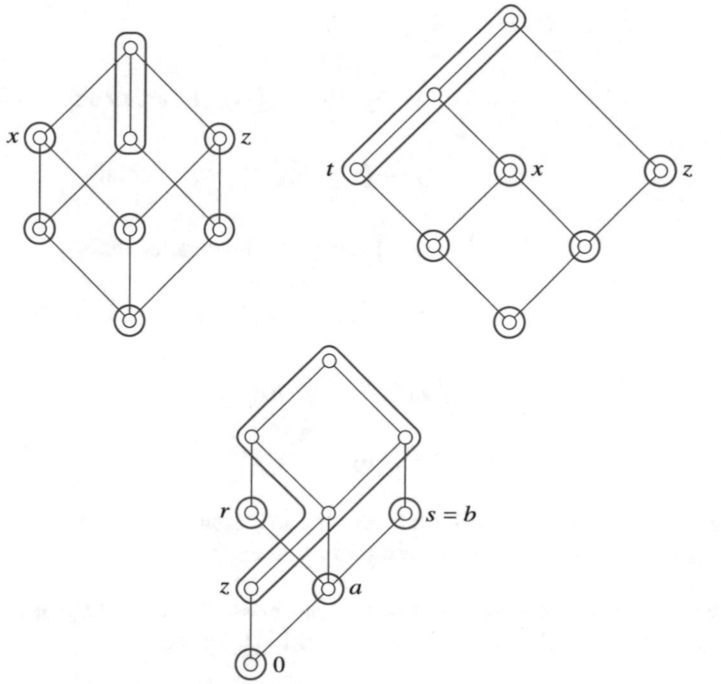

Fig. 3.3 Lattices with preclops that fail (‡)

with one set of operators H at a time, we can suppress that in the notation. The next lemma gives some basic technical properties of $\mathbb{DMO}$.

Lemma 3.10 *Let H be a monoid of operators on an algebraic lattice* **S**. *Suppose K_j ($j \in J$) is a down-directed collection of subsets of H, i.e., for all $i, j \in J$ there exists $\ell \in J$ such that $K_i \cap K_j \supseteq K_\ell$. Define new operators by*

$$h_j(x) = \bigwedge_{g \in K_j} g(x)$$

$$\hat{h}(x) = \bigvee_{j \in J} h_j(x).$$

Then

(1) *$\{h_j(x) : j \in J\}$ is up-directed for any $x \in S$.*
(2) *If A is an H-closed algebraic subset of* **S** *and $x \in A$, then $\hat{h}(x) \in A$.*
(3) *$\hat{h}(\bigwedge X) = \bigwedge_{x \in X} \hat{h}(x)$ for any finite subset $X \subseteq S$.*
(4) *If $\{d_\ell : \ell \in L\}$ is an up-directed subset of S, then $\hat{h}(\bigvee d_\ell) = \bigvee \hat{h}(d_\ell)$.*

Proof Items (1), (2), and (4) are straightforward. For (3), we calculate as follows:

$$\begin{aligned}\hat{h}(x) \wedge \hat{h}(y) &= \bigvee_j h_j(x) \wedge \bigvee_k h_k(y) \\ &= \bigvee_k ((\bigvee_j h_j(x)) \wedge h_k(y)) \quad \text{by continuity} \\ &= \bigvee_k \bigvee_j (h_j(x) \wedge h_k(y)) \quad \text{by continuity} \\ &\leq \bigvee_\ell (h_\ell(x) \wedge h_\ell(y)) \quad \text{by directedness} \\ &= \bigvee_\ell h_\ell(x \wedge y) \\ &\leq (\bigvee_\ell h_\ell(x)) \wedge (\bigvee_\ell h_\ell(y)) \\ &= \hat{h}(x) \wedge \hat{h}(y).\end{aligned}$$

Applying Lemma 3.10(3), plus a more obvious calculation for meets of maps and $1_{\mathbf{S}}$, yields that the maps in $\mathbb{DMO}$ are finitary operators.

Theorem 3.11 *Every $k \in \mathbb{DMO}$ preserves finite meets and the largest element $1_{\mathbf{S}}$, that is, $k(x_1 \wedge \ldots \wedge x_n) = kx_1 \wedge \ldots \wedge kx_n$ and $k(1_{\mathbf{S}}) = 1_{\mathbf{S}}$.*

Examples 3.15 and 3.16 show that we cannot do much better than that.

The next lemma deals with the situation when $x \in \mathrm{Ag}(y)$.

Lemma 3.12 *Let H be a monoid of operators on an algebraic lattice $\mathbf{S}$. Suppose the following situation occurs.*

(1) *$x, y \in S$,*
(2) *$x = \bigvee_{j \in J} d_j$ where $\{d_j : j \in J\}$ is up-directed,*
(3) *for each $j \in J$, $d_j = \bigwedge_{g \in G_j} g(y)$ with $G_j \subseteq H$.*

For each $j \in J$, let $K_j = \{f \in H : f(y) \geq d_j\}$. Define $h_j = \bigwedge K_j$ and $\hat{h} = \bigvee_{j \in J} h_j$ as in Lemma 3.10. *Then*

(4) *$\{K_j : j \in J\}$ is down-directed,*
(5) *$h_j(y) = d_j$ for all j,*
(6) *$\hat{h}(y) = x$.*

Proof Clearly $d_i \leq d_j$ implies $K_i \supseteq K_j$. On the other hand, $G_j \subseteq K_j$, so for each j we have $d_j \leq \bigwedge_{f \in K_j} f(y) \leq \bigwedge_{g \in G_j} g(y) = d_j$. This gives (4) and (5), whence (2) and (5) yield (6). □

Since $x = h(y)$ clearly implies $x \in \mathrm{Ag}(y)$, Lemma 3.12 has the following crucial consequence.

Theorem 3.13 *For $x, y \in S$, we have $x \in \mathrm{Ag}(y)$ if and only if $x = h(y)$ for some $h \in \mathbb{DMO}$.*

Let us illustrate how the results are used by proving that (‡) from (K9) must hold in a representable pair $(\mathbf{L}, \gamma)$. (The proof of the full condition (K9) in [92] involves complications that mask the simplicity of the argument.)

Theorem 3.14 *If* $(\mathbf{L}, \gamma)$ *can be represented as* $\mathrm{S_p}(\mathbf{S}, H)$ *with* $\mathbf{S}$ *algebraic and* H *a monoid of operators, then*

$$\gamma[x \wedge \tau(x \vee z)] \geq x \wedge \tau(z). \tag{‡}$$

Proof Let X, Z be H-closed algebraic sets in $\mathrm{S_p}(\mathbf{S}, H)$, and let x_0 and z_0 denote the least elements of X and Z, respectively. Consider an element $x \in X \wedge \tau(Z)$. Then $x \in \mathrm{Ag}(z_0)$, whence $x \in \mathbb{DMO}(z_0)$. By Lemma 3.12 there is a map $\hat{h} \in \mathbb{DMO}$ such that $x = \hat{h}(z_0)$.

Now $\hat{h}(x_0 \wedge z_0) = \hat{h}(x_0) \wedge x$, so $\hat{h}(x_0 \wedge z_0) \in X \wedge \tau(X \vee Z)$ and $\hat{h}(x_0 \wedge z_0) \leq x$. Hence $x \in \Gamma(X \wedge \tau(X \vee Z))$, as desired. □

Unfortunately, maps in $\mathbb{DMO}$ need not preserve infinite meets nor directed joins of elements.

Example 3.15 To see that meets need not be preserved, let $\mathbf{S} = (\omega + 1)^d$ with elements $0 > -1 > -2 > \cdots > -\omega$. For $j \in \omega$, define maps

$$h_j(x) = \begin{cases} 0 & \text{if } x \geq -j \\ -\omega & \text{if } x < -j. \end{cases}$$

Note $h_0 \leq h_1 \leq h_2 \leq \cdots$, so that if we let $K_j = \{h_k : k \geq j\}$, then $h_j = \bigwedge K_j$. Let $\hat{h} = \bigvee h_j$ and consider $X = \omega^d = \{0, -1, -2, \ldots\}$. Then $\hat{h}(\bigwedge X) = \hat{h}(-\omega) = -\omega$, while $\bigwedge_{x \in X} \hat{h}(x) = 0$ since $h_j(x) = 0$ eventually for every $x > -\omega$. Thus a directed join of maps in H need not preserve arbitrary meets.

Example 3.16 Likewise, an infinite meet of maps in H need not preserve directed joins. Let $\mathbf{S} = \omega + 1$, and for $j \in \omega$ define

$$k_j(x) = \begin{cases} x & \text{if } x > j, \\ 0 & \text{if } x \leq j \end{cases}$$

and let $m = \bigwedge_{j \in \omega} k_j$. Then $m(\bigvee j) = m(\omega) = \omega$ while $\bigvee m(j) = 0$. Thus a meet of infinitely many maps in H need not preserve directed joins.

Lemma 9.3 shows that a map h on an algebraic lattice $\mathbf{S}$ preserves directed joins if and only if it is determined by its values on compact elements, i.e., $h(s) = \bigvee\{h(c) : c \leq s,\ c \text{ compact}\}$.

Straightforward arguments show that finite meets of maps in H preserve arbitrary meets and directed joins of elements. Table 3.1 summarizes the information on what preserves what: rows correspond to maps, columns to elements. But there is a lot more to be understood about the role of $\mathbb{DMO}$.

Maps\elements	Finite meets	Arbitrary meets	Directed joins
Fin meets	✓	✓	✓
Arb meets	✓	✓	X
Dir joins	✓	X	✓

Table 3.1 Whether operations on maps in H (rows) preserve operations on elements of $\mathbf{S}$ (columns)

3.3 Companion Lattices

Like any closure operator, an equaclosure operator produces a lattice of closed sets. Given a pair $(\mathbf{L}, \gamma)$ the lattice $\gamma(\mathbf{L})$ is called the *companion* lattice. It is useful to think of the elements of $\gamma(\mathbf{L})$ as classes of the equapartition, which will be denoted by $[x]$. So $[x] \leq [y]$ iff $\gamma x \leq \gamma y$. To confuse matters further, for purposes that will become apparent later, it is useful to choose τx as the canonical representative of $[x]$. Thus we label the class as $[\tau x]$.

Worse, as we shall see below, we need the dual lattice $\gamma^d(\mathbf{L})$. Denote its elements by $\hat{x}$ with $x \in \tau(\mathbf{L}) = \{\tau(x) : x \in L\}$. The order on $\gamma^d(\mathbf{L})$ is given by $\hat{x} \leq \hat{y}$ iff $\gamma x \geq \gamma y$. Since $\tau(x \vee y) \neq \tau x \vee \tau y$ in general, we will have occasion to use the map

$$\hat{\tau}(w) = \widehat{\tau(w)}.$$

These conventions are illustrated in Fig. 3.4.

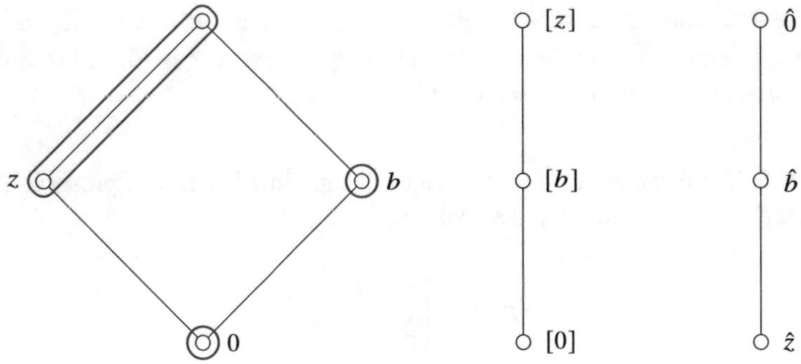

Fig. 3.4 Conventions for $\mathbf{L}, \gamma(\mathbf{L}), \gamma^d(\mathbf{L})$

The next two lemmas tell us how to calculate in the companion lattices.

Lemma 3.17 *Let γ be an equaclosure operator on a dually algebraic lattice* $\mathbf{L}$, *and let $\{x_i : i \in I\} \subseteq L$. Then in $\gamma(\mathbf{L})$:*

(1) $[0]$ *is the least element and* $[\tau(1)]$ *is the greatest element,*
(2) $\bigvee_{\gamma(\mathbf{L})}[x_i] = [\bigvee_{\mathbf{L}} x_i]$,

(3) $\bigwedge_{\gamma(\mathbf{L})}[x_i] = [\bigwedge_{\mathbf{L}} \gamma x_i]$.

Lemma 3.18 *Let γ be an equaclosure operator on a dually algebraic lattice* $\mathbf{L}$, *and let $\{x_i : i \in I\} \subseteq \tau(\mathbf{L})$, so that $\tau(x_i) = x_i$ for every i. Then in $\gamma^d(\mathbf{L})$:*

(1) $\hat{\tau}(1)$ *is the least element and* $\hat{0}$ *is the greatest element,*
(2) $\bigvee_{\gamma^d(\mathbf{L})} \hat{x}_i = \hat{\tau}(\bigwedge_{\mathbf{L}} \gamma x_i)$,
(3) $\bigwedge_{\gamma^d(\mathbf{L})} \hat{x}_i = \hat{\tau}(\bigvee_{\mathbf{L}} x_i)$.

The proofs are straightforward. Let us illustrate this with the proof of the last one, $\hat{x} \wedge \hat{y} = \hat{\tau}(x \vee y)$. Assume $x = \tau x$, $y = \tau y$, $u = \tau u$. We calculate:

$$\begin{aligned} \hat{u} \leq \hat{x} \;\&\; \hat{u} \leq \hat{y} &\text{ iff } \gamma u \geq \gamma x \;\&\; \gamma u \geq \gamma y \\ &\text{ iff } \gamma u \geq x \;\&\; \gamma u \geq y \\ &\text{ iff } \gamma u \geq x \vee y \\ &\text{ iff } \gamma u \geq \gamma(x \vee y) = \gamma(\tau(x \vee y)) \\ &\text{ iff } \hat{u} \leq \hat{\tau}(x \vee y). \end{aligned}$$

Obvious modifications yield Lemma 3.18(3) for an arbitrary index set.

We are of course primarily interested in the case when $\mathbf{L} = \mathrm{S_p}(\mathbf{S}, H)$ where $\mathbf{S}$ an algebraic lattice and H a monoid of operators, with Γ the natural weak equaclosure operator given by $\Gamma(X) = \uparrow x_0$ where x_0 is the least element of X.

For an element u of $\mathbf{S}$, $\mathrm{Ag}(u)$ denotes the H-closed algebraic set generated by u. As long as H is a monoid, containing the identity map and closed under composition, the least element of $\mathrm{Ag}(u)$ is $u_0 = \bigwedge_{h \in H} h(u)$, since for this u_0 the filter $\uparrow u_0$ is an H-closed algebraic subset containing u.

The next lemma tells us how to interpret the abstract notions in the pair $(\mathrm{S_p}(\mathbf{S}, H), \Gamma)$.

Lemma 3.19 *Let $\mathbf{L} = \mathrm{S_p}(\mathbf{S}, H)$ and let Γ be the natural weak equaclosure operator. Let X, Y be H-closed algebraic subsets of* $\mathbf{S}$ *with least elements x_0, y_0, respectively.*

(1) $\Gamma X = \uparrow x_0$.
(2) $\tau X = \mathrm{Ag}(x_0)$.
(3) $\Gamma X \leq \Gamma Y$ *iff* $x_0 \geq y_0$.
(4) $\tau X \leq \tau Y$ *iff* $x_0 = h(y_0)$ *for some* $h \in \mathbb{DMO}$.

This brings us to the main result linking $\Gamma(\mathrm{S_p}(\mathbf{S}, H))$ and $\mathbf{S}$. Recall from Theorem 1.22 that the fully invariant elements of an algebraic lattice $\mathbf{L}$ with operators form a complete sublattice $\mathrm{Fi}(\mathbf{S})$. The fully invariant elements are precisely the least elements of H-closed algebraic subsets, and two algebraic subsets are in the same class when they have the same least element, i.e., $[X] = [Y]$ if and only if $x_0 = y_0$. The next results state these facts in terms of companion lattices.

Theorem 3.20 *Let $\mathbf{L} = \mathrm{S_p}(\mathbf{S}, H)$ and let Γ be the natural weak equaclosure operator. Define $f : \Gamma(\mathbf{L}) \to \mathbf{S}$ via $f[X] = x_0$. Then f is a dual complete lattice embedding. The range of f is the set* $\mathrm{Fi}(\mathbf{S})$ *of fully invariant elements of* $\mathbf{S}$.

Proof Clearly f is one-to-one and order-reversing.

Let X_i $(i \in I)$ be Γ-closed algebraic subsets, say $X_i = \uparrow x_{i0}$ for each $i \in I$. The equation $f(\bigwedge_{\Gamma(\mathbf{L})}[X_i]) = \bigvee_{\mathbf{S}} x_{i0}$ says that the least element of $\bigcap(\uparrow x_{i0})$ is $\bigvee x_{i0}$. Similarly, the equation $f(\bigvee_{\Gamma(\mathbf{L})}[X_i]) = \bigwedge_{\mathbf{S}} x_{i0}$ says that the least element of $\bigvee(\uparrow x_{i0}) = \mathrm{Ag}(\bigcup(\uparrow x_{i0}))$ is $\bigwedge x_{i0}$. □

Corollary 3.21 *Let* $\mathbf{L} = \mathrm{S_p}(\mathbf{S}, H)$ *and let* Γ *be the natural weak equaclosure operator. The map* $f^d : \Gamma^d(\mathbf{L}) \to \mathbf{S}$ via $f(\hat{X}) = x_0$ *is a complete lattice embedding.*

Corollary 3.22 *Let* $\mathbf{L} = \mathrm{S_p}(\mathbf{S}, H)$ *and let* Γ *be the natural weak equaclosure operator. The companion* $\Gamma^d(\mathbf{L})$ *is algebraic, and hence* $\Gamma(\mathbf{L})$ *is dually algebraic.*

This despite the fact that $\Gamma(\mathbf{L})$ need not be a dually algebraic subset of $\mathbf{L}$; generally, it is only a complete meet subsemilattice of $\mathbf{L}$. However, the companion $\Gamma^d(\mathbf{L})$ can be *any* algebraic lattice, due to the following observation of Adaricheva and Gorbunov [16].

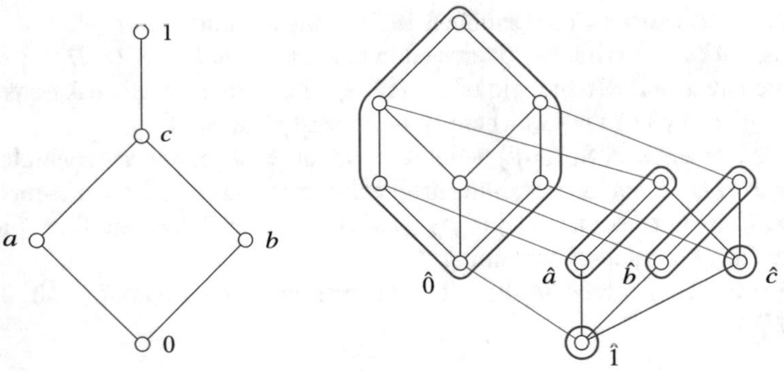

Fig. 3.5 Example of Theorem 3.23 that $\mathbf{K} \cong \Gamma^d(\mathrm{S_p}(\mathbf{K}))$

Theorem 3.23 *Let* $\mathbf{K}$ *be an algebraic lattice and* $\mathbf{S} = \mathrm{S_p}(\mathbf{K})$ *(with no operators). For the natural equaclosure operator,* $\Gamma^d(\mathbf{S}) \cong \mathbf{K}$.

Theorem 3.23 is illustrated for a small lattice in Fig. 3.5.

To recapitulate, if $\mathbf{L} = \mathrm{S_p}(\mathbf{S}, H)$ with $\mathbf{S}$ algebraic, H a monoid of operators and Γ the natural weak equaclosure operator, then

$$\mathbf{G} = f(\Gamma(\mathbf{L})) = \{x_0 : [X] \in \Gamma(\mathbf{L})\}$$

is a complete sublattice of $\mathbf{S}$ dually isomorphic to $\Gamma(\mathbf{L})$. Note that $\mathbf{L}$ and $\Gamma(\mathbf{L})$ are dually algebraic, and thus generated by their completely join irreducible elements, while $\mathbf{S}$ and $\mathbf{G}$ are algebraic, and hence generated by their completely meet irreducibles. The next lemma gives a little more information about this relation.

Lemma 3.24 *Let* $\mathbf{L} = \mathrm{S_p}(\mathbf{S}, H)$ *and let* Γ *be the natural weak equaclosure operator. Let* $f : \Gamma(\mathbf{L}) \to \mathbf{S}$ via $f[X] = x_0$ *and let* $\mathbf{G} = f(\Gamma(\mathbf{L}))$. *For an algebraic subset* X *in* $\mathbf{L}$, $f[X] = x_0$ *is meet irreducible in* $\mathbf{G}$ *if and only if* τX *is join prime in* $\mathbf{L}$.

Proof First suppose that τX is join prime in $\mathbf{L}$ and that $x_0 = y_0 \wedge z_0$ in $\mathbf{G}$. This implies both $\tau X \le Y \vee Z$ and $Y, Z \le \Gamma X$. Thus either $\tau X \le Y \le \Gamma X$ or $\tau X \le Z \le \Gamma X$, so that $[X] = [Y]$ or $[X] = [Z]$, whence $x_0 = y_0$ or $x_0 = z_0$. Therefore, x_0 is meet irreducible in $\mathbf{G}$.

Conversely, assume that x_0 is meet irreducible in $\mathbf{G}$. Suppose $\tau X \le Y \vee Z$. Then

$$\tau X \le \Gamma X \wedge (Y \vee Z) = (\Gamma X \wedge Y) \vee (\Gamma X \wedge Z) \le \Gamma X.$$

Let $Y' = \Gamma X \wedge Y$ and $Z' = \Gamma X \wedge Z$. From the above, $[Y' \vee Z'] = [X]$, so that $y_0' \wedge z_0' = x_0$. Since x_0 is meet irreducible in $\mathbf{G}$, either $y_0' = x_0$ or $z_0' = x_0$. Hence either $\tau X = \tau Y' \le Y' \le Y$ or similarly $\tau X \le Z$. Thus τX is join prime in $\mathbf{L}$. □

Besides $\mathbf{G}$, the lattice $\mathbf{S}$ must contain elements from $P \setminus P_*$ for each completely join irreducible $P \in \mathbf{L}$.

Lemma 3.25 *Let* $\mathbf{L} = \mathrm{S_p}(\mathbf{S}, H)$ *and let* P *be completely join irreducible in* $\mathbf{L}$.

(1) $P = \mathrm{Ag}(p)$ *for each* $p \in P \setminus P_*$.
(2) *If* $p, p_1 \in P \setminus P_*$, *then* $p = h(p_1)$ *for some* $h \in \mathbb{DMO}$.
(3) *If* $p \in P \setminus P_*$ *and* $x_0 \in \mathbf{G}$, *then* $p \ge x_0$ *iff* $\Gamma P \le \Gamma X$.

We would like to know more about $\mathbf{S} \setminus \mathbf{G}$. It turns out that if $\mathbf{L}$ is a finite lower bounded lattice and $\mathrm{J}(\mathbf{L}) \subseteq \tau(\mathbf{L}) = \{\tau x : x \in L\}$ and $(\mathbf{L}, \gamma)$ can be represented as $(\mathrm{S_p}(\mathbf{S}, H), \Gamma)$ for some pair $(\mathbf{S}, H)$, then it can be represented using $\mathbf{S} = \mathbf{G} = \gamma^d(\mathbf{L})$ (Theorem 5.6). Section 5.2 gives an algorithm for determining whether a finite lower bounded lattice with $\mathrm{J}(\mathbf{L}) \subseteq \tau(\mathbf{L})$ has such a representation; we just need to figure out when it gives an affirmative answer! It may even be that if $\mathrm{J}(\mathbf{L}) \subseteq \tau(\mathbf{L})$ and $(\mathbf{L}, \gamma)$ satisfies all the known restrictions, then it is representable as $(\mathrm{S_p}(\mathbf{S}, H), \Gamma)$. That would be nice. (This is Problem 2 in Sect. 10.1.)

3.4 A New Condition for Equaclosure Operators

This section presents a new restriction on equaclosure operators, called (K10). It was motivated by our desire to show that the pair $(\mathbf{L}, \gamma)$ in Fig. 3.6 is not representable as $(\mathrm{S_p}(\mathbf{S}, H), \Gamma)$. That same lattice, with a different preclop, is represented as a subquasivariety lattice in Example 7.21.

Theorem 3.26 *Suppose we are given* $(\mathbf{L}, \gamma)$ *with elements satisfying*

(1) $\tau a \le \tau b$,
(2) $\gamma s \le \gamma b \le \gamma(r \vee s)$.

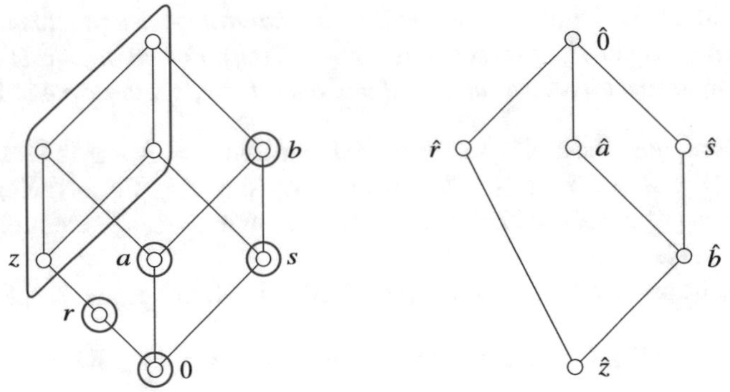

Fig. 3.6 $\mathbf{L}$ and $\gamma^d(\mathbf{L})$ for the pair $(\mathbf{L}, \gamma)$ motivating (K10). This pair satisfies (I1)–(I8) and (K9)

If $(\mathbf{L}, \gamma)$ *can be represented as* $(\mathrm{S_p}(\mathbf{S}, H), \Gamma)$ *with* $\mathbf{S}$ *algebraic, H a monoid of operators, and* Γ *the natural weak equaclosure operator, then*

(3) $a \le \gamma(\tau(r \vee s) \wedge (\tau r \vee (\tau s \wedge \gamma a)))$.

Proof Recall Corollary 1.20, the definition of Γ, and Lemma 3.19(4) for the least elements of H-closed algebraic subsets R, S, A, B:

- The least element of $R \vee S$ is $r_0 \wedge s_0$,
- $\Gamma S \le \Gamma B$ iff $s_0 \ge b_0$,
- $\tau A \le \tau B$ iff $a_0 = h(b_0)$ for some $h \in \mathbb{DMO}$.

Condition (1) says that $a_0 = h(b_0)$ for some $h \in \mathbb{DMO}$, while (2) says that $s_0 \ge b_0 \ge r_0 \wedge s_0$. Applying h yields $h(s_0) \ge a_0 \ge h(r_0 \wedge s_0)$. This shows that $h(s_0)$ is in $\tau s \wedge \Gamma a$ and also that a_0 is in Γe, where $e = h(r_0 \wedge s_0)$. Thinking of e as $h(r_0 \wedge s_0)$, it is in $\tau(r \vee s)$. Thinking of e as $h(r_0) \wedge h(s_0)$, it is in $\tau r \vee (\tau s \wedge \Gamma a)$. Thus e is in both. Combining this with $a_0 \in \Gamma e$ gives (3). □

Thus we obtain condition (K10) for an equaclosure operator:

(K10) If $\tau a \le \tau b$ and $\gamma s \le \gamma b \le \gamma(r \vee s)$, then $a \le \gamma(\tau(r \vee s) \wedge (\tau r \vee (\tau s \wedge \gamma a)))$.

Note that the condition can be written in terms of r and s, or in terms of τr and τs, with the strongest conclusion being when $r = \tau r$ and $s = \tau s$. Using just r and s, the conclusion looks like

(3′) $a \le \gamma(\tau(r \vee s) \wedge (r \vee (s \wedge \gamma a)))$.

It is straightforward to check that the pair $(\mathbf{L}, \gamma)$ in Fig. 3.6 fails (K10), but satisfies (I1)–(I8), (K9), and (K11). Thus (K10) is independent of the other ten conditions for an equaclosure operator.

Example 3.27 To see how (K10) works, consider the familiar example in Fig. 3.7 with the indicated labeling. Note that the hypotheses of (K10) are satisfied. The pair $(\mathbf{K}, \gamma)$ in the figure has the property that $x = \tau x$ for every join irreducible element x. This property, written as $\mathrm{J}(\mathbf{K}) \subseteq \tau(\mathbf{K})$, is the subject of Chap. 5. Theorem 5.6 says that if a finite pair $(\mathbf{L}, \gamma)$ has the property $\mathrm{J}(\mathbf{L}) \subseteq \tau(\mathbf{L})$ and it can be represented as $(\mathrm{S_p}(\mathbf{S}, H), \Gamma)$ for some $\mathbf{S}$ and H, then in fact it can be represented with $\mathbf{S} = \gamma^d(\mathbf{L})$. Moreover, in that representation the elements of $\mathbf{L}$ correspond to meet subsemilattices of $\gamma^d(\mathbf{L})$ with 1, and the question of representation is whether there exists a collection H of meet homomorphisms such that the meet subsemilattices corresponding to elements of $\mathbf{L}$ are the *only* H-closed subsemilattices.

For the pair $(\mathbf{K}, \gamma)$ of Fig. 3.7, a representation as $(\mathrm{S_p}(\gamma^d(\mathbf{K}), H), \Gamma)$ would have

$$A = \{\hat{a}, \hat{0}\}$$
$$B = \{\hat{b}, \hat{a}, \hat{0}\}$$
$$R = \{\hat{r}, \hat{a}, \hat{0}\}$$
$$Z = \{\hat{z}, \hat{0}\}.$$

Since $a \leq b$ in $\mathbf{K}$, we must have $A \leq B = \mathrm{Ag}(\hat{b})$. Thus, as in Lemma 3.19(4), we need a map $h \in H$ with $h(\hat{b}) = \hat{a}$. Alas, $\hat{r} \wedge \hat{b} = \hat{z}$, whence $h(\hat{z}) = h(\hat{r}) \wedge h(\hat{b}) = h(\hat{r}) \wedge \hat{a}$. It is also required that $h(\hat{z}) \in Z$, and with $h(\hat{r}) \in R$ and $h(\hat{z}) = h(\hat{r}) \wedge \hat{a}$, that is impossible. Thus $(\mathbf{K}, \gamma)$ has no representation in $\gamma^d(\mathbf{K})$, and hence by Theorem 5.6 no representation at all. (Yes, we already knew that this lattice is not representable, since it fails (‡), but see the next proposition.)

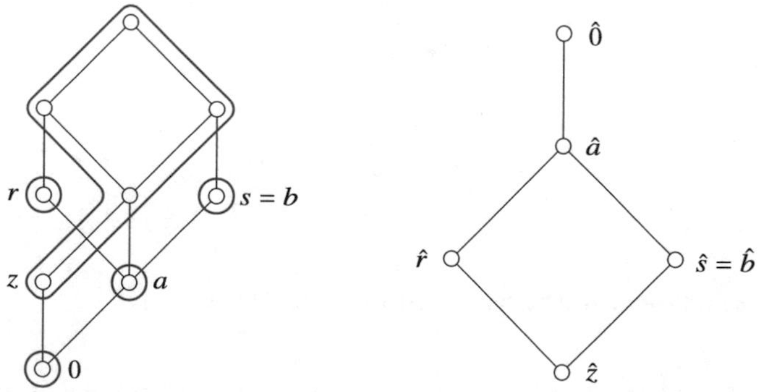

Fig. 3.7 $\mathbf{K}$ and $\gamma^d(\mathbf{K})$ failing (K10). This one also fails (K9)

Next, we note that (K10) implies the older version of (†), i.e., the original from [20], before refinement to (‡) in [92].

Proposition 3.28 *If* $(\mathbf{L}, \gamma)$ *is a pair where the preclop* γ *satisfies* (K10), *then it satisfies*

$$x \wedge \tau z \leq \gamma(\tau(x \vee z) \wedge \gamma x). \qquad (\dagger)$$

Proof Into (K10), substitute

$$a \mapsto x \wedge \tau z \qquad b \mapsto z \qquad r \mapsto x \qquad s \mapsto \tau z\,.$$

The hypotheses of (K10) hold, so we conclude

$$\begin{aligned} x \wedge \tau z &\leq \gamma(\tau(x \vee \tau z) \wedge (\tau x \vee (\tau z \wedge \gamma(x \wedge \tau z)))) \\ &\leq \gamma(\tau(x \vee z) \wedge \gamma x) \end{aligned}$$

as desired. □

For comparison, the new version of (K9) for finite index sets is

$$x \wedge \tau(z) \leq \gamma(\tau(x \vee z) \wedge x) \tag{‡}$$

with no γ on the last x. The pair $(\mathbf{L}, \gamma)$ in Fig. 3.8 satisfies (K10), and hence (†), but fails (‡). Thus (‡) is indeed stronger than (†).

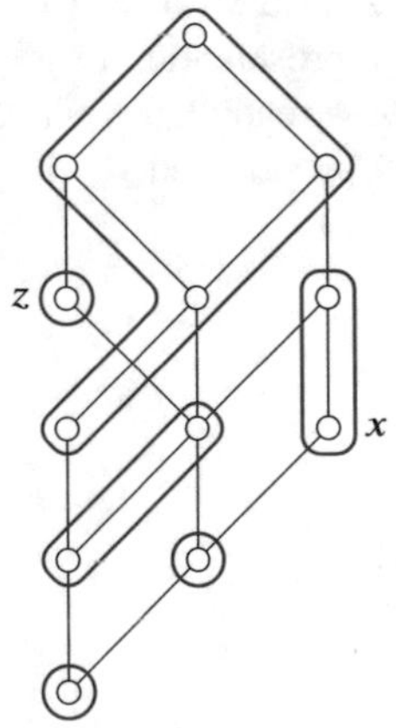

Fig. 3.8 Lattice with preclop that satisfies (K10), and hence (†), but fails (‡)

The algorithm from Sect. 5.2 suggests a refinement of (K10), which may or may not be useful, but is true.

Theorem 3.29 *Suppose we are given* $(\mathbf{L}, \gamma)$ *with elements satisfying*

(1) $\tau a \leq \tau b$,
(2) $\gamma s \leq \gamma t \leq \gamma b \leq \gamma(r \vee s)$.

If $(\mathbf{L}, \gamma)$ *can be represented as* $(\mathrm{S_p}(\mathbf{S}, H), \Gamma)$ *with* $\mathbf{S}$ *algebraic,* H *a monoid of operators, and* Γ *the natural weak equaclosure operator, then*

(3) $a \leq \gamma(\tau(r \vee s) \wedge (\tau r \vee (\tau s \wedge \gamma(\tau t \wedge \gamma a))))$.

Proof As in the beginning of the proof of Theorem 3.26, condition (1) says that $a_0 = h(b_0)$ for some $h \in \mathbb{DMO}$, while (2) says that $s_0 \geq t_0 \geq b_0 \geq r_0 \wedge s_0$. Applying h yields $h(s_0) \geq h(t_0) \geq a_0 \geq h(r_0 \wedge s_0)$. Thus $h(t_0)$ is in $\tau t \wedge \Gamma a$, whence $h(s_0)$ is in $\tau s \wedge \Gamma(\tau t \wedge \Gamma a)$. Also a_0 is in Γe, where $e = h(r_0 \wedge s_0)$. Thinking of e as $h(r_0 \wedge s_0)$, it is in $\tau(r \vee s)$. Thinking of e as $h(r_0) \wedge h(s_0)$, it is in $\tau r \vee (\tau s \wedge \Gamma(\tau t \wedge \Gamma a))$. Thus e is in both. Combining this with $a_0 \in \Gamma e$ gives (3). □

Indeed, we can make apparently stronger versions of (K10) by increasing the length of the chains in its hypothesis (2). Whether or not these conditions are really stronger remains open for investigation (Problem 14 in Chap. 10).

We can relate (K10) to the meet semidistributive law ($\mathrm{SD}_\wedge$) in the companion lattice $\gamma^d(\mathbf{L})$. Recall that a lattice is *meet semidistributive* if it satisfies the law

$$a \wedge b \approx a \wedge c \text{ implies } a \wedge b \approx a \wedge (b \vee c). \qquad (\mathrm{SD}_\wedge)$$

Corollaries 3.30 and 3.31 below say that, with an additional hypothesis, ($\mathrm{SD}_\wedge$) holds for certain triples in $\gamma^d(\mathbf{L})$. However, there must be a condition referring to the lattice $\mathbf{L}$ itself, because $\gamma^d(\mathbf{L})$ can be any algebraic lattice by Theorem 3.23.

To avoid confusion with substitutions, let us rewrite (K10) with different letters:

$$\text{If } \tau x \leq \tau y \text{ and } \gamma s \leq \gamma y \leq \gamma(r \vee s), \text{ then } x \leq \gamma(\tau(r \vee s) \wedge (\tau r \vee (\tau s \wedge \gamma x))). \quad (\mathrm{K10}^*)$$

Also, we must recall from Lemma 3.18 how to calculate in $\gamma^d(\mathbf{L})$:

$$\begin{aligned} \hat{x} \leq \hat{y} &\text{ iff } \gamma x \geq \gamma y \\ \hat{x} \wedge \hat{y} &= \hat{\tau}(x \vee y) \\ \hat{x} \vee \hat{y} &= \hat{\tau}(\gamma x \wedge \gamma y). \end{aligned}$$

Now we consider a couple types of instances of ($\mathrm{SD}_\wedge$) in $\gamma^d(\mathbf{L})$.

Corollary 3.30 *Assume* $(\mathbf{L}, \gamma)$ *can be represented by* $(\mathrm{S_p}(\mathbf{S}, H), \Gamma)$ *with* $\mathbf{S}$ *algebraic,* H *a monoid of operators, and* Γ *the natural weak equaclosure operator. For elements* $a, b, c \in L$, *if*

(1) $\tau b \leq \tau(b \vee c)$,
(2) $\hat{a} \wedge \hat{c} \leq \hat{a} \wedge \hat{b}$ *in* $\gamma^d(\mathbf{L})$,

then

(3) $\hat{a} \wedge \hat{b} = \hat{a} \wedge (\hat{b} \vee \hat{c})$ *in* $\gamma^d(\mathbf{L})$.

In particular, ($\mathrm{SD}_\wedge$) *holds for the triple* $(\hat{a}, \hat{b}, \hat{c})$ *whenever* $\tau b \leq \tau(b \vee c)$.

Proof The hypothesis (2) translates into $b \leq \gamma(a \vee c)$. Into ($\mathrm{K10}^*$), substitute

$$x \mapsto b \qquad y \mapsto b \vee c \qquad r \mapsto a \qquad s \mapsto c.$$

The hypotheses of (K10) are satisfied, so we conclude that

$$b \le \gamma(\tau(a \vee c) \wedge (\tau a \vee (\tau c \wedge \gamma b))) \le \gamma(\tau a \vee (\gamma c \wedge \gamma b))).$$

This in turn translates into $\hat{a} \wedge (\hat{b} \vee \hat{c}) \le \hat{b}$, which gives the desired equality $\hat{a} \wedge \hat{b} = \hat{a} \wedge (\hat{b} \vee \hat{c})$. □

A slight twist yields another version.

Corollary 3.31 *Assume* $(\mathbf{L}, \gamma)$ *can be represented by* $(\mathrm{S_p}(\mathbf{S}, H), \Gamma)$ *with* **S** *algebraic, H a monoid of operators, and* Γ *the natural weak equaclosure operator. For elements* $a, b, c \in L$, *if*

(1) $\tau b \le \tau(b \vee c)$,
(2) $\hat{c} \wedge \hat{a} = \hat{c} \wedge \hat{b}$ *in* $\gamma^d(\mathbf{L})$,

then

(3) $\hat{c} \wedge \hat{b} = \hat{c} \wedge (\hat{a} \vee \hat{b})$ *in* $\gamma^d(\mathbf{L})$.

Thus $(\mathrm{SD}_\wedge)$ *holds for the triple* $(\hat{c}, \hat{a}, \hat{b})$ *whenever* $\tau b \le \tau(b \vee c)$.

Proof The hypothesis (2) yields both $a \le \gamma(b \vee c)$ and $b \le \gamma(a \vee c)$. Into (K10*), substitute

$$x \mapsto b \qquad y \mapsto b \vee c \qquad r \mapsto c \qquad s \mapsto a.$$

The hypotheses of (K10) are satisfied, so we obtain that

$$b \le \gamma(\tau(c \vee a) \wedge (\tau c \vee (\tau a \wedge \gamma b))) \le \gamma(\tau c \vee (\gamma a \wedge \gamma b))).$$

This translates into $\hat{c} \wedge (\hat{a} \vee \hat{b}) \le \hat{b}$, which gives $\hat{c} \wedge \hat{b} = \hat{c} \wedge (\hat{a} \vee \hat{b})$, as desired. □

The lattice in Fig. 3.9 has the same non-semidistributive (both ways) companion as that in Fig. 3.6 but is easily representable as $(\mathrm{S_p}(\mathbf{S}, H), \Gamma)$ by the method of Chap. 5, as we only need a map with $\hat{z} \to \hat{a}$. (We have not yet tried to represent it as $\mathrm{L_q}(\mathcal{K})$.)

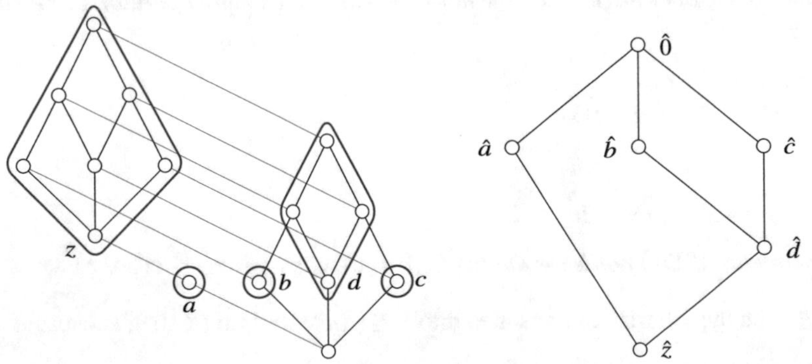

Fig. 3.9 Lattice with $\gamma^d(\mathbf{L})$ not meet semidistributive, but easily representable as $(\mathrm{S_p}(\mathbf{S}, H), \Gamma)$

3.5 Directed Meets in $\tau(\mathbf{L})$

Our last condition for equaclosure operators is comparatively simple. Recall that the map τ need not be order preserving. For arbitrary joins, we have Lemma 3.1: $\tau(\bigvee X) \leq \bigvee_{x \in X} \tau x$. Property (K11) is an analogue for down-directed meets.

Theorem 3.32 *The natural (weak) equaclosure operator on* $\mathrm{S_p}(\mathbf{S}, H)$ *satisfies*

(K11) *If* x *and* c_i $(i \in I)$ *are such that*

> (a) $\gamma c_i = c_i$,
> (b) $\{c_i : i \in I\}$ *is down-directed,*
> (c) $\tau c_i \leq x$ *for all* $i \in I$,

then $\tau(\bigwedge_i c_i) \leq x$.

Proof In $\mathrm{S_p}(\mathbf{S}, H)$, let X be any H-closed algebraic set, and let C_i for $i \in I$ be a down-directed collection of Γ-closed sets. Thus we can write $C_i = \uparrow c_{i0}$ for each i. If property (c) holds, $\tau(C_i) = \mathrm{Ag}(c_{i0}) \leq X$, then each c_{i0} is in X, and these least elements form an up-directed set in $\mathbf{S}$. Hence $\bigvee_i c_{i0}$ is in X. Moreover, it is clearly the least element of $\bigwedge_i C_i$. Therefore, $\tau(\bigwedge_i C_i) = \mathrm{Ag}(\bigwedge_i c_{i0}) \leq X$, as desired. □

Lemma 3.1 and Theorem 3.32 can be useful; see, e.g., the proof of Theorem 5.13.

Example 3.33 To show that (K11) is independent of the other ten properties of equaclosure operators, we need a pair $(\mathbf{L}, \gamma)$ that fails (K11) but satisfies the remaining properties. Let $\omega + 1$ be the ordered set $0 < 1 < 2 < \cdots < \omega$. The lattice $\mathbf{L}$ is the lattice of all subsets of $\omega + 1$, which is (algebraic and) dually algebraic. The closure operator γ uses the order on $\omega + 1$: for $S \subseteq \omega + 1$,

$$\gamma(S) = \uparrow S = \uparrow \min(S) .$$

One can check that γ satisfies (I1)–(I8) and (K9), (K10). For (I8), recall that the Boolean lattice $\mathbf{B}_\omega$ is isomorphic to $\mathrm{S_p}(\omega + 1)$.

Now, into (K11) substitute $C_j = \uparrow j$ for $j \in \omega + 1$ and $X = \omega = \{0, 1, 2, \dots\}$. We calculate that $\bigwedge_{j \in \omega} C_j = \{\omega\} = C_\omega$ while $\tau(C_j) = \{j\}$. Thus $\tau(C_j) \leq X$ for $j \in \omega$, but $\tau(\bigwedge_{j \in \omega} C_j) = \{\omega\} \nleq X$, so that (K11) fails.

3.6 More Properties of Equaclosure Operators

3.6.1 Three Properties of Weak Equaclosure Operators

In this subsection, we present some additional necessary conditions for representability, denoted (E1)–(E3), that turned out to be consequences of (I1)–(I8) and (K9)–(K11).

Recall the case $|I| = 1$ of condition (K9):

$$\gamma(x \wedge \tau(x \vee z)) \geq x \wedge \tau z. \tag{$\ddagger$}$$

Proposition 3.34 *Let γ be an equaclosure operator on a lattice* **L**. *Then the following property holds.*

(E1) *If $\tau a \leq \tau b$, then for every $s \in L$ there exists $s' \leq \tau s$ such that $\tau(a \vee s') \leq \tau(b \vee s)$.*

Indeed, (E1) is equivalent to ($\ddagger$).

Proof First assume that **L** satisfies ($\ddagger$). Assume $\tau a \leq \tau b$ and without loss of generality $\tau a = a$, $\tau b = b$, $\tau s = s$. (Since $\gamma(x \vee y) = \gamma(\tau x \vee \tau y)$ holds in general, then $\tau(x \vee y) = \tau(\tau x \vee \tau y)$.) Into ($\ddagger$) substitute $x \mapsto \tau a \vee s$ and $z \mapsto \tau b$, yielding

$$\gamma((\tau a \vee s) \wedge \tau(b \vee s)) \geq (\tau a \vee s) \wedge \tau b \geq \tau a.$$

Put $r = (\tau a \vee s) \wedge \tau(b \vee s)$, and note that $\tau r \leq \tau(b \vee s)$ and $r \leq \tau a \vee s$. Moreover, since $\tau a \leq \gamma r$ by the displayed inequality, $r' = \tau a \vee r$ has $\gamma r' = \gamma r$, whence $\tau r' = \tau r$. Thus $\tau r' \leq \tau(b \vee s)$ and $\tau a \leq r' \leq \tau a \vee s$. Consider that

$$r' \leq \gamma r' \wedge (\tau a \vee s) = \tau a \vee (\gamma r' \wedge s)$$

by meet distributivity, and put $s' = \gamma r' \wedge s$. Then $s' \leq s = \tau s$ and $r' \leq \tau a \vee s' \leq \gamma r'$, so that $\gamma(a \vee s') = \gamma r'$ and $\tau(a \vee s') = \tau r' \leq \tau(b \vee s)$.

Conversely, assume that **L** satisfies (E1). Into (E1) substitute $a \mapsto x \wedge \tau z$, $b \mapsto z$, and $s \mapsto x$. Then $\tau a \leq \tau b$, so there exists $s' \leq \tau x$ such that $\tau((x \wedge \tau z) \vee s') \leq \tau(x \vee z)$. Moreover, $\tau((x \wedge \tau z) \vee s') \leq (x \wedge \tau z) \vee s' \leq x$, whence $\tau((x \wedge \tau z) \vee s') \leq x \wedge \tau(x \vee z)$. Therefore, $x \wedge \tau z \leq \gamma((x \wedge \tau z) \vee s') \leq \gamma(x \wedge \tau(x \vee z))$, as desired. □

The next two conditions follow from condition (I6), the meet distributivity of γx.

Proposition 3.35 *Let γ be an equaclosure operator on a lattice* **L**. *Then the following property holds.*

(E2) *If $\tau a < \tau b$ and $\gamma b = \gamma(r \vee s)$, then $\tau a \leq \tau(\tau r \wedge \gamma a) \vee \tau(\tau s \wedge \gamma a)$.*

Proof We calculate:

$$\begin{aligned} \tau a &\leq \gamma a \wedge \tau b \\ &\leq \gamma a \wedge (\tau r \vee \tau s) \\ &= (\gamma a \wedge \tau r) \vee (\gamma a \wedge \tau s) \\ &\leq \gamma a. \end{aligned}$$

So $\tau a = \tau((\gamma a \wedge \tau r) \vee (\gamma a \wedge \tau s)) \leq \tau(\gamma a \wedge \tau r) \vee \tau(\gamma a \wedge \tau s)$ by Lemma 3.1. □

The meet distributivity of $\gamma(x)$ also gives a condition on minimal join covers in $\gamma(\mathbf{L})$.

Proposition 3.36 *Let γ be an equaclosure operator on a lattice* $\mathbf{L}$. *Then the following property holds.*

(E3) *If* $[a] \vee [b] = [z] > [c]$ *is a minimal join cover in* $\gamma(\mathbf{L})$, *then* $\tau z \ngeq \tau c$.

Proof For notational convenience assume $a = \tau a$, etc. Then

$$\gamma c \wedge (a \vee b) = (\gamma c \wedge a) \vee (\gamma c \wedge b).$$

We cannot have $\gamma c \geq a$ and $\gamma c \geq b$, else $[c] = [a] \vee [b]$. Hence, by the minimality of the join cover, the join on the right is some $[w]$ with $[w] \ngeq [c]$. So the class of the left-hand side is not above $[c]$ either, whence $a \vee b \ngeq c$, *a fortiori* $z \ngeq c$. □

3.6.2 An Almost Old Observation

We will expand on the following lemma proved in Adaricheva and Gorbunov [16, Lemma 1.6].

Lemma 3.37 *For any dually algebraic lattice* $\mathbf{L}$ *and a preclop* γ, *if* $a = \gamma(a)$ *and* $a < b$, *then* $\mathbf{L}_1 = [a, b]$ *is a dually algebraic lattice with preclop* $\gamma_1(x) = \gamma(x) \wedge b$ *for* $x \in L_1$.

The preclop γ_1 on $\mathbf{L}_1$ is called the *restriction* of γ to $\mathbf{L}_1$. The new version of the statement is the following.

Theorem 3.38 *Let* $\mathbf{L} = \mathrm{S_p}(\mathbf{S}, H)$ *where* $\mathbf{S}$ *is algebraic and* H *is a monoid of operators. Let* $A \in \mathbf{L}$ *with* $\Gamma(A) = A$, *so that* $A = \uparrow a_0$. *For any* $B \in L$ *with* $A \subseteq B$, *there exists a set of operations* H_1 *such that the interval* $\mathbf{L}_1 = [A, B]$ *of* $\mathbf{L}$ *is represented as* $\mathrm{S_p}(B, H_1)$. *(The operations in* H_1 *preserve nonempty meets and nonempty directed joins, but not necessarily* 1.*) Moreover, the natural equaclosure operator on* $\mathbf{L}_1$ *is the restriction of the natural equaclosure operator on* $\mathbf{L}$.

Proof Note that B, as an algebraic subset of $\mathbf{L}$, is itself an algebraic lattice. One needs to add operations to H to guarantee that A is the smallest element of $\mathrm{S_p}(B, H_1)$. For each $a \in A$ define the operation $h_a(x) = x \wedge a$. These operations preserve nonempty meets and nonempty directed joins (by upper continuity). Let $H_1 = H \cup \{h_a : a \in A\}$. Apparently, $A = \uparrow a_0$ is the smallest H_1-closed algebraic subset of B.

Note that every H-closed algebraic subset C of B that contains A will be closed under each h_a. Thus there is a one-to-one correspondence between the interval $\mathbf{L}_1 = [A, B]$ of $\mathbf{L}$ and $\mathrm{S_p}(B, H_1)$. Moreover, the Γ-closed subsets of $\mathrm{S_p}(B, H_1)$ are of the form $\uparrow c_0 \cap B$ for an H-closed principal filter with $b_0 \leq c_0 \leq a_0$. Thus the natural equaclosure operator on $\mathrm{S_p}(B, H_1)$ is the restriction to B of the natural equaclosure operator on $\mathrm{S_p}(\mathbf{S}, H)$. □

Chapter 4
Preclops on Finite Lattices

It takes at least five years of rigorous training to be spontaneous. – Martha Graham

Après cela, il y aura, j'espère, des gens qui trouveront leur profit à déchiffrer tout ce gâchis. – Evariste Galois

Our central problem for the next few chapters is this:

Given a finite lattice* L*, determine whether* $\mathbf{L} \cong \mathrm{L_q}(\mathcal{K})$ *for some quasivariety* $\mathcal{K}$*.

We divide the problem into three parts.

(1) *Does* **L** *support an equaclosure operator?* If not, the answer is NO, and we are done. In Chap. 4, we consider how to find all the preclops on **L**. Those preclops can then be tested for (K9) and (K10) to see if we have an equaclosure operator, the remaining properties (I8) and (K11) being trivial when **L** is finite.

If **L** has one or more equaclosure operators, then we consider pairs $(\mathbf{L}, \gamma)$.

(2) *Can the pair* $(\mathbf{L}, \gamma)$ *be represented as* $(\mathrm{Sub}(\mathbf{S}, \wedge, 1, H), \Gamma)$ *for* **S** *a finite semilattice with operators and its natural weak equaclosure operator? or more generally, as* $(\mathrm{S_p}(\mathbf{S}, H), \Gamma)$ *with* **S** *an algebraic lattice?* Chapters 5 and 6 address this topic.

If $(\mathbf{L}, \gamma)$ can be represented as $(\mathrm{Sub}(\mathbf{S}, \wedge, 1, H), \Gamma)$, then the pair is $(\mathrm{L_q}(\mathcal{K}_0), \Gamma)$ for a quasivariety of structures in a language without equality, with Γ the natural equaclosure operator for quasivarieties.

(3) *If* $(\mathbf{L}, \gamma)$ *can be represented as* $(\mathrm{Sub}(\mathbf{S}, \wedge, 1, H), \Gamma)$*, can we convert the representation* $(\mathrm{L_q}(\mathcal{K}_0), \Gamma)$ *to one* $(\mathrm{L_q}(\mathcal{K}_1), \Gamma)$ *with* $\mathcal{K}_1$ *in a language with equality?* Methods that sometimes achieve this are described in Chap. 7 under the names longstyle, shortstyle, and mediumstyle.

We have only partial answers to all three questions, to which we might add two more unanswered questions:

K. Adaricheva et al., *A Primer of Subquasivariety Lattices*, CMS/CAIMS Books in Mathematics 3, https://doi.org/10.1007/978-3-030-98088-7_4

(3a) *When can we represent* $(\mathbf{L}, \gamma)$ *with a quasivariety* $\mathcal{K}_2$ *of algebras?* That is, without using predicates.

(3b) *When can we represent* $(\mathbf{L}, \gamma)$ *with a locally finite quasivariety?*

With this itinerary, let us begin the journey.

4.1 Meet Distributive Elements and Preclops

Recall that a *preclop* satisfies properties (I1)–(I7). This chapter explores how to find preclops on finite join semidistributive lattices. Since property (I7) always holds for closure operators on a finite lattice, it will not be a concern in this chapter.

We consider join semidistributive lattices because every lattice $\mathrm{S_p}(\mathbf{S}, H)$ with $\mathbf{S}$ algebraic satisfies $\mathrm{SD}_\vee$. When the lattice $\mathbf{S}$ is finite, then $\mathrm{S_p}(\mathbf{S}, H) = \mathrm{Sub}(\mathbf{S}, \wedge, 1, H)$, which has the stronger property of being a lower bounded lattice. However, in Sect. 8.1 we will construct subquasivariety lattices $\mathrm{L_q}(\mathcal{K})$ that are finite and join semidistributive, but not lower bounded.

Call an element $a \in L$ *meet distributive* if $a \wedge (x \vee y) = (a \wedge x) \vee (a \wedge y)$ for all $x, y \in L$. Let $\mathrm{MD}(\mathbf{L})$ denote the set of all meet distributive elements of $\mathbf{L}$. Note that $\mathrm{MD}(\mathbf{L})$ is a meet subsemilattice of $\mathbf{L}$ containing 0 and 1. With each 0-1-meet subsemilattice K of $\mathrm{MD}(\mathbf{L})$ we can associate the closure operator $\kappa(x) = \bigwedge(\uparrow x \cap K)$, that is, the least element of K above x. Each such closure operator on $\mathbf{L}$ satisfies conditions (I1)–(I4) and (I6), while none of them need to satisfy (I5). We can interpret (I5) for finite lattices as: $\kappa(x) = \kappa(y)$ implies $\kappa(x) = \kappa(x \wedge y)$, or the subsemilattice form

$$\uparrow x \cap K = \uparrow y \cap K \text{ implies } \uparrow x \cap K = \uparrow(x \wedge y) \cap K. \qquad (*)$$

Theorem 4.1 *Let* $\mathbf{L}$ *be a finite lattice. There is a one-to-one correspondence between preclops on* $\mathbf{L}$ *and* 0-1*-meet subsemilattices* K *of* $\mathrm{MD}(\mathbf{L})$ *satisfying* $(*)$*. Given a preclop* κ*, the corresponding subsemilattice is* $K = \kappa(L)$*, the set of* κ*-closed elements. Given* K*, the corresponding closure operator is* $\kappa(x) = \bigwedge(\uparrow x \cap K)$*.*

Proof Note that $x \le \kappa(x) \in K$, so that for $k \in K$ we have $k \ge x$ iff $k \ge \kappa(x)$. Thus $\kappa(x) = \kappa(y)$ iff $\uparrow x \cap K = \uparrow y \cap K$. With this observation, $(*)$ translates as $\kappa(x) = \kappa(y) \to \kappa(x) = \kappa(x \wedge y)$, which is (I5). □

Let $\mathrm{Pre}(\mathbf{L})$ denote the set of all preclops on a lattice $\mathbf{L}$. Note that $\mathrm{Pre}(\mathbf{L})$ may well be empty; indeed, that is the case for most finite lattices. When $\mathrm{Pre}(\mathbf{L})$ is nonempty, the preclops are ordered pointwise: $\alpha \le \beta$ iff $\alpha(x) \le \beta(x)$ for all $x \in L$. As usual for closure operators, $\alpha \le \beta$ implies that every β-closed set is α-closed. In other words, if α corresponds to $\mathbf{A} \le \mathrm{MD}(\mathbf{L})$ and β to $\mathbf{B}$, then $\alpha \le \beta$ iff $\mathbf{A} \supseteq \mathbf{B}$.

The next theorem reproduces the result from Adaricheva and Gorbunov [16] that with this order the preclops form a meet semilattice.

Theorem 4.2 *Let* $\mathbf{L}$ *be a finite lattice. If* $\mathrm{Pre}(\mathbf{L})$ *is nonempty, then it is a meet semilattice.*

Proof We show that if A and B are 0-1-subsemilattices of MD(**L**) satisfying (∗), then $A \vee B$ also satisfies (∗). Note that since 1 is in A and B, we have $A \vee B = \mathrm{Sg}(A \cup B) = \{a \wedge b : a \in A,\ b \in B\}$.

Corresponding to $A \vee B$, let $\xi(x) = \alpha(x) \wedge \beta(x)$. Note that $\alpha(x) = \alpha(\xi(x))$ because $x \le \xi(x) \le \alpha(x)$, and likewise $\beta(x) = \beta(\xi(x))$. Thus $\xi(x) = \xi(y)$ implies that $\alpha(x) = \alpha(\xi(x)) = \alpha(\xi(y)) = \alpha(y)$, and similarly that $\beta(x) = \beta(y)$. But then

$$\xi(x \wedge y) = \alpha(x \wedge y) \wedge \beta(x \wedge y) = \alpha(x) \wedge \beta(x) = \xi(x),$$

verifying that (I5) holds for $\xi = \alpha \wedge \beta$. □

In fact, it is shown in [16] that if **L** is join continuous and Pre(**L**) is nonempty, then it is closed under arbitrary *nonempty* meets. This leads us to ask: *Is* Pre(**L**) *closed under directed joins?* If so, that would ensure the existence of maximal preclops in the infinite case.

Throughout what follows we will let μ denote the map on L

$$\mu(x) = \bigwedge(\uparrow x \cap \mathrm{MD}(\mathbf{L})),$$

which may or may not satisfy (I5). If μ satisfies (I5), then it is the least member of Pre(**L**).

Figure 4.1 shows all the preclops on a lattice **W**, and their order. Since we will use this example again, it is named **W** for *Will*, in honor of William the Bastard, who had to conquer England (1066) in order to change his nickname.

Given a finite lattice **L**, *how do we go about determining whether* **L** *supports a preclop? If it does, how do we find* Pre(**L**)*?*

This is a two-part question: first, we must find the set of meet distributive elements MD(**L**), and then decide which subsemilattices of that satisfy (I5). The next few results address the first part by characterizing meet distributive elements in a way that is easily tested. Later, Sect. 4.2 presents an algorithm that terminates with no possible preclop or it constructs the least preclop on **L** if Pre(**L**) is nonempty.

Recall these basics from Alan Day's theory of finite lower bounded lattices, i.e., lower bounded homomorphic images of a free lattice [35]. (A more complete summary is in Sect. 1.8.)

(1) $x \mathrm{D} y$ if some minimal nontrivial join cover of x contains y.
(2) $\mathrm{D}_0(\mathbf{L})$ is the set of join-prime elements.
(3) Recursively, $x \in \mathrm{D}_{k+1}(\mathbf{L})$ if every nontrivial join cover of x refines to one contained in $\mathrm{D}_k(\mathbf{L})$.
(4) A finite lattice is lower bounded if and only if $\mathrm{D}_n(\mathbf{L}) = \mathbf{L}$ for some n.

Let $\overline{\mathrm{D}}$ denote the reflexive, transitive closure of the D relation on J(**L**), the set of nonzero join irreducible elements of **L**, regarded as an ordered set. On finite lattices in general, $\overline{\mathrm{D}}$ is a quasi-order, but on finite lower bounded lattices it is a partial order. (In fact, this is equivalent to lower boundedness.)

Finding $(\mathrm{J}(\mathbf{L}), \overline{\mathrm{D}})$ for a finite lattice is relatively straightforward. Minimal nontrivial join covers are pairs (x, U) with $x \in \mathrm{J}(\mathbf{L})$, $U \subseteq \mathrm{J}(\mathbf{L})$ an antichain, $x \notin U$, such

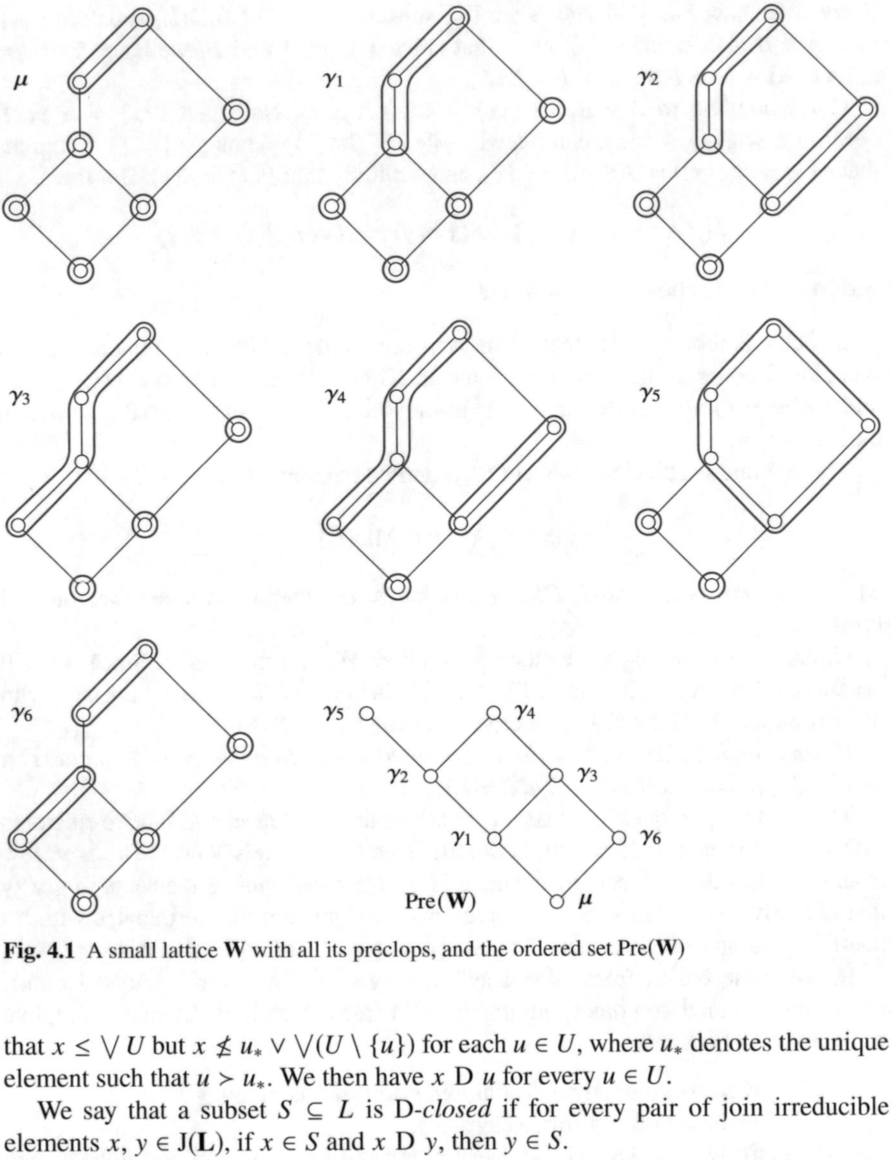

Fig. 4.1 A small lattice **W** with all its preclops, and the ordered set Pre(**W**)

that $x \leq \bigvee U$ but $x \nleq u_* \vee \bigvee(U \setminus \{u\})$ for each $u \in U$, where u_* denotes the unique element such that $u \succ u_*$. We then have x D u for every $u \in U$.

We say that a subset $S \subseteq L$ is D-*closed* if for every pair of join irreducible elements $x, y \in \mathrm{J}(\mathbf{L})$, if $x \in S$ and x D y, then $y \in S$.

Theorem 4.3 *Let* **L** *be a finite lattice and $a \in L$. Then $a \in \mathrm{MD}(\mathbf{L})$ if and only if $\downarrow a$ is* D*-closed. In other words, for any $x \in L$, $\mu(x)$ is the least element m of L such that $m \geq x$ and $\downarrow m$ is* D*-closed. Consequently, $a \in \mathrm{MD}(\mathbf{L})$ if and only if $\downarrow a \cap \mathrm{J}(\mathbf{L})$ is a filter in* $(\mathrm{J}(\mathbf{L}), \overline{\mathrm{D}})$.

Proof Assume $a \in \mathrm{MD}(\mathbf{L})$, and let $b \leq a$ be join irreducible. If $b \leq \bigvee U$ is a minimal nontrivial join cover, then $b \leq a \wedge (\bigvee U) = \bigvee_{u \in U} a \wedge u$, whence $u \leq a$ for all $u \in U$. Thus $\downarrow a$ is D-closed.

Conversely, assume that $\downarrow a$ is D-closed. We want to show that $a \wedge (x \vee y) = (a \wedge x) \vee (a \wedge y)$ for any $x, y \in L$. If suffices to show that for any join irreducible element p, if $p \leq a \wedge (x \vee y)$, then $p \leq (a \wedge x) \vee (a \wedge y)$.

If $p \leq a \wedge (x \vee y)$, then $p \leq a$ and $p \leq x \vee y$. If $p \leq x$ or $p \leq y$, then $p \leq a \wedge x$ or $p \leq a \wedge y$, whence $p \leq (a \wedge x) \vee (a \wedge y)$. If not, then $p \leq x \vee y$ refines to a minimal nontrivial join cover $p \leq \bigvee U$ with $U \ll \{x, y\}$, meaning that for all $u \in U$, either $u \leq x$ or $u \leq y$. Now $p \leq a$, and for each $u \in U$ we have $u \in \mathrm{J}(\mathbf{L})$ and $p \mathrel{\mathrm{D}} u$. Since $\downarrow a$ is D-closed that implies $u \leq a$, whence either $u \leq a \wedge x$ or $u \leq a \wedge y$. Therefore, $p \leq \bigvee U \leq (a \wedge x) \vee (a \wedge y)$, as desired. □

Corollary 4.4 *If* $\downarrow a \cap \mathrm{J}(\mathbf{L}) \subseteq \mathrm{D}_0(\mathbf{L})$, *then* $a \in \mathrm{MD}(\mathbf{L})$.

Proof A join-prime element has no nontrivial join cover. If $\downarrow a \cap \mathrm{J}(\mathbf{L}) \subseteq \mathrm{D}_0(\mathbf{L})$, then $\downarrow a$ is trivially D-closed. □

The next corollary says that in a finite join semidistributive lattice, only a few elements need be tested for meet distributivity. Recall that in a finite join distributive lattice, every element has a canonical join representation $a = \bigvee C$; see Sect. 1.8. Let $\mathrm{CJ}(a)$ denote the set of canonical joinands of an element a.

Corollary 4.5 *Let* $\mathbf{L}$ *be a finite join semidistributive lattice. If* $a \in \mathrm{MD}(\mathbf{L})$, *then* $\mathrm{CJ}(a) \subseteq \mathrm{D}_0(\mathbf{L})$.

Proof When a is meet distributive, then by Theorem 4.3, every minimal nontrivial join cover of an element in $\downarrow a$ is contained in $\downarrow a$. The Jónsson-Kiefer Property [69] says that in a finite join semidistributive lattice, the canonical joinands of any element a are join prime in $\downarrow a$, that is, $\mathrm{CJ}(a) \subseteq \mathrm{D}_0(\downarrow a)$. Thus the elements of $\mathrm{CJ}(a)$ for $a \in \mathrm{MD}(\mathbf{L})$ are join prime in $\mathbf{L}$. □

Theorem 4.3 gives a practical method to find $\mathrm{MD}(\mathbf{L})$. From that, we can determine the operator μ and $\mathrm{Pre}(\mathbf{L})$ for small lattices, and finding $\mathrm{MD}(\mathbf{L})$ is the first step of the more general algorithm of Sect. 4.2 to decide whether a finite lattice supports any preclop. Let us go through the details and then an example.

Suppose we are given a finite join semidistributive lattice $\mathbf{L}$. We begin by finding $\mathrm{J}(\mathbf{L})$ and all the minimal nontrivial join covers, from which we get the D relation on $\mathrm{J}(\mathbf{L})$. If $\mathbf{L}$ is lower bounded, then the transitive closure $\overline{\mathrm{D}}$ is a partial order. If $\mathbf{L}$ is not lower bounded, then $\overline{\mathrm{D}}$ is only a quasi-order, and we must factor out by the equivalence relation induced by cycles $x_1 \mathrel{\mathrm{D}} x_2 \mathrel{\mathrm{D}} \dots \mathrel{\mathrm{D}} x_n \mathrel{\mathrm{D}} x_1$. (This is relevant only when we are seeking an infinite representation of a non-lower bounded lattice, as in Sect. 8.1.) Let us assume that has been done. This enables us to draw the ordered set $(\mathrm{J}(\mathbf{L}), \overline{\mathrm{D}})$. The join-prime elements will be its maximal elements, since they have no nontrivial join cover.

It is useful, though not strictly necessary, at this point to identify the set C of candidates for $\mathrm{MD}(\mathbf{L})$, which are the join-prime elements and their joins. This includes 0, 1, and the canonical joinands of 1. If $\mathbf{L}$ is also meet semidistributive, then its atoms are join prime, but otherwise they need not be. Those members of C that satisfy the property that $\downarrow x$ is D-closed constitute $\mathrm{MD}(\mathbf{L})$.

Here is the algorithm for finding MD(**L**) and the operator μ for a finite join semidistributive lattice, starting at the bottom of the lattice and working our way up.

(0) Find the minimal nontrivial join covers in **L** and form the ordered set (J(**L**), $\overline{\mathrm{D}}$).
(1) Index the elements of L with a linear extension of the order of **L**: $L = \langle x_0, x_1, \ldots, x_n \rangle$ so that $x_i \leq x_j$ implies $i \leq j$. (Thus $x_0 = 0$ and $x_n = 1$.)
(2) Set $i = 0$, $M = \{0\}$, and $\mu(x_0) = 0$.
(3) Set $i \to i + 1$.
(4) If $i > n$ stop, return μ, and return M as MD(**L**).
(5) If $x_j < x_i \leq \mu(x_j)$ for some $j < i$, set $\mu(x_i) = \mu(x_j)$ and go to step (3).
(6) Else, compute $\mu(x_i)$ as follows.

   (a) $\mu_0(x_i) = x_i$
   (b) Recursively, set

$$\mu_{k+1}(x_i) = \bigvee \{q \in \mathrm{J}(\mathbf{L}) : p\,\overline{\mathrm{D}}\,q \text{ for some } p \in \downarrow \mu_k(x_i) \cap \mathrm{J}(\mathbf{L})\}.$$

   (c) When $\mu_{k+1}(x_i) = \mu_k(x_i)$, return $\mu(x_i) = \mu_k(x_i)$.

(7) Set $M \to M \cup \{\mu(x_i)\}$.
(8) Go to step (3).

The subroutine in step (6) calculates $\mu(x)$ as the least element $m \geq x$ such that $\downarrow m$ is D-closed, per Theorem 4.3.

We can prove that the algorithm is correct by applying the following straightforward lemma.

Lemma 4.6 *Let* **L** *be a finite lattice. For* $a \in L$, *let*

$$n(a) = \bigvee \{q \in \mathrm{J}(\mathbf{L}) : p\,\overline{\mathrm{D}}\,q \textit{ for some } p \in \downarrow a \cap \mathrm{J}(\mathbf{L})\}$$

so that $\mu_{k+1}(x) = n(\mu_k(x))$. *Then*

(1) $a \leq n(a)$,
(2) *if* $a \leq m \in$ MD(**L**), *then* $n(a) \leq m$,
(3) $n(a) = a$ *if and only if* $a \in$ MD(**L**).

Once we have found μ, it remains to check whether property (I5) holds, to see whether μ is a preclop. (Properties (I1)–(I4) and (I6) always hold for μ.) Sometimes a lattice **L** supports no preclop. But in Fig. 4.9, there is an example of a lattice that admits a real equaclosure operator, satisfying (I1)–(I8) and (K9)–(K11), even though μ is not a preclop. And in Example 4.13 we will see a lattice where μ is a preclop, but not an equaclosure operator, because it fails the condition (‡) from (K9).

For small lattices, once we have MD(**L**) and μ, it is not hard to find all of Pre(**L**). Preclops on **L** correspond to 0-1-subsemilattices of MD(**L**) that satisfy ($*$) on page 94. The order on Pre(**L**) is of course reversed: if α corresponds to **A** and β to **B**, then $\alpha \leq \beta$ iff $\mathbf{A} \supseteq \mathbf{B}$.

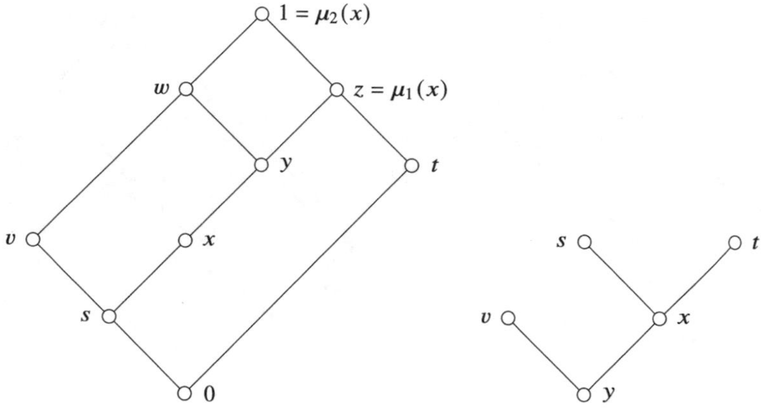

Fig. 4.2 Lattice **G** for Example 4.7 and (J(**G**), $\overline{\mathrm{D}}$)

Example 4.7 Consider the lattice **G** on the left in Fig. 4.2. The minimal nontrivial join covers in **G** are

$$x \leq s \vee t$$
$$y \leq s \vee t$$
$$y \leq x \vee v$$

yielding (J(**G**), $\overline{\mathrm{D}}$) as on the right in the figure. The join-prime elements are s, t, and v, making the candidates for meet distributivity $C = \{0, s, t, v, z, 1\}$. Thus it makes sense to list G as $\langle 0, s, t, v, x, y, z, w, 1\rangle$ in step (1) of the algorithm.

We easily get $\mu(0) = 0$, $\mu(s) = s$, $\mu(t) = t$, and $\mu(v) = v$. The crucial step in the algorithm is to calculate $\mu(x)$:

$$\mu_0(x) = x$$
$$\mu_1(x) = \bigvee\{x, s, t\} = z$$
$$\mu_2(x) = \bigvee\{x, s, t, y, v\} = 1$$

whence $\mu(x) = 1$. It follows of course that $\mu(a) = 1$ for all $a \geq x$, and thus we obtain MD(**G**) $= \{0, s, t, v, 1\}$. This gives μ as indicated on the left in Fig. 4.3, and this indeed satisfies (I5), so μ is a preclop.

Suppose γ is another preclop on **G**, whence $\gamma > \mu$. Since $x \wedge t = 0$ and $\gamma(x) = 1$, we must have $\gamma(t) = t$. If $\gamma(v) > v$, then $\gamma(v) = 1$, so that by (I5), $\gamma(s) = 1$. This is a valid possibility, and we obtain the second preclop γ drawn in Fig. 4.3. Thus Pre(**G**) consists of the two preclops in that figure.

In Sect. A.1 we will represent **G** as Sub(**T**, $\wedge$, 1, H) for a finite semilattice with operators.

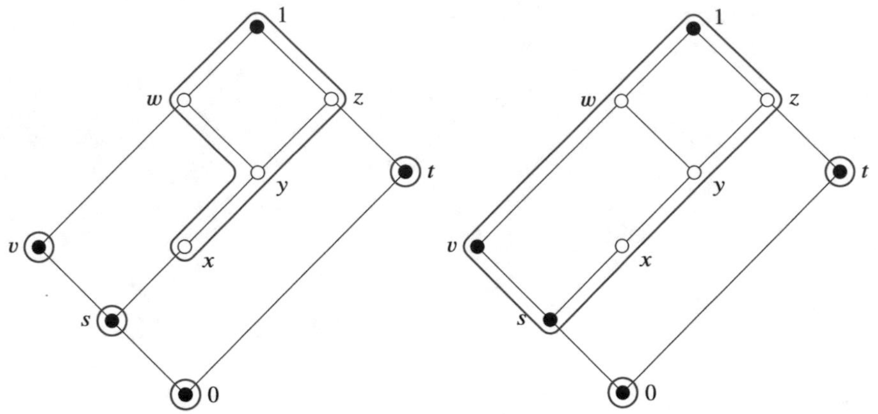

Fig. 4.3 The two preclops on **G** from Example 4.7. Meet distributive elements are marked with solid dots

It is a good exercise to replicate Fig. 4.1 by applying the algorithm to the lattice **W**.

4.1.1 Some Finite Join Semidistributive Lattices That Support No Preclop

The lattices in Fig. 4.4 are both upper and lower bounded, but admit no preclop. To see that there is no preclop, we repeatedly use the following argument, which we record for future reference.

Lemma 4.8 *If γ is a preclop on* **L**, $x \in \mathrm{J}(\mathbf{L})$, *and* $x \mathrel{\mathrm{D}} y$, *then* $\gamma(x) \geq x \vee y$.

This is because $\mu(x) \in \mathrm{MD}(\mathbf{L})$ and $\mu(x) \geq x \vee y$ by Theorem 4.3, and $\gamma \geq \mu$.

Suppose γ were a preclop on the hexagon **H** on the left of Fig. 4.4. Since $a \leq b \vee d$ is a minimal nontrivial join cover, we have $a \mathrel{\mathrm{D}} d$, whence $\gamma(a) \geq a \vee d = 1$, so $\gamma(a) = 1$. Similarly $\gamma(c) = 1$. But then property (I5) implies $\gamma(0) = \gamma(a \wedge c) = 1$, contradicting (I4). Thus there is no preclop on the hexagon.

Suppose γ were a preclop on the lattice **J** on the right of Fig. 4.4. In this lattice, $p \leq q \vee r$ and $r \leq s \vee t$ are minimal nontrivial join covers. In particular, $p \mathrel{\mathrm{D}} r$ and $r \mathrel{\mathrm{D}} t$. Therefore, $\gamma(t) \geq \gamma(p) \geq r$, so $\gamma(t) = 1$. Likewise, $\gamma(r) \geq r \vee t$, whence $\gamma(r) = 1$. Thus by (I5), $\gamma(0) = \gamma(r \wedge t) = 1$, contradicting (I4). So again we conclude that the lattice supports no preclop.

An alternative argument, really just a variation, works for the hexagon. By Corollary 4.5, every $\gamma(x)$ is a join of join-prime elements. Thus $\gamma(L) \subseteq \{0, b, d, 1\}$. This yields $\gamma(a) = \gamma(c) = 1$ and the same contradiction as before. But $t \in \mathrm{D}_0(\mathbf{L})$ for the second lattice, so the weaker argument does not suffice for it.

The following result from Adaricheva and Gorbunov [16] is a good source of examples. Recall that the lattice $\mathrm{Co}(\mathbf{P})$ of convex subsets of a finite ordered set **P**

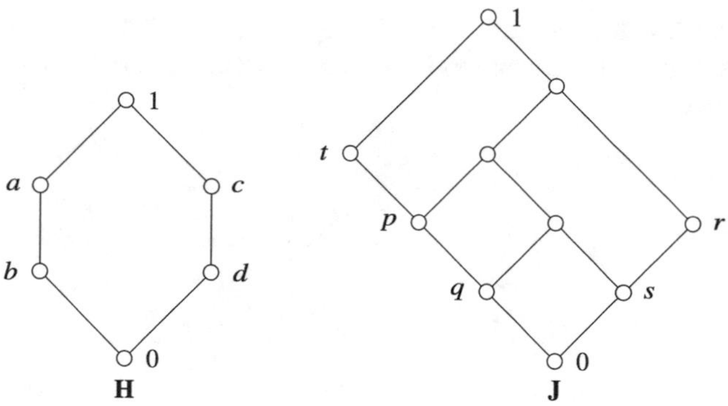

Fig. 4.4 Examples of lower bounded lattices that support no preclop. These two lattices are also upper bounded

is atomistic and join semidistributive; it is lower bounded if and only if **P** does not contain a 4-element chain. Let $\mathbf{P}_5$ be the ordered set with $a < b < c > d > e$.

Theorem 4.9 *The following properties are equivalent for a finite ordered set* **P**.

(1) $\mathrm{Co}(\mathbf{P}) \cong \mathrm{L_q}(\mathcal{K})$ *for some quasivariety* $\mathcal{K}$.
(2) $\mathrm{Co}(\mathbf{P})$ *admits a preclop.*
(3) $\mathrm{Co}(\mathbf{P})$ *admits an equaclosure operator.*
(4) **P** *contains none of* $\mathbf{4}$, $\mathbf{2}^2$, $\mathbf{P}_5$, *and* $\mathbf{P}_5^d$.

An easy part of Theorem 4.9, that (2) is equivalent to (4), fits into the current discussion. Note that $\mathrm{Co}(\mathbf{P})$ is atomistic, with 1-element subsets being the atoms; let us denote $\overline{a} = \{a\}$. Arguments similar to those used for the lattices in Fig. 4.4 show that if **P** contains one of the four forbidden subposets, then $\mathrm{Co}(\mathbf{P})$ does not support a preclop. For example, assume that **P** contains $\mathbf{P}_5$, and suppose that γ is a preclop on $\mathrm{Co}(\mathbf{P})$. Note that $\{a, d\}$ and $\{b, e\}$ are convex, since the elements in these sets are incomparable. Moreover, in $\mathrm{Co}(\mathbf{P})$ we have $\overline{b}\ \mathrm{D}\ \overline{a}, \overline{c}$ and $\overline{d}\ \mathrm{D}\ \overline{c}, \overline{e}$. Thus $\gamma(\{a, d\})$ contains $\{c, e\}$, hence by convexity also $\{b\}$, so that $\gamma(\{a, d\}) \supseteq \{a, b, c, d, e\} \supseteq \{b, e\}$. Similarly, $\gamma(\{b, e\}) \supseteq \{a, b, c, d, e\} \supseteq \{a, d\}$. Thus $\gamma(\{a, d\}) = \gamma(\{b, e\})$, while $\{a, d\} \cap \{b, e\} = \varnothing$, contradicting (I5). The other cases are similar.

On the other hand, if **P** contains none of the forbidden four subposets, then it is not hard to see that $\mathrm{Co}(\mathbf{P})$ is a direct product of copies of $\mathrm{Co}(\mathbf{3})$, which does admit a preclop, and the lattice $\mathbf{2}$. It is straightforward that a direct product of lattices that support a preclop itself supports a preclop. So for any ordered set that does not contain $\mathbf{4}$, $\mathbf{2}^2$, $\mathbf{P}_5$, or $\mathbf{P}_5^d$, the lattice $\mathrm{Co}(\mathbf{P})$ does admit a preclop.

See also Theorem 4.11 below with regard to the equivalence of (1) and (3).

The class of all finite lattices that admit a preclop is not even closed under taking 0-1-sublattices. Let $\mathbf{K} = \mathbf{3}\times\mathbf{3}$, with the two corner points labeled c and d. Form **L** by doubling both c and d. The lattice **L** is bounded and admits the equaclosure operator γ with $\gamma(c_1)$ its upper cover, $\gamma(d_1)$ its upper cover, and $\gamma(x) = x$ otherwise. See

Fig. 4.5. Indeed, this lattice can easily be represented as $\mathrm{Sub}(\mathbf{S}, \wedge, 1, H)$ using the methods of Chap. 7. But $\mathbf{L}$ contains the hexagon, which does not admit a preclop, as a 0-1-sublattice.

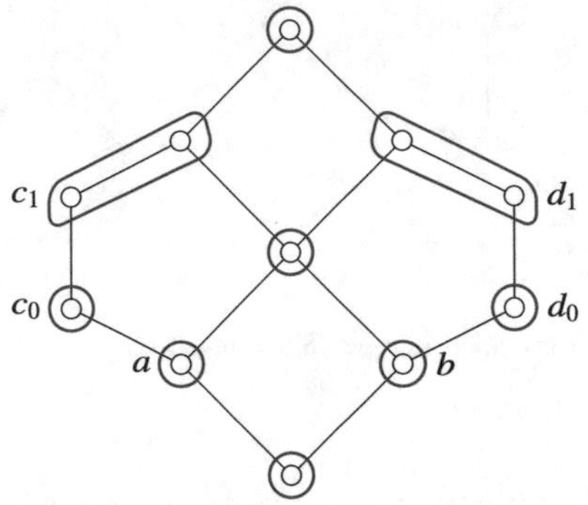

Fig. 4.5 Preclop on lattice containing a hexagon

In fact, every finite lower bounded lattice $\mathbf{K}$ can be 0-1-embedded into a lattice $\mathbf{L} = \mathrm{Sub}(\mathbf{S}, \wedge, 1)$ for some finite lattice $\mathbf{S}$, due to Adaricheva [10]; see Theorem 4.30 below. Every such $\mathbf{L}$ admits an equaclosure operator, while $\mathbf{K}$ may not.

The one positive result on sublattices is Lemma 3.37 above, from [16]: *For any dually algebraic lattice* $\mathbf{L}$ *and preclop* γ, *if* $a = \gamma(a)$ *and* $a < b$, *then the interval* $\mathbf{S} = [a, b]$ *is a dually algebraic lattice with preclop* $\sigma(x) = \gamma(x) \wedge b$ *for* $x \in S$. The nontrivial part of the proof is checking (I5), where we use the observation that $x \le \sigma(x) \le \gamma(x)$, whence $\sigma(x) = \sigma(y)$ implies $\gamma(x) = \gamma(y)$.

4.1.2 Examples and Sufficient Conditions

There are a few sufficient conditions to ensure that a finite lattice supports a preclop.

(1) On a finite distributive lattice, the identity map $i(x) = x$ is a preclop.

(2) A linear sum $\mathbf{1} + \mathbf{L}$ always has the preclop

$$\gamma(x) = \begin{cases} 0 & \text{if } x = 0 \\ 1 & \text{otherwise.} \end{cases}$$

This is bad news in a sense: any condition necessary for a lattice to admit a preclop must involve the least element 0. Indeed, Chap. 8 will feature examples of lattices such that $\mathbf{L}$ does not admit a preclop, but $\mathbf{1} + \mathbf{L}$ is isomorphic to $\mathrm{L_q}(\mathcal{K})$ for a quasivariety $\mathcal{K}$.

(3) It is easy to see that if $\mathbf{L}_1$ and $\mathbf{L}_2$ admit a preclop, then so does $\mathbf{L}_1 \times \mathbf{L}_2$.

(4) The next observation is that if **L** has a least join-irreducible-but-not-join-prime element, then it supports a preclop.

Theorem 4.10 *Let* **L** *be a finite lattice. If* $\mathrm{J}(\mathbf{L}) \setminus \mathrm{D}_0(\mathbf{L})$ *contains a least element* u*, then* **L** *admits the preclop*

$$\kappa(x) = \begin{cases} x & \text{if } x \ngeq u \\ 1 & \text{if } x \geq u. \end{cases}$$

Proof We need to know that if $x \ngeq u$, then $x \in \mathrm{MD}(\mathbf{L})$. Since every join irreducible below x will be join prime, this is so, as in Corollary 4.4. Property (I5) is immediate.□

The lattice **G** of Example 4.7 illustrates Theorem 4.10; see the first preclop in Fig. 4.3.

The variety $\mathcal{LB}(k)$ of lower bounded lattices of rank k is generated by all finite lattices **L** with $\mathrm{J}(\mathbf{L}) \subseteq \mathrm{D}_k(\mathbf{L})$; see [89]. Theorem 4.10 includes all subdirectly irreducible lattices in $\mathcal{LB}(1)$, which have exactly one non-join-prime join irreducible element. In fact, Theorem 8.7 will show that *every finite, subdirectly irreducible, lower bounded lattice of rank* 1 *is isomorphic to* $\mathrm{Sub}(\mathbf{S}, \wedge, 1, H)$ *for a finite semilattice with operators.*

The lattices in Fig. 4.4 show that this does not extend. The hexagon is in $\mathcal{LB}(1)$ with 2 join irreducibles in $\mathrm{D}_1(\mathbf{H})$, while the second lattice **B** is in $\mathcal{LB}(2)$ with 1 join irreducible in each of $\mathrm{D}_1(\mathbf{B})$ and $\mathrm{D}_2(\mathbf{B}) \setminus \mathrm{D}_1(\mathbf{B})$.

4.1.3 Finite Atomistic Lattices

Every finite atomistic lattice that admits an equaclosure operator is a subquasivariety lattice. This result is due to Adaricheva, Dziobiak, and Gorbunov [13]; cf. Theorem 2.65.

Theorem 4.11 *For a finite atomistic lattice* **L***, the following are equivalent.*

(1) $\mathbf{L} \cong \mathrm{L}_{\mathrm{q}}(\mathcal{K})$ *for some quasivariety* $\mathcal{K}$.
(2) **L** *is isomorphic to* $\mathrm{Sub}(\mathbf{S}, \wedge, 1)$ *for some finite semilattice* **S**.
(3) **L** *admits an equaclosure operator.*

Note that a finite lattice that admits an equaclosure operator cannot contain a D-cycle of atoms. For if there are atoms a, b such that $a\ \mathrm{D}^m\ b\ \mathrm{D}^n\ a$, then $b \leq \mu(a)$ and $a \leq \mu(b)$ by Theorem 4.3. Thus $\mu(a) = \mu(b)$, and since $\gamma \geq \mu$ for every equaclosure operator γ, this implies $\gamma(a) = \gamma(b) = \gamma(a \wedge b) = \gamma(0) = 0$, a contradiction.

As a consequence, every finite atomistic lattice of subquasivarieties is lower bounded, but not conversely. For example, the lattice $\mathbf{B}_4^-$ that is a Boolean algebra on 4 atoms with one coatom removed, is lower bounded and atomistic, but does not support an equaclosure operator. Finite non-atomistic subquasivariety lattices need not be lower bounded [20].

Theorem 4.11 extends the finite case of Theorem 2.64, and in turn is extended to algebraic atomistic lattices with appropriate finiteness conditions added in Theorem 2.65 [14, 53]. In part (3) of the theorem, it is important that **L** admits an equaclosure operator, not just a preclop, as it is required that **L** be biatomic, which is a consequence of property (K9) (Theorem 3.9).

Lattices of subsemilattices of a meet semilattice with 1 (without operators for now) provide a rich source of examples.

Let $\mathbf{S} = (S, \wedge, 1)$ be a finite meet semilattice with 1. There are at least two (not necessarily distinct) preclops on Sub **S**, μ and the natural equaclosure operator Γ.

Again, for $x \neq 1$ let $\overline{x}$ denote the subsemilattice $\{x, 1\}$; these are the atoms (and the only join irreducibles) of Sub **S**.

Consider the operator μ on Sub **S**. For an element $p < 1$, $\mu(\overline{p})$ is the least subsemilattice $\mathbf{X} \leq \mathbf{S}$ such that

(1) $p \in X$,
(2) if $x \in X$ and $x = y \wedge z$ properly, then $y, z \in X$.

Note with respect to (2) that if $x = y \wedge z$ properly, then $\overline{x} \leq \overline{y} \vee \overline{z}$ is a minimal nontrivial join cover in Sub **S**, whence $\overline{x}\ \mathrm{D}\ \overline{y}$ and $\overline{x}\ \mathrm{D}\ \overline{z}$. For any $Y \in$ Sub **S**, we then have $\mu(Y) = \bigcup\{\mu(\overline{p}) : p \in Y\}$. This is because the right-hand side is D-closed.

It is a nontrivial exercise that μ on Sub **S** satisfies (I5), and thus is a preclop [16].

There is also the natural equaclosure operator Γ on Sub **S**, which has $\Gamma(\mathbf{X}) = \uparrow x_0$ where x_0 is the least element of **X**. Note that $\tau(\mathbf{X}) = \overline{x}_0$ is an atom. Thus Γ can have no proper extension, and it is maximal in Pre(Sub **S**) [16].

These observations can be summarized as follows.

Lemma 4.12 *Let* $\mathbf{S} = (S, \wedge, 1)$ *be a finite meet semilattice with* 1. *On the lattice* Sub **S** *of* 1*-subsemilattices,* μ *is the minimal preclop and the natural equaclosure operator* Γ *is a maximal preclop.*

It is easy to design a semilattice **S** for which $\mu < \Gamma$, which we leave as an exercise.

Example 4.13 The preclop μ on Sub **S** might not be a real equaclosure operator, as it could fail one of (K9) or (K10). As an example, consider the semilattice $\mathbf{S}^* = \mathbf{2}\times\mathbf{3}$ labeled as on the left in Fig. 4.6. The right side of that figure gives the ordered set $(\mathrm{J}(\mathrm{Sub}\ \mathbf{S}^*), \overline{\mathrm{D}})$. Evidently μ defined on Sub $\mathbf{S}^*$ satisfies $\mu(\overline{c}) = S^*$, $\mu(\overline{b}) = \{b, d, e, 1\}$, and $\mu(\overline{x}) = \{x, 1\}$ otherwise. Substitute $X = \{b, d, 1\}$ and $Z = \{a, d, 1\}$ into (‡), which is (K9) for finite index sets. Then $\mu(Z) = Z = \tau(Z)$, while $c = a \wedge d \in X \vee Z$ whence $\mu(X \vee Z) = S^*$, so that $\tau(X \vee Z) = \overline{c}$. Thus

$$X \wedge \tau(Z) = \{d, 1\} \nleq \{1\} = \mu(X \wedge \tau(X \vee Z))$$

so that (‡) fails.

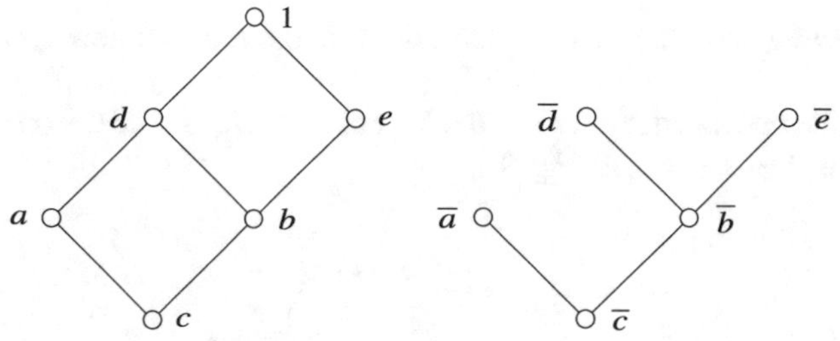

Fig. 4.6 Semilattice $\mathbf{S}^* = \mathbf{2} \times \mathbf{3}$ and the join irreducibles of Sub $\mathbf{S}^*$ ordered by the D relation

4.1.4 Necessary Conditions

Roughly speaking, most lattices do not support a preclop. Let us record some of the necessary conditions for one to do so.

Dziobiak's original condition from [39] is still useful.

Lemma 4.14 *Let* $\mathbf{L}$ *be a finite lattice that admits a preclop. If an element* u *is the join of* n *atoms of* $\mathbf{L}$*, then* $\downarrow u$ *contains at most* $2^n - 1$ *atoms.*

Proof Assume $\mathbf{L}$ admits a preclop γ, and let u be an element that is a join of atoms. Let the atoms below u be $p_1, \ldots, p_m$ with $u = p_1 \vee \ldots \vee p_n$. For each $i \leq m$ we have, by meet distributivity (I6),

$$\begin{aligned}\gamma(p_i) \wedge u &= \gamma(p_i) \wedge (p_1 \vee \ldots \vee p_n) \\ &= (\gamma(p_i) \wedge p_1) \vee \ldots \vee (\gamma(p_i) \wedge p_n). \qquad (\sharp)\end{aligned}$$

Since each p_j is an atom, $\gamma(p_i) \wedge p_j$ is either p_j or 0. Moreover, we have $\gamma(p_i) \wedge u \geq p_i$, so for any given $i \leq m$ the right-hand side of ($\sharp$) cannot be 0. Thus $\gamma(p_i) \wedge u$ can take on one of at most $2^n - 1$ values, *viz.*, the join of a nonempty subset of $\{p_1, \ldots, p_n\}$.

Suppose $m \geq 2^n$. Then for some $j \neq k$ we have $\gamma(p_j) \wedge u = \gamma(p_k) \wedge u$. However, since $p_i \leq \gamma(p_i) \wedge u \leq \gamma(p_i)$ for every index, we have $\gamma(\gamma(p_i) \wedge u) = \gamma(p_i)$. Hence

$$\gamma(p_j) = \gamma(\gamma(p_j) \wedge u) = \gamma(\gamma(p_k) \wedge u) = \gamma(p_k).$$

Then by (I5), $\gamma(p_j) = \gamma(p_j \wedge p_k) = \gamma(0) = 0$, a contradiction. Therefore, $m < 2^n$. □

The next condition, which we have used before, is an easy consequence of the fact that if $\mathbf{L}$ admits a preclop κ, then $\kappa \geq \mu$.

Theorem 4.15 *Let* $\mathbf{L}$ *be a finite lattice. If* $\mathbf{L}$ *contains elements* $x, y > 0$ *such that* $x \wedge y = 0$ *and* $\mu(x) = \mu(y)$*, then* $\mathbf{L}$ *does not admit a preclop.*

Proof Suppose that x, y are as in the statement and that γ is a preclop on $\mathbf{L}$. Note that $x \leq \mu(x) \leq \gamma(x)$, whence $\gamma(\mu(x)) = \gamma(x)$. Thus $\mu(x) = \mu(y)$ implies $0 < \gamma(x) = \gamma(y) = \gamma(x \wedge y) = \gamma(0)$, a contradiction. □

The following condition is a bit esoteric. It is offered as a new type of result to pursue.

Let $\mathbf{B}$ be the *bowtie* with $a_1, a_2 < b_1, b_2$. For $k \geq 2$ let $\mathbf{Y}_k$ be the ordered set with $a < b < c_j$ for $1 \leq j \leq k$. See Fig. 4.7.

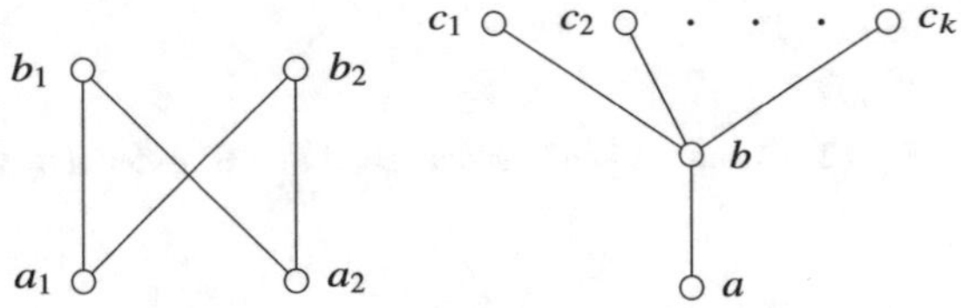

Fig. 4.7 Ordered sets $\mathbf{B}$ and $\mathbf{Y}_k$

Lemma 4.16 *Let* $\mathbf{L}$ *be a finite lower bounded lattice. If either*

(1) $\mathbf{B}$ *is an order filter in* $(\mathrm{J}(\mathbf{L}), \overline{\mathrm{D}})$ *and* $a_1 \wedge a_2 = 0$ *in* $\mathbf{L}$ *or*
(2) *some* $\mathbf{Y}_k$ $(k \geq 2)$ *is an order filter in* $(\mathrm{J}(\mathbf{L}), \overline{\mathrm{D}})$ *and* $a \wedge b = 0$ *in* $\mathbf{L}$,

then $\mathbf{L}$ *does not admit an equaclosure operator.*

Proof For case (1), both $a_1 \leq b_1 \vee b_2$ and $a_2 \leq b_1 \vee b_2$ are minimal nontrivial join covers. Thus a_i D b_j for all pairs. By Theorem 4.3, we get $\mu(a_i) \geq b_1 \vee b_2 \geq a_i$. Hence $\mu(a_1) = \mu(b_1 \vee b_2) = \mu(a_2)$, and we can apply Theorem 4.15.

For case (2), let $C = \{c_1, \ldots, c_k\}$. The minimal nontrivial join covers of b are $b \leq \bigvee U$ for various subsets $V \subseteq C$, while there is a minimal nontrivial join cover $a \leq b \vee \bigvee V$ for some $V \subseteq C$ (with $b \nleq \bigvee V$). Therefore, as in the first case, $\mu(a) = \mu(\bigvee C) = \mu(b)$ and we can again apply Theorem 4.15. □

4.2 Algorithm to Determine Whether a Lattice has a Preclop

In this section we build an algorithm that determines whether a given finite lattice has a preclop. If it does, the algorithm produces the minimal preclop on the lattice. This extends and formalizes our earlier discussions of how to find Pre($\mathbf{L}$) and how to determine whether μ is a preclop.

The authors thank David Casperson for his contributions to this section.

We start from the description of preclops in terms of equapartitions suggested in [16].

Theorem 4.17 *Let* $\mathbf{L}$ *be a finite lattice, and let* $\mathrm{MD}(\mathbf{L})$ *be its semilattice of meet distributive elements. Then* $\mathbf{L}$ *admits a preclop iff there exists a subset* $H \subseteq \mathrm{MD}(\mathbf{L})$ *and a map* $\beta : H \to L$ *such that*

(1) $0 \in H$,
(2) $b(h) \leq h$ *for all* $h \in H$,
(3) $L = \bigcup_{h \in H}[b(h), h]$,
(4) *if* $b(h) \leq k$ *for some* $h, k \in H$, *then* $h \leq k$.

If the conditions hold, the map $\kappa(x) = h$ *if* $x \in [b(h), h]$ *is a preclop.*

Let **L** be a finite lattice with meet distributive elements MD(**L**). A pair (S, β) with S a meet subsemilattice of **L** contained in MD(**L**) and $\beta : S \to \mathbf{L}$ is a *preclop setup* if it satisfies

(Q1) $0 \in S$,
(Q2) $\beta(x) \leq x$,
(Q3) $L = \bigcup_{s \in S}[\beta(s), s]$,
(Q4) for all preclops $\gamma \in \text{Pre}(\mathbf{L})$ we have $\gamma(L) \subseteq S$,
(Q5) for all preclops $\gamma \in \text{Pre}(\mathbf{L})$ and $s \in S$, $\gamma(\beta(s)) \geq \gamma(s)$.

Notice that (Q5) implies that for all preclops $\gamma \in \text{Pre}(\mathbf{L})$ and $s \in S$, $\gamma(\beta(s)) = \gamma(s)$ as $\beta(s) \leq s$ and γ respects order.

Given a preclop setup (S, β), a pair $(m, n) \in S^2$ is called a *conflict pair* if $\beta(m) \leq n$ but $m \nleq n$.

Immediate consequences of these definitions are that 1 is in S and it is a $(0, 1)$-meet-subsemilattice of **L**. If $\beta(m) = 0$ for some $m \in S$ with $m \neq 0$, then $(m, 0)$ is a conflict pair.

Recall that $\mu : L \to \text{MD}(\mathbf{L})$ is given by $\mu(x)$ is the least meet distributive element above x. Define $\beta_0 : \text{MD}(L) \to L$ by $\beta_0(m) = \bigwedge\{x \in L : \mu(x) = m\}$.

Lemma 4.18 *The pair* $(\text{MD}(\mathbf{L}), \beta_0)$ *is a preclop setup.*

Proof It is sufficient to show that (Q5) holds. By (I6) any preclop γ has the property that $\gamma(x)$ is in MD(**L**) for every $x \in L$. Set $G_m = \{y \in L : \mu(y) = m\}$, so that for $m \in \text{MD}(\mathbf{L})$ we have $\beta_0(m) = \bigwedge G_m$. Consider an element $x \in G_m$. As $\gamma(x)$ is meet distributive and above x, while m is the least meet distributive element above x, we have $m \leq \gamma(x)$ which implies $\gamma(m) \leq \gamma(x)$. However, γ is order preserving and $x \leq m$, whence $\gamma(x) \leq \gamma(m)$. Thus $\gamma(\beta_0(m)) = \gamma(\bigwedge G_m) = \gamma(m)$ by (I5). □

Lemma 4.19 *If a preclop setup* (S, β) *has no conflict pairs, then* $L = \bigcup_{s \in S}[\beta(s), s]$ *is a partition.*

Proof Suppose $x \in [\beta(m), m] \cap [\beta(n), n]$ with $m \neq n$. Then either $m \nleq n$ or $n \nleq m$, say the former. That makes (m, n) a conflict pair as $\beta(m) \leq x \leq n$. □

A conflict pair $(m, n) \in S^2$ is *maximal* if every pair $(a, b) \in S^2$ with either $n < b$ or $n = b$ and $m < a$ is not a conflict pair. This is lexicographical order from right to left.

Given a preclop setup (S, β) that is not a partition, choose a conflict pair $(m, n) \in S^2$. Such a pair exists by Lemma 4.19.

Set $S' = S \setminus \{n\}$. Let $a = m \oplus_S n$ be the least element in S above m and n; this makes sense because S is a meet subsemilattice of the finite lattice **L**. Define $\beta' : S' \to L$ by $\beta'(a) = \beta(a) \wedge \beta(n)$ and $\beta'(x) = \beta(x)$ otherwise. The pair (S', β') is called a *(m, n)-refinement* of (S, β).

Lemma 4.20 *Let (S, β) have (m, n)-refinement (S', β') with (m, n) a maximal conflict pair. Then either $\beta'(x) = 0$ for some $x \neq 0$ or (S', β') is a preclop setup and $|S'| = |S| - 1$.*

Proof To see that S' is meet closed, assume for contradiction that $n = u \wedge v$ with $u, v > n$. This implies $\beta(m) \leq u$ and $\beta(m) \leq v$ as $\beta(m) \leq n$. As (m, n) is a maximal conflict pair, the pairs (u, m) and (v, m) are not conflict pairs. Thus $m \leq u$ and $m \leq v$, which implies $m \leq n$, a contradiction.

Assume that $\beta'(x) \neq 0$ for $x \neq 0$; otherwise, the conclusion on the lemma holds. Note this implies $0 \in S'$. Since $\beta'(a) = \beta(a) \wedge \beta(n)$ and $a \geq n$ the interval $[\beta'(a), a]$ contains both $[\beta(a), a]$ and $[\beta(n), n]$, so (Q3) holds for S'. Clearly $\beta'(a) \leq \beta(a) \leq a$ so (Q2) holds for S'.

Assume that there is a preclop $\gamma : L \longrightarrow L$. To see that (Q4) holds, for contradiction pick $u \in L$ with $\gamma(u) = n$. Note $\gamma(n) = n$. As $\beta(m) \leq n$ and γ is a preclop, we have $\gamma(\beta(m)) \leq \gamma(n) = n$. By (Q5), $\gamma(\beta(m)) \geq \gamma(m)$. Thus $\gamma(m) \leq \gamma(n)$. That is, $m \leq \gamma(m) \leq \gamma(n) = n$ which contradicts $m \nleq n$.

To see that (Q5) holds for S', it suffices to show that $\gamma(\beta'(a)) \geq \gamma(a)$. This is done by showing $\gamma(\beta(a)) = \gamma(\beta(n)) = \gamma(a)$, whereby (I5) implies $\gamma(\beta'(a)) = \gamma(a)$. From $\beta(m) \leq n$ we have $\gamma(m) = \gamma(\beta(m)) \leq \gamma(n)$ and we obtain $n \vee m \leq \gamma(n) \vee \gamma(m) = \gamma(n) = \gamma(\beta(n))$. The fact that $\gamma(\beta(n))$ is in S by (Q4) means that $\gamma(\beta(n))$ is larger than the least element in S above $m \vee n$, which is a. Thus $\gamma(\beta(n)) \geq \gamma(a)$. However, $\beta(n) \leq n \leq a$ implies $\gamma(\beta(n)) \leq \gamma(a)$. Thus $\gamma(\beta'(a)) = \gamma(a)$. □

The *Refinement Algorithm* for finding a minimal preclop, if one exists, is found in the next theorem.

Theorem 4.21 *Given a finite lattice* **L** *with meet distributive elements* MD(**L**) *and map $\mu : L \longrightarrow$* MD(**L**) *where $\mu(x)$ is the least meet distributive element above x. Set $S_0 =$* MD(**L**) *and $\beta_0(m) = \bigwedge\{x \in L : \mu(x) = m\}$ for $m \in$* MD(**L**).

Recursively do the following.

(1) *If there is an $m \in S_i$ with $\beta_i(m) = 0$ but $m \neq 0$, then stop.*
(2) *If $L = \bigcup_{s \in S_i} [\beta_i(s), s]$ is a partition, then stop.*
(3) *If neither of the above holds, then find a maximal conflict pair (m, n) and construct (S_{i+1}, β_{i+1}) the (m, n)-refinement of (S_i, β_i).*

This process terminates under either Case (1) with no possible preclop or under Case (2) with preclop setup (S_n, β_n). In the latter case $\kappa : L \longrightarrow$ MD(**L**) *given by $\kappa(x) = m$ for $x \in [\beta_n(m), m]$ is the minimal preclop.*

Proof By Lemma 4.20 either the algorithm stops at Case (1) or the size of the semilattice $|S_i|$ decreases at each step. Since L is finite the process must stop.

Assume that γ is a preclop and that during the running of this process some β_i and $m \neq 0$ satisfies $\beta_i(m) = 0$. Then a contradiction arises as $0 = \gamma(0) = \gamma(\beta(m)) \geq \gamma(m) \geq m > 0$. Thus the existence of any preclop implies the refinement process stops in Case (2).

Now assume the algorithm stops in Case (2) with (S_n, β_n). The map κ is a preclop by Theorem 4.17 and it is minimal by (Q5). □

Figure 4.8 illustrates the $\overline{\mathrm{D}}$ relation on a lattice, the meet distributive elements, and the minimal preclop found by the Refinement Algorithm.

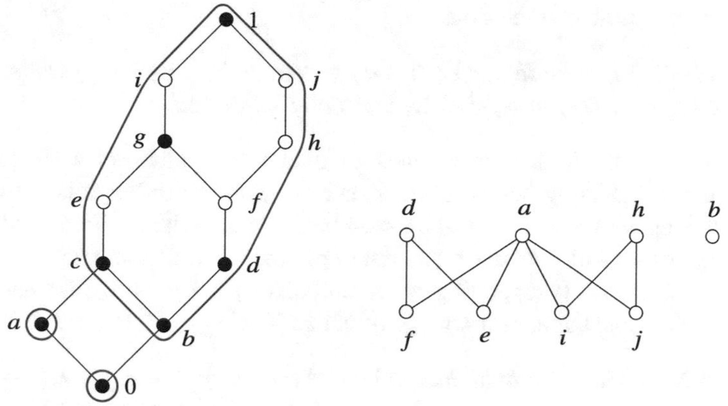

Fig. 4.8 A lattice and its $\overline{\mathrm{D}}$ relation on the join irreducibles. Solid dots indicate the meet distributive elements. Red curves indicate the minimal preclop

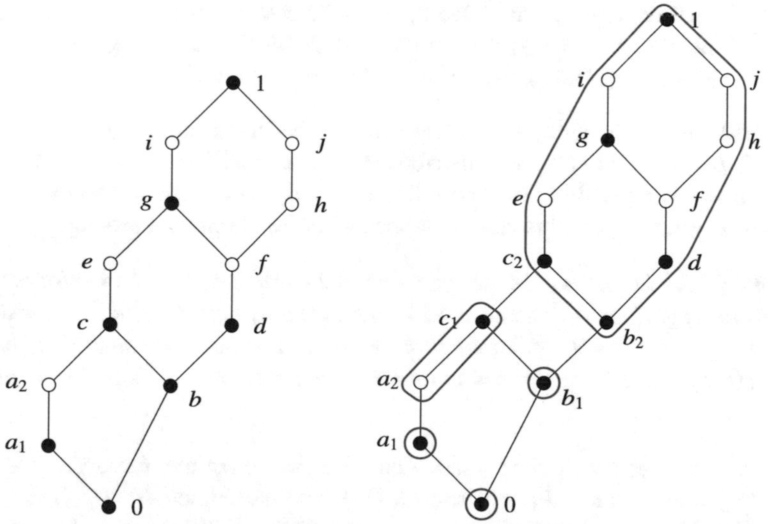

Fig. 4.9 The first lattice does not admit a preclop. The second lattice admits a preclop and has a 0-separating homomorphism onto the first. Solid dots indicate the meet distributive elements. Red curves indicate the minimal preclop

4.2.1 0-Separating Homomorphisms

The lattices in Fig. 4.9 provide a counterexample to the following conjecture, which seemed quite plausible at the time.

Conjecture 4.22 *If finite lattice* $\mathbf{L}_1$ *has a preclop, and* $f : \mathbf{L}_1 \to \mathbf{L}_2$ *is a surjective* 0*-separating homomorphism, then* $\mathbf{L}_2$ *has a preclop as well.*

By extending the lattices in Fig. 4.9 we obtain an example of a lattice with a preclop and 0-separating homomorphism such that the algorithm fails in the image on the nth step with $n > 1$. Corollary 4.24 below is a partial positive result saying that the algorithm will not fail on the first step. First we need a lemma.

Given a surjective 0-separating homomorphism $f : \mathbf{L}_1 \to \mathbf{L}_2$, for an element $a \in L_2$ let a' denote the least preimage of a, that is, $a' = \bigwedge f^{-1}(a)$ in $\mathbf{L}_1$.

Lemma 4.23 *Suppose the finite lattice* $\mathbf{L}_1$ *has a preclop* ν*, and* $f : \mathbf{L}_1 \to \mathbf{L}_2$ *is a surjective* 0*-separating homomorphism. If* $\mu(b_1) = \mu(b_2)$ *in* $\mathbf{L}_2$*, then* $\nu(b_1') = \nu(b_2')$ *in* $\mathbf{L}_1$.

Proof First note that $f(\mathrm{MD}(\mathbf{L}_1)) \subseteq \mathrm{MD}(\mathbf{L}_2)$, by a straightforward argument.

Let $c = \mu(b_1) = \mu(b_2)$. Apparently, $b_i' \leq c'$, for $i = 1, 2$. Let $m \geq b_i'$ be any meet distributive element above b_i' in L_1, for some fixed $i \in \{1, 2\}$. Then $f(m) \geq f(b_i') = b_i$ is a meet distributive element above b_i in L_2. Therefore, $f(m) \geq c$, which implies $m \geq c'$. Thus we get $\mu(b_i') \geq c'$ for both $i = 1, 2$. By a standard argument, we have $b_i' \leq c' \leq \mu(b_i') \leq \nu(b_i')$, which implies $\nu(c') = \nu(b_i')$ for both i. □

The first step of algorithm of Theorem 4.21 applied to the lattice L_2 builds intervals $[b, m] \subseteq L_2$, where m is a meet distributive element and $b = \beta_0(m) = \bigwedge\{x \in L_2 : \mu(x) = m\}$. The algorithm fails at the first step if $\beta_0(m) = 0$ for some $m \in \mathrm{MD}(\mathbf{L}_2)$. The next result says that this cannot happen when $\mathbf{L}_1$ supports a preclop.

Corollary 4.24 *Let* $\mathbf{L}_1$*,* $\mathbf{L}_2$ *be as in* Lemma 4.23*, and let* $[b, m]$ *be one of intervals built by the algorithm of* Theorem 4.21 *on* $\mathbf{L}_2$ *on the first step. Then there exists an interval* $[c, \nu(c)]$ *from the equapartition of* $\mathbf{L}_1$ *with respect to* ν *such that* $[b, m] \subseteq [f(c), f(\nu(c))]$*. In particular,* $b \neq 0$*, and the algorithm never fails on* $\mathbf{L}_2$ *on the first step.*

Proof Let $M = \{z \in L_2 : \mu(z) = m\}$. According to the lemma, for any $x, y \in M$ we have $\nu(x') = \nu(y') = a \in L_1$. By property (I5) for ν, the element $b^* = \bigwedge\{z' : z \in M\}$ of L_1 also has the properties $\nu(b^*) = a$ and $f(b^*) = \beta_0(m) = b$. In particular, $b^* \geq c = \bigwedge\{t \in L_1 : \nu(t) = a\}$.

Also, according to proof of Lemma 4.23, $m' \leq \mu(z') \leq \nu(z') = a$ for any $z' \in M$, therefore, $f(m') = m \leq f(a)$. Thus, we have $c \leq b^* \leq a$, whence we get $f(c) \leq b \leq m \leq f(a) = f(\nu(c))$, as claimed. □

Let us do more analysis of the consecutive steps of the algorithm. If for all built intervals $[b, m] \subseteq L_2$, and for every $y \in [b, m]$ it holds that $\mu(y) = m$, then intervals represent the partition of L_2, and the algorithm finishes on the first step.

If algorithm does not finish, then there exists an interval $[b, m]$ and $y \in [b, m]$ such that $\mu(y) \wedge m_1 < m$. This would require resolution by taking $b := b \wedge b_1$ and removing $[b_1, m_1]$. So the goal is to show that $b \wedge b_1 \neq 0$. If we have the same interval $[c, \nu(c)]$ serving both $[b, m]$ and $[b_1, m_1]$ as per Corollary 4.24, then this resolution still satisfies the conditions of Corollary 4.24. Note that this situation will always occur, if all elements x with $\mu(x) = m_1$ are in $[b, m]$, so that $[b_1, m_1] \subseteq [b, m]$.

Thus, for the nontrivial case, we should have $p \notin [b, m]$ with $\mu(p) = m_1$. In such a nontrivial case, we have two intervals: $[c_1, \nu(c_1)]$ and $[c, \nu(c)]$ so that $y' \in [c_1, \nu(c_1)]$ and another $y^* \in [c, \nu(c)]$ such that $f(y') = f(y^*) = y$. It is easy to show that $\nu(c_1) \leq \nu(c)$.

4.3 Embedding a Lower Bounded Lattice into Sub S

Our eventual goal (in later chapters) is the following scheme.

- Start with a finite lower bounded lattice **L**.
- If possible, represent **L** as Sub(**S**, $\wedge$, 1, H) for some finite meet semilattice and set of operators.
- Again if possible, represent **L** as $\mathrm{L_q}(\mathcal{K})$ for a quasivariety $\mathcal{K}$.

If **L** is join semidistributive but not lower bounded, then we would use $\mathrm{S_p}(\mathbf{S}, H)$ for an infinite algebraic lattice in the second step. Toward that end, in this section we consider in detail embeddings of a finite lower bounded lattice into Sub(**S**, $\wedge$, 1), with the operators to be added later to convert the embedding into an isomorphism with Sub(**S**, $\wedge$, 1, H), when possible.

It was shown in [10] that any finite lower bounded lattice can be embedded into Sub **S** for some finite semilattice **S**, which we reproduce as Theorem 4.30 below. This construction yields a (0, 1)-embedding, but it need not be atom-preserving. See also Theorem 2-2.8 of [21], and the independent proof of Repnitskiĭ [102]. The same sort of construction embeds a finite join semidistributive (but not lower bounded) lattice **L** into Sub **S** with **S** infinite.

Example 4.25 The lower bounded (and biatomic) lattice Co($\mathbf{P}_5$) of convex subsets of partially ordered set $\mathbf{P}_5 = \{a, b, c, d, e\}$ with $a < b < c > d > e$ cannot be embedded into Sub **S** atom-preservingly for any finite semilattice **S**. It is also known from [16] that Co($\mathbf{P}_5$) does not have a preclop. These two observations about Co($\mathbf{P}_5$) might be connected, but it is still vague.

At one point, it was conjectured that if a finite lower bounded lattice can be embedded into Sub **S** atom-preservingly, then it has a preclop. However, this is false, as shown by the next example.

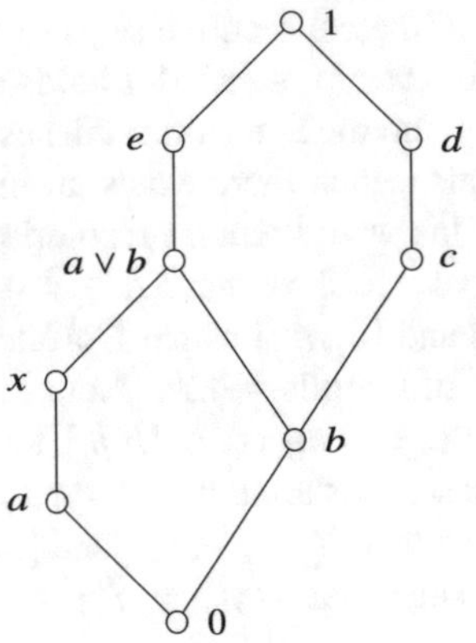

Fig. 4.10 Lattice **L** of Example 4.26

Example 4.26 Let **S** be freely generated by $\{a, b, c_1, c_2\}$ and identify the elements

$$x = a \wedge b \qquad d = a \wedge c_1 \qquad e = a \wedge c_2.$$

The lattice in Fig. 4.10 can be embedded atom-preservingly into Sub **S**, but it does not admit a preclop. An embedding of **L** into Sub$(S, \wedge, 1)$ is given by

$$\begin{aligned} a &= \{a, 1\} & x &\mapsto \{a, x, 1\} \\ b &\mapsto \{b, 1\} & a \vee b &\mapsto \{a, b, x, 1\} \\ c &\mapsto \{b, c_1, c_2, 1\} & d &\mapsto \{b, c_1, c_2, d, 1\} \\ e &\mapsto \{a, x, b, e, 1\} & 1 &\mapsto S. \end{aligned}$$

However, exactly as in the first lattice in Fig. 4.9, **L** does not admit a preclop.

We collect some easy facts about preclops on sublattices of Sub **S**. Again it is convenient to consider meet semilattices with a constant 1, so that the bottom element of Sub **S** is $\{1\}$. The problem then becomes: *Given a* $(0, 1)$*-sublattice* $\mathbf{K} \leq$ Sub$(\mathbf{S}, \wedge, 1)$*, find the preclops in* Pre$(\mathbf{K})$.

Let ε be a quasi-order (reflexive, transitive) on **S**. We will call this quasi-order *distributive* if

(i) $(a \wedge b)\, \varepsilon\, c$ implies there are $a', b' \in S$ such that $a\, \varepsilon\, a'$, $b\, \varepsilon\, b'$ and $c = a' \wedge b'$,
(ii) $1\, \varepsilon\, x$ implies $x = 1$.

Denote by Sub$(\mathbf{S}, \varepsilon)$ the lattice of ε-closed subsemilattices of **S**, i.e., subsemilattices $X \in$ Sub **S** such that $a \in X$ and $a\, \varepsilon\, b$ implies $b \in X$.

The next lemma shows that distributive quasi-orders determine sublattices of Sub **S**, and *vice versa*. The idea goes back to Gorbunov and Tumanov [54]; this version is from Section 4-3.1 of [22].

Lemma 4.27 *For any distributive quasi-order ε on a finite semilattice* **S**, Sub(**S**, ε) *is a* (0, 1)*-sublattice of* Sub **S**.

Conversely, for any (0, 1)*-sublattice* $\mathbf{T} \leq$ Sub **S**, *there is a distributive quasi-order* ρ *on* **S** *such that* $\mathbf{T} =$ Sub(**S**, ρ). *Moreover,* ρ *can be chosen to also satisfy*

(iii) *If* $c \rho d, e$ *then* $c \rho (d \wedge e)$.
(iv) *For all* $c \in S$, $c \rho 1$.

Proof Let ε be a distributive quasi-order on a semilattice **S**. It is clear that the intersection of ε-closed subsemilattices is ε-closed. Condition (i) is designed to ensure that the join of ε-closed subsemilattices is ε-closed, while (ii) guarantees that $\{1\}$ is ε-closed.

Conversely, suppose we are given a (0,1)-sublattice $\mathbf{T} \leq$ Sub **S**. For the quasi-order ρ in the second part, let $c \rho d$ if for all $X \in \mathbf{T}$, $c \in X$ implies $d \in X$.

For each subset $A \subseteq S$, let $\mathrm{Tg}(A) = \bigcap\{X \in \mathbf{T} : A \subseteq X\}$, that is, the least subsemilattice in **T** containing A. Then

- $\mathrm{Tg}(\{a\}) = \{b \in S : a \rho b\}$,
- $\mathrm{Tg}(A) = A$ iff $A \in \mathbf{T}$.

Thus a subset $A \subseteq \mathbf{S}$ is a ρ-closed subsemilattice if and only if $A \in \mathbf{T}$, so that $\mathbf{T} =$ Sub(**S**, ρ).

It is straightforward to check that ρ has properties (ii), (iii), and (iv). For (i), assume that $(a \wedge b) \rho c$. Now $a \wedge b \in \mathrm{Tg}(a) \vee \mathrm{Tg}(b) \in \mathbf{T}$, so $c \in \mathrm{Tg}(a) \vee \mathrm{Tg}(b)$. Thus $c = a' \wedge b'$ for some $a' \in \mathrm{Tg}(a)$ and $b' \in \mathrm{Tg}(b)$. □

Remember, though, not every sublattice of Sub **S** admits a preclop.

Let us call an element $x \in S$ an *ε-element*, if $x \varepsilon y$ implies $y \geq x$.

The following property of a quasi-order was introduced in [51]. We say that the quasi-order ε is *filterable*, if for any ε-element $x \in S$ it follows from $y > x$ and $y \varepsilon z$ that $z \geq x$. An example of filterable quasi-order is any quasi-order ε such that $a \varepsilon b$ implies $b \geq a$.

Lemma 4.28 *Let* ε *be a distributive filterable quasi-order on* **S**. *Then there is a preclop on* Sub(**S**, ε).

Proof If $X \in$ Sub(**S**, ε), then its least element $x_0 = \bigwedge X$ is an ε-element. It follows from the filterability of ε that $\uparrow x_0 \in$ Sub(**S**, ε). Hence, for any $X \in$ Sub(**S**, ε) and for the natural preclop Γ on Sub **S**, we have $\Gamma(X) = \uparrow x_0 \in$ Sub(**S**, ε). Therefore, the restriction of Γ to Sub(**S**, ε) provides a preclop on Sub(**S**, ε). □

Problem 4.29 *Can every finite lower bounded lattice with a preclop be presented as* Sub(**S**, ε) *for suitable* **S** *and distributive and filterable quasi-order* ε *on* **S**? (This is Question 6 of [1].)

The prime model for a distributive filterable quasi-order comes from semilattices with operators. Let H be a monoid of operators on a semilattice $\mathbf{S} = (S, \wedge, 1)$. Then $\mathbf{T} = \mathrm{Sub}(S, \wedge, 1, H)$ is a (0,1)-sublattice of Sub **S**. Moreover, ρ as defined in the proof of Lemma 4.27 is a filterable quasi-order: the ρ-elements are the least elements of H-closed subsemilattices, that is, fully invariant elements.

Historically, this is backwards, as filterable quasi-orders were invented first, but that matters not. And now we see a refinement in our scheme: given a lattice $\mathbf{L}$ to represent, we first find a semilattice $\mathbf{S}$ and filterable quasi-order ε such that $\mathbf{L} \cong \mathrm{Sub}(\mathbf{S}, \varepsilon)$, and then try to realize the quasi-order ε in terms of operators.

As the first step toward such a description, we present a construction that represents every lower bounded lattice in the form $\mathrm{Sub}(\mathbf{P}, \varepsilon)$, by constructing a semilattice $\mathbf{P}$ and a distributive quasi-order ε on $\mathbf{P}$. (This quasi-order may not be filterable, though.) The construction presented here combines elements of the original in Adaricheva [10] and a method employing colored trees in Semenova [108]. As a matter of fact, our current construction borrows the definition of the relation ε from [108], but uses it on the original construction of [10], which seems to be simpler than colored trees.

Theorem 4.30 *For every finite lower bounded lattice* $\mathbf{L}$ *one can find a finite semilattice* $\mathbf{P}$ *and a distributive quasi-order* ε *on* P *such that* $\mathbf{L} = \mathrm{Sub}(\mathbf{P}, \varepsilon)$.

Proof Let us use the meet symbol $\sqcap$ for the operation in $\mathbf{P}$, thinking of it as a meet semilattice, as we already have $\wedge$ in $\mathbf{L}$ and concatenation for sequences.

The semilattice $\mathbf{P} = \langle P, \sqcap \rangle$ is generated by the set $\overline{P}$ of all sequences $\overline{d} = \langle d_1, \ldots, d_n \rangle$ of join irreducible elements $d_1, \ldots, d_n \in \mathrm{J}(\mathbf{L})$, $n \geq 1$, such that $d_i \mathrel{\mathrm{D}} d_{i+1}$ and d_1 is a minimal element with respect to the relation D, i.e., there is no $c \in \mathrm{J}(\mathbf{L})$ such that $c \mathrel{\mathrm{D}} d_1$. We also allow the empty sequence ι as an element of $\mathbf{P}$. The semilattice $\mathbf{P}$ is then defined as freely generated by $\overline{P}$ modulo the relations

$$\overline{d} = \iota \sqcap \overline{d} = \overline{d} \sqcap \iota,$$
$$\overline{d} = \overline{d} \sqcap \overline{f} = \overline{f} \sqcap \overline{d},$$

when $\overline{d}$ is an initial segment of $\overline{f}$, for nonempty sequences $\overline{d}$ and $\overline{f}$, and

$$\langle \overline{d}a \rangle = \langle \overline{d}ab_1 \rangle \sqcap \ldots \sqcap \langle \overline{d}ab_n \rangle,$$

whenever $a \leq b_1 \vee \ldots \vee b_n$ is a nontrivial minimal join cover of a in $\mathbf{L}$.

In particular, $\mathbf{P}$ has a top element $1_{\mathbf{P}}$, which is represented by the empty sequence ι.

It was proved in [10] that every element of $\mathbf{P}$ can uniquely be presented as a "reduced" $\sqcap$-product of elements of P, meaning that in that form none of the relations above can be applied to the product.

For $a \in \mathbf{L}$, we let $\overline{P}_a$ be the set of all $\overline{d} \in \overline{P}$ with $l(\overline{d}) \leq a$. Here $l(\overline{d})$ denotes the last term of $\overline{d}$, that is, $l(a_1 a_2 \ldots a_n) = a_n$. For the empty sequence $\iota = 1_{\mathbf{P}}$ we define $l(\iota) = 0_{\mathbf{L}}$. In particular, $1_{\mathbf{P}} \in \overline{P}_a$ for every $a \in L$. It was proved in [10] that the mapping Φ from $\mathbf{L}$ that sends any element $a \in L$ to the $\sqcap$-subsemilattice of $\mathbf{P}$ generated by $\overline{P}_a$, is an embedding of $\mathbf{L}$ into the lattice of subsemilattices (with $1_{\mathbf{P}}$) of $\mathbf{P}$.

Now we want to introduce the quasi-order ε on $\mathbf{P}$.

Let $p = \overline{p_1} \sqcap \ldots \sqcap \overline{p_n}$ and $q = \overline{q_1} \sqcap \ldots \sqcap \overline{q_s}$ are elements of **P** in their canonical reduced form. Then we define $p \, \varepsilon \, q$, if $l(q_1) \vee \ldots \vee l(q_s) \leq_{\mathbf{L}} l(p_1) \vee \ldots \vee l(p_n)$. In particular, $p \, \varepsilon \, \iota$ for every $p \in \mathbf{P}$.

First, we prove that this quasi-order is distributive.

Suppose that $p \sqcap r \, \varepsilon \, q$, where $p = \overline{p_1} \sqcap \ldots \sqcap \overline{p_n}$, $r = \overline{r_1} \sqcap \ldots \sqcap \overline{r_m}$, and $q = \overline{q_1} \sqcap \ldots \sqcap \overline{q_s}$ are canonical forms of elements p, r, and q. We bring $p \sqcap r$ to its canonical form $\overline{u_1} \sqcap \ldots \sqcap \overline{u_k}$. It is straightforward to check that $l(\overline{u_1}) \vee \ldots \vee l(\overline{u_k}) \leq_{\mathbf{L}} l(\overline{p_1}) \vee \ldots \vee l(\overline{p_n}) \vee l(\overline{r_1}) \vee \ldots \vee l(\overline{r_m})$. Hence, $l(\overline{q_1}) \vee \ldots \vee l(\overline{q_s}) \leq_{\mathbf{L}} l(\overline{p_1}) \vee \ldots \vee l(\overline{p_n}) \vee l(\overline{r_1}) \vee \ldots \vee l(\overline{r_m})$.

Every $l(\overline{q_i})$ is either below some $l(\overline{p_j})$ or $l(\overline{r_k})$, or has a nontrivial minimal join cover $Q_i \ll \{l(\overline{p_1}), \ldots, l(\overline{p_n}), l(\overline{r_1}), \ldots, l(\overline{r_m})\}$. In the first case, $\overline{p_j} \, \varepsilon \, \overline{q_i}$ or $\overline{r_k} \, \varepsilon \, \overline{q_i}$. In the second case, we consider elements $\langle \overline{q_i} q_{ij} \rangle$, for all $q_{ij} \in Q_i$. By the definition, $\langle \overline{q_i} \rangle = \sqcap \{\langle \overline{q_i} q_{ij} \rangle : q_{ij} \in Q_i\}$, and $p_i \, \varepsilon \, \langle \overline{q_i} q_{ij} \rangle$ or $r_k \, \varepsilon \, \langle \overline{q_i} q_{ij} \rangle$, for each $q_{ij} \in Q_i$.

It follows that q can be presented as a product $\overline{v_1} \sqcap \ldots \sqcap \overline{v_d}$, where, for each $\overline{v_k}$, either $\overline{p_i} \, \varepsilon \, \overline{v_k}$, for some $\overline{p_i}$, or $\overline{r_j} \, \varepsilon \, \overline{v_k}$, for some $\overline{r_j}$. If for every k there is i such that $\overline{p_i} \, \varepsilon \, \overline{v_k}$, then $p \, \varepsilon \, q$ and $q = q \sqcap 1_{\mathbf{P}}$. Similarly, when for every k there is j such that $\overline{r_j} \, \varepsilon \, \overline{v_k}$. Otherwise, we form the product of all $\overline{v_k}$ for which $\overline{p_i} \, \varepsilon \, \overline{v_k}$, for some i, and call that element q_1, we form the product of all $\overline{v_k}$ for which $\overline{r_j} \, \varepsilon \, \overline{v_k}$, for some j and call that element q_2. Notice that $p \, \varepsilon \, q_1$ and $r \, \varepsilon \, q_2$, and also $q = q_1 \sqcap q_2$, as is required for distributivity.

We now want to show that **L** is isomorphic to Sub(**P**, ε). Evidently, $\Phi(a)$ is an ε-closed subsemilattice, hence, we have an embedding of **L** into Sub(**P**, ε).

It remains to show that Φ is surjective. Take any $A \in$ Sub(**P**, ε), and let $a = \bigvee\{l(\overline{p}) : \overline{p} \in A\}$. We notice that if $p = \overline{p_1} \sqcap \ldots \sqcap \overline{p_n}$ is in A, then $p \, \varepsilon \, \overline{p_i}$, thus $\overline{p_i}$ are also in A, since A is ε-closed. It follows that, for every such element $p \in A$, we have $\bigvee l(\overline{p_i}) \leq_L a$. It remains to show that $A = \Phi(a)$. Clearly, $A \subseteq \Phi(a)$. Now, take any $q = \overline{q_1} \sqcap \ldots \sqcap \overline{q_n} \in \Phi(a)$ in its reduced form. For every $\overline{q_i}$ we have either $l(\overline{q_i}) \leq_{\mathbf{L}} l(\overline{p})$ for some $\overline{p} \in A$ or we can find for $l(\overline{q_i})$ a nontrivial minimal cover $Q \ll \bigvee\{l(\overline{p}) : \overline{p} \in A\}$. In the first case, $\overline{p} \, \varepsilon \, \overline{q_i}$, hence $\overline{q_i}$ must be in A. In the second case, $\overline{q_i} = \sqcap\{\langle \overline{q_i} q \rangle : q \in Q\}$. Every $\langle \overline{q_i} q \rangle$ must be in A, since $\overline{p} \, \varepsilon \, \langle \overline{q_i} q \rangle$ for some $\overline{p} \in A$, and $\overline{q_i}$ must be in A, since A is $\sqcap$-subsemilattice. Thus, every $\overline{q_i}$ is in A, hence $q \in A$, and we are done. □

Example 4.31 The lattice **J** in Fig. 4.11 illustrates how the construction and embedding work. The minimal nontrivial join covers in **J** are

$$a \leq b \vee c \qquad b \leq d \vee e$$

giving the ordered set (J(**J**), $\overline{\mathrm{D}}$) shown in the figure. Therefore, we take **P** to be the semilattice generated by the sequences $\{a, ab, ac, abd, abe\}$ subject to the relations

$$a = ab \sqcap ac \qquad ab = abd \sqcap abe$$

as illustrated in Fig. 4.12.

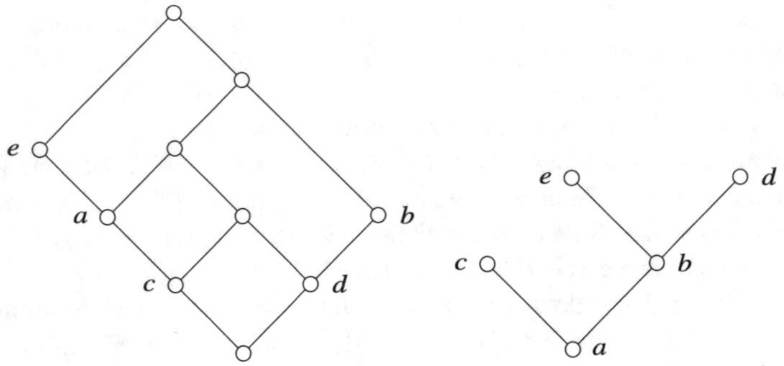

Fig. 4.11 Lattice **J** of Example 4.31 and (J(**J**), $\overline{\text{D}}$)

Define a mapping $\lambda : \mathbf{P} \to \mathbf{J}$ such that $\lambda(\overline{x}) = \ell(\overline{x})$, e.g., $\lambda(abe) = e$. Then $\lambda(\overline{x} \sqcap \overline{y}) = \lambda\overline{x} \vee \lambda\overline{y}$, so that $\lambda(ac \sqcap abd) = c \vee d$ and $\lambda(ac \sqcap abe) = e$. By default, $\lambda(1_P) = 0_{B^2}$.

Then the embedding $\zeta : \mathbf{J} \to \text{Sub}(\mathbf{P}, \sqcap, 1)$ is given by $\zeta(u) = \{p \in P : \lambda p \leq u\}$. Explicitly, ζ is the following embedding:

$$\begin{aligned}
\zeta(0_J) &= \{1\} \\
\zeta(c) &= \{ac, 1\} \\
\zeta(a) &= \{a, ac, 1\} \\
\zeta(e) &= \{abe, a, ac, ac \sqcap abe, 1\} \\
\zeta(c \vee d) &= \{ac, abd, ac \sqcap abd, 1\} \\
\zeta(a \vee d) &= \{a, ac, abd, ac \sqcap abd, 1\} \\
\zeta(d) &= \{abd, 1\} \\
\zeta(b) &= \{ab, abd, 1\} \\
\zeta(a \vee b) &= \{a, ab, ac, abd, ac \sqcap abd, 1\} \\
\zeta(1_J) &= P.
\end{aligned}$$

We want to look closer at the sublattices of Sub **S**, where **S** is a meet semilattice with 1, that might have a preclop. Again we use the convention that a_0 denotes the least element of a subsemilattice $\mathbf{A} \leq \mathbf{S}$.

Lemma 4.32 *Let* **S** *be a finite* $(\wedge, 1)$*-semilattice and* **L** *a finite (lower bounded) lattice. Let* $\phi : \mathbf{L} \leq \text{Sub}\,\mathbf{S}$ *be a* $(0, 1)$*-embedding. For any* $a \in L$*, let* $\eta(a) = \bigvee\{x \in \mathbf{L} : (\phi(x))_0 \geq_{\mathbf{S}} (\phi(a))_0\}$. *Then* η *on* **L** *satisfies properties* (I1)–(I5) *of an equaclosure operator.*

Proof For (I5), we need the observation that in Sub **S**, we have $A \vee B = \{a \wedge b : a \in A,\ b \in B\}$. The rest is easy. □

It remains to decide when the map η also satisfies (I6), in order to obtain a preclop.

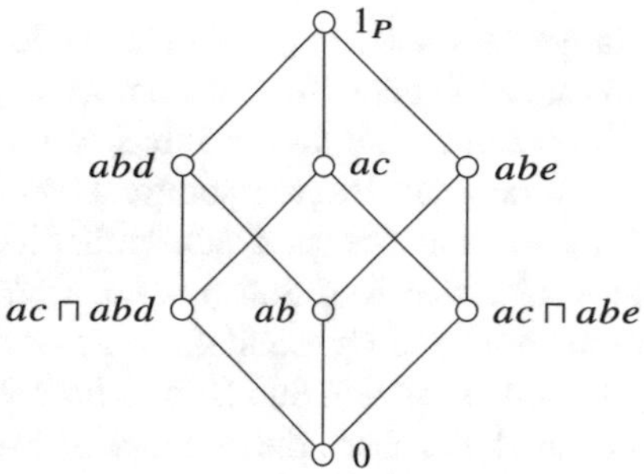

Fig. 4.12 Semilattice **P** of Example 4.31 with $0 = a$

Fix some (0, 1)-embedding ϕ of a lower bounded lattice **L** into Sub **S**. Define a map $\lambda : S \to L$ by $\lambda(p) = \bigwedge_{\mathbf{L}}\{x \in L : p \in \phi(x)\}$ for any $p \in S$. Note that $\lambda(p)$ is defined for every $p \in S$ because $\phi(\mathbf{L})$ is a 1-sublattice of Sub **S**.

Consider the following property of ϕ for any $a \in L$:

$$\text{If } p \geq \phi(a)_0 \text{ in } \mathbf{S}, \text{ then } \phi(\lambda(p)) \leq \uparrow\phi(a)_0 \text{ in Sub } \mathbf{S}. \qquad (\diamond)$$

The property ($\diamond$) says that $\phi(a)_0$ mimics a property of fully invariant elements in Sub($\mathbf{S}, \wedge, 1, H$): if a_0 is fully invariant and $p \geq a_0$, then $\mathrm{Sg}(p) \leq \uparrow a_0$.

Lemma 4.33 *If* $\phi : \mathbf{L} \to$ Sub **S** *is an embedding of a lower bounded lattice* **L** *into* Sub **S** *that satisfies* ($\diamond$) *for each* $a \in L$, *then* η *on* **L** *defined in* Lemma 4.32 *is a preclop.*

Proof We need to check property (I6).

Let $A, V, W \in \phi(L)$. If $t \in \eta(A) \wedge (V \vee W)$, then $t \in \eta(A)$ and $t = v \wedge w$ for some $v \in V$, $w \in W$. Thus $v, w \geq t \geq a_0$. By ($\diamond$), this implies that $\phi(\lambda(v)) \leq \uparrow a_0$ and $\phi(\lambda(w)) \leq \uparrow a_0$ in Sub **S**. It follows from the definition of η that $\phi(\lambda(v)), \phi(\lambda(w)) \leq \eta(A)$. Hence, v and w are in $\eta(A)$, whence $t \in (\eta(A) \wedge V) \vee (\eta(A) \wedge W)$. □

Example 4.34 Recall that $\mathbf{P}_5$ is the ordered set with $a < b < c > d > e$, which appears in Theorem 4.9. The embedding of $\mathbf{L} = \mathrm{Co}(\mathbf{P}_5)$ into Sub **S** that uses the construction of Theorem 4.30 does not satisfy ($\diamond$). Every atom of Co($\mathbf{P}_5$), except c, is mapped to a 2-element sublattice $\{1, x\}$ of **S**, and c is mapped to the 4-element sublattice $\{1, c_1, c_2, c_3, \}$, with $c_3 = c_1 \wedge c_2$. We also have $b = c_1 \wedge a$ and $d = c_2 \wedge d$ in **S**. Thus, we have b and d that are least elements for which ($\diamond$) fails: $c_1 \geq b$ and $\lambda(c_1) = \{1, c_1, c_2, c_3\}$, so that $\lambda(c_1) \not\leq \uparrow b$, and analogously for d.

Lemma 4.32 raises the question:

Question 4.35 *Let* **L** *be a finite lower bounded lattice with a preclop, and let* ϕ *be some* (0, 1)*-embedding into* Sub **S**, *where* ($\diamond$) *fails for some* $a \in \mathbf{L}$. *Can this embedding be fixed at* a*?*

Namely, can we build a larger semilattice $\mathbf{S}_1 \geq \mathbf{S}$ such that $\mathbf{L}$ is still (0, 1)-embedded into Sub $\mathbf{S}_1$, and ($\diamond$) holds at $a \in \mathbf{L}$ under this new embedding?

For a semilattice with operators, the lattice Sub($\mathbf{S}, \wedge, 1, H$) is a (0, 1)-sublattice of Sub($\mathbf{S}, \wedge, 1$), and the natural equaclosure operator Γ on Sub($\mathbf{S}, \wedge, 1, H$) has the property that $\Gamma(A) = \uparrow a_0$ for any H-closed subsemilattice A. Generalizing this situation, a preclop κ on a sublattice $\mathbf{K} \leq$ Sub $\mathbf{S}$ is *filterable* if $\kappa(A) \leq \uparrow a_0$ for any $A \in K$. The idea of Lemma 4.32 and Question 4.35 is that any preclop on a lower bounded lattice $\mathbf{L}$ could be represented in Sub $\mathbf{S}_1$ as a filterable preclop.

The natural first step would be to check the preclops on Sub $\mathbf{S}$. There was a long-standing hypothesis that for every preclop κ on Sub $\mathbf{S}$ one could find a semilattice $\mathbf{S}^*$ such that Sub $\mathbf{S}$ is isomorphic to Sub $\mathbf{S}^*$ and κ is filterable on Sub $\mathbf{S}^*$. The following example shows that this is not true.

Example 4.36 Let $\mathbf{S}$ be the semilattice in Fig. 4.13. According to results of [8], if Sub $\mathbf{P}$ isomorphic to Sub $\mathbf{S}$ for some semilattice $\mathbf{P}$, then $\mathbf{P}$ and $\mathbf{S}$ are isomorphic.

Define a preclop on Sub $\mathbf{S}$ as the minimal preclop κ satisfying $b \in \kappa(\{c_1, c_2\})$. The preclop μ on Sub $\mathbf{S}$ has $\mu(X)$ being the least subsemilattice Y such that $X \subseteq Y$ and whenever $y \in Y$ and $y = z \wedge t$ properly, then $z, t \in Y$. Thus, for example, the subsemilattice $\{1, c_1, c_2\}$ has $\mu(C) = \uparrow c_1$. Our κ extends this by making $\kappa(C) = \uparrow b$, and more generally, $b \in \kappa(Y)$ whenever $C \leq Y$. One can check that κ is indeed a preclop on Sub $\mathbf{S}$, but it is not filterable as witnessed by $\kappa(C)$.

It is interesting to note that κ also fails (‡) with $X = \{b, c_1, 1\}$ and $Z = \{b, c_2, 1\}$.

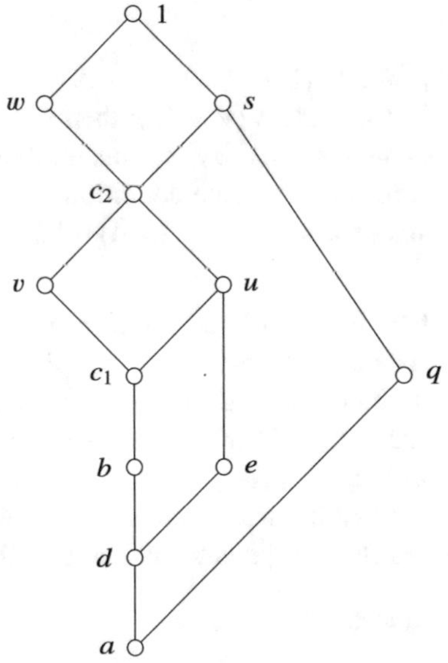

Fig. 4.13 Semilattice $\mathbf{S}$ of Example 4.36

4.4 A General Embedding Method

Let **L** be a lattice and $\mathcal{V}$ a variety of algebras. We seek a construction that will embed **L** into Sub **A** for some algebra $\mathbf{A} \in \mathcal{V}$, if possible. In our applications, $\mathcal{V}$ will be semilattices or semilattices with operators, but the approach is quite general.

Suppose we can find $\mathbf{A} \in \mathcal{V}$ and a map $\lambda : \mathbf{A} \to \mathbf{L}$ with the following properties.

(a) $a \in \mathrm{Sg}(B)$ implies $\lambda a \leq \bigvee \lambda B$.
(b) $\lambda c = 0_L$ iff c is in the subalgebra generated by the constants of **A** (the least subalgebra).
(c) If $\lambda a \leq x \vee y$, then there exists a finite set $B \subseteq A$ such that $\lambda B \ll \{x, y\}$ and $a \in \mathrm{Sg}(B)$.
(d) $\lambda(A)$ $\bigvee$-generates **L**.

Define $\zeta : \mathbf{L} \to \mathcal{P}(A)$ by

$$\zeta x = \{a \in A : \lambda a \leq x\}$$

so that $a \in \zeta x$ iff $\lambda a \leq x$. Note that ζ is order preserving.

Theorem 4.37 *Let* $\mathbf{A} \in \mathcal{V}$ *and* $\lambda : \mathbf{A} \to \mathbf{L}$. *If* λ *satisfies* (a)–(d), *then* $\zeta : \mathbf{L} \leq \mathrm{Sub}\,\mathbf{A}$ *is an embedding such that* $\zeta(0_L)$ *is the least subalgebra of* **A**.

Proof Properties (a), (b) guarantee that ζx is a subalgebra of **A**, and that $\zeta(0_L)$ the least subalgebra of **A**. It is straightforward to verify that $\zeta(x \wedge y) = \zeta x \wedge \zeta y$. Property (c) is a straightforward translation of ζ being join-preserving.

To complete the proof of Theorem 4.37, we must show that ζ is one-to-one, using (d). Let $x \not\leq y$ in **L**. There exists a subset $B \subseteq A$ such that $x = \bigvee \lambda(B)$. For some $b_0 \in B$ we have $\lambda b_0 \leq x$, $\lambda b_0 \not\leq y$. Then $\lambda b_0 \in \zeta x \setminus \zeta y$, so that $\zeta x \not\leq \zeta y$, as desired.□

For the converse, let us assume that **L** is finite.

Theorem 4.38 *Let* **L** *be a finite lattice,* $\mathbf{A} \in \mathcal{V}$ *and* $\xi : \mathbf{L} \leq \mathrm{Sub}\,\mathbf{A}$ *a lattice embedding. Assume that* $\xi(0_L) = 0_{\mathrm{Sub}\,\mathbf{A}}$ *and (without loss of generality) that* $\xi(1_L) = \mathbf{A}$. *Define a map* $\lambda : \mathbf{A} \to \mathbf{L}$ *by*

$$\lambda a = \bigwedge \{x \in L : a \in \xi x\}.$$

Then λ *satisfies* (a)–(d).

Proof Note that $a \in \xi \lambda a$ as ξ preserves meets. Thus

$$a \in \xi x \quad \text{iff} \quad \lambda a \leq x.$$

For condition (a), let $a \in \mathrm{Sg}(B)$ where say $B = \{b_1, \ldots, b_k\}$. Since $b_j \in \xi \lambda b_j$ for each j, in Sub **A** we have

$$a \in \xi \lambda b_1 \vee \ldots \vee \xi \lambda b_k = \xi(\lambda b_1 \vee \ldots \vee \lambda b_k).$$

Therefore, $\lambda a \leq \lambda b_1 \vee \ldots \vee \lambda b_k$, as desired.

Condition (b) holds by assumption.

For (c), assume $\lambda a \leq x \vee y$. Then $a \in \xi(x \vee y) = \xi x \vee \xi y$, so there exists a finite set $B \subseteq \xi x \cup \xi y$ such that $a \in \mathrm{Sg}(B)$. If $b \in B$ and $b \in \xi x$, then $\lambda b \leq x$, and similarly for y. Thus $\lambda B \ll \{x, y\}$.

For (d), let $x \in \mathrm{J}(\mathbf{L})$, and let x_* denote its lower cover. Choose $a \in \xi x \setminus \xi x_*$, so that $\lambda a = x$. □

Aside Though we will mostly use the method when **L** is finite, the most natural setting for Theorem 4.37 is to assume that **L** is algebraic, and that $\lambda : \mathbf{A} \to \mathbf{L}^c$, where $\mathbf{L}^c$ denotes the join semilattice of compact elements of **L**. In this case, ζ is a complete lattice embedding. Theorem 4.39 below, regarding when ζ is an isomorphism, is unchanged.

The converse, Theorem 4.38, goes through with the weaker hypotheses that ξ preserves complete meets and that **L** is *spatial*, meaning that every element is a join of completely join irreducible elements.

Modifying the construction of Theorem 4.30, one can use Theorem 4.37 to prove that *every lattice* **L** *can be embedded into* Sub **S** *for some semilattice* **S**. But the relevant structure when **S** is infinite is the lattice of algebraic subsets $\mathrm{S_p}(\mathbf{S})$, and not every lattice can be embedded into one of those, because $\mathrm{S_p}(\mathbf{S})$ is join semidistributive (whereas Sub **S** need not be when **S** is infinite).

4.4.1 A Semilattice Example

Some easy examples will show how this works. (These particular examples might be done more easily, but we are trying to illustrate a method.)

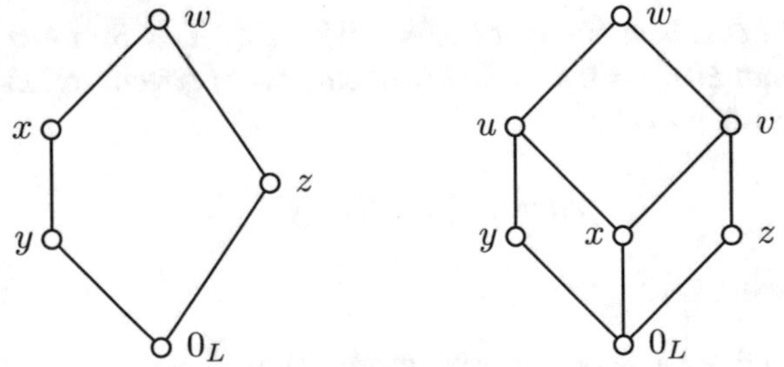

Fig. 4.14 Pentagon $\mathbf{N}_5$ and $\mathbf{L}_1 = \mathrm{L}(x \wedge (y \vee z))$

Consider the pentagon as labeled in Fig. 4.14, and the variety $\mathcal{S}$ of meet semilattices with 1. The minimal nontrivial join cover in $\mathbf{N}_5$ we have to take care of is $x \leq y \vee z$.

In $\mathcal{S}$ take the 4-element semilattice $\mathbf{S}$ with elements $1, a, b, a \wedge b$. Let λ be the map $1 \mapsto 0_L$, $a \mapsto y$, $b \mapsto z$, $a \wedge b \mapsto x$. This satisfies the properties of Theorem 4.37 and gives us the embedding

$$\begin{aligned} \zeta(0_L) &= \{1\} \\ \zeta(x) &= \{1, a, a \wedge b\} \\ \zeta(y) &= \{1, a\} \\ \zeta(z) &= \{1, b\} \\ \zeta(w) &= S \end{aligned}$$

of $\mathbf{N}_5$ into Sub $\mathbf{S}$.

The second lattice $\mathbf{L}_1$ of Fig. 4.14 is quite similar, in that it is generated by x, y, z with the minimal nontrivial join cover $x \leq y \vee z$. We can use the same 4-element semilattice $\mathbf{S}$ and map λ, but since x and y are now incomparable, the embedding becomes

$$\begin{aligned} \zeta(0_L) &= \{1\} \\ \zeta(x) &= \{1, a \wedge b\} \\ \zeta(y) &= \{1, a\} \\ \zeta(z) &= \{1, b\} \\ \zeta(u) &= \{1, a, a \wedge b\} \\ \zeta(v) &= \{1, b, a \wedge b\} \\ \zeta(w) &= S \end{aligned}$$

which is in fact an isomorphism $\mathbf{L}_1 \cong$ Sub $\mathbf{S}$.

Recall that for any finite lower bounded lattice $\mathbf{L}$, Theorem 4.30 constructs a finite semilattice $\mathbf{P}$ such that $\mathbf{L} \leq$ Sub $\mathbf{P}$. The construction in the proof of that theorem fits directly into this scheme. The generators of $\mathbf{P}$ are sequences $d_1 \ldots d_k$ with each $d_i \in \mathrm{J}(\mathbf{L})$, d_1 minimal with respect to the D-relation order, and $d_i \mathrel{D} d_{i+1}$. The order on the generators is that $\overline{d} \leq \overline{e}$ if $\overline{d}$ is an initial segment of $\overline{e}$, and meets are determined by the minimal nontrivial join covers of $\mathbf{L}$ (see the proof). The empty sequence ι is the largest element of $\mathbf{P}$. The map λ is given by

$$\begin{aligned} \lambda(\iota) &= 1_{\mathbf{L}} \\ \lambda(d_1 \ldots d_k) &= d_k \\ \lambda(\overline{s}_1 \wedge \cdots \wedge \overline{s}_m) &= \lambda\overline{s}_1 \vee \cdots \vee \lambda\overline{s}_m \\ &\qquad \text{if } \overline{s}_1 \wedge \cdots \wedge \overline{s}_m \text{ is in reduced form.} \end{aligned}$$

In general the requirement is only that $\lambda(\overline{s}_1 \wedge \cdots \wedge \overline{s}_m) \leq \lambda\overline{s}_1 \vee \cdots \vee \lambda\overline{s}_m$, and indeed that is what happens if $\bigwedge_i \overline{dab_i} = \overline{da}$ in $\mathbf{P}$ because of the defining relations, i.e., when $a \le \bigvee_i b_i$ is a minimal nontrivial join cover. Then $\zeta x = \{p \in P : \lambda p \le x\}$ for $x \in L$ embeds $\mathbf{L}$ into Sub $\mathbf{P}$.

In the terminology of the proof of Theorem 4.30, $\lambda p = \ell(p)$ and $\zeta x = \mathbf{P}_x$.

We use this construction in Chap. 6. Sometimes it will be necessary to enlarge $\mathbf{P}$ with additional elements to accommodate the operators. In that case, we will extend λ to a map λ' on the larger semilattice to obtain the desired embedding.

4.4.2 Group Examples

Now let $\mathcal{V}$ be the variety generated by the 8-element dihedral group $\mathbf{D}_4$. We rework the examples from Fig. 4.14. Let us use the representation that $\mathbf{D}_4$ is the group generated by a, b satisfying $a^2 = b^2 = (ab)^4 = 1$. Its elements are 1, a, b, ab, ba, aba, bab, $abab = baba$. The subgroup lattice Sub $\mathbf{D}_4$ is in Fig. 4.15. Note that the center of $\mathbf{D}_4$ is $\{1, abab\}$.

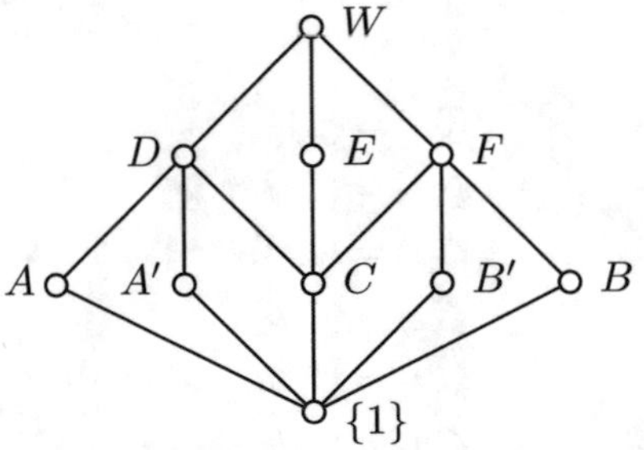

Fig. 4.15 Subgroup lattice of $\mathbf{D}_4$, where $A = \{1, a\}$, $A' = \{1, bab\}$, $C = \{1, abab\}$, $B' = \{1, aba\}$, $B = \{1, b\}$, $D = \{1, a, bab, abab\}$, $E = \{1, ab, abab, ba\}$, $F = \{1, b, aba, abab\}$

Now let us represent the pentagon in Sub $\mathbf{D}_4$. Again the only minimal nontrivial join cover is $x \le y \vee z$. This time we take $\lambda : \mathbf{D}_4 \to \mathbf{N}_5$ to be the map

$$
\begin{aligned}
1 &\mapsto 0_L & ba &\mapsto w \\
a &\mapsto y & aba &\mapsto w \\
b &\mapsto z & bab &\mapsto x \\
ab &\mapsto w & abab &\mapsto x.
\end{aligned}
$$

This yields the embedding

$$
\begin{aligned}
\zeta(0_L) &= \{1\} \\
\zeta(x) &= \{1, a, bab, abab\} \\
\zeta(y) &= \{1, a\} \\
\zeta(z) &= \{1, b\} \\
\zeta(w) &= D_4
\end{aligned}
$$

of $\mathbf{N}_5$ into Sub $\mathbf{D}_4$.

Next consider the second lattice $\mathbf{L}_1$ of Fig. 4.14. The pentagon is a sublattice of $\mathbf{L}_1$, and indeed a join-semilattice retract with $x \mapsto u$ and $v \mapsto w$. To embed $\mathbf{L}_1$ into Sub $\mathbf{D}_4$, take $\lambda : \mathbf{D}_4 \to \mathbf{L}_1$ to be the map

$$
\begin{aligned}
1 &\mapsto 0_L & \qquad ba &\mapsto w \\
a &\mapsto y & \qquad aba &\mapsto v \\
b &\mapsto z & \qquad bab &\mapsto u \\
ab &\mapsto w & \qquad abab &\mapsto x.
\end{aligned}
$$

This yields the embedding

$$
\begin{aligned}
\zeta(0_L) &= \{1\} \\
\zeta(x) &= \{1, abab\} \\
\zeta(y) &= \{1, a\} \\
\zeta(z) &= \{1, b\} \\
\zeta(u) &= \{1, a, bab, abab\} \\
\zeta(v) &= \{1, b, aba, abab\} \\
\zeta(w) &= D_4
\end{aligned}
$$

of $\mathbf{L}_1$ into Sub $\mathbf{D}_4$.

4.4.3 Isomorphism

Theorem 4.37 gives an embedding of $\mathbf{L}$ into Sub $\mathbf{A}$. With an extra condition, we obtain an isomorphism.

Theorem 4.39 *Let* $\mathbf{A} \in \mathcal{V}$ *and* $\lambda : \mathbf{A} \to \mathbf{L}$. *If* λ *satisfies* (a)–(d) *and has the additional property*

(e) $\lambda b \le \lambda a$ *iff* $b \in \mathrm{Sg}(a)$,

then $\zeta : \mathbf{L} \cong \mathrm{Sub}\,\mathbf{A}$ *is an isomorphism.*

Proof Given the fact that ζ is an embedding, we need that $\zeta\lambda(a) = \mathrm{Sg}(a)$ for each $a \in A$. Since $\zeta\lambda(a) = \{b \in A : \lambda b \le \lambda a\}$, condition (e) translates that property. □

In practical terms, condition (e) means that whenever $a, b \in A$ and $0_L < \lambda b \leq \lambda a$, then there should be a term operation h on $\mathbf{A}$ such that $b = h(a)$. For our "semilattice with operator" constructions, we will be seeking to find operators to make this happen.

Chapter 5
Finite Lattices as Sub(S, ∧, 1, H): The Case J(L) ⊆ τ(L)

Woman has a double clutch, but man must shift for himself. – Walt Kelley

Женская интуиция позволяет предугадать то, что еще не случилось, но мешает видеть то, что уже произошло. – Яна Джангирова

In this chapter and the next, we address the question: *Given a pair* $(\mathbf{L}, \gamma)$ *consisting of a finite lower bounded lattice and a preclop, when can we find a finite semilattice* **S** *and a set H of operators such that* $\mathbf{L} \cong \mathrm{Sub}(\mathbf{S}, \wedge, 1, H)$ *with* γ *corresponding to the natural equaclosure operator* Γ? That is, γ-closed sets should map to Γ-closed sets under the isomorphism. Then, in Chap. 7, we consider when such a representation can be converted to one of the form $\mathbf{L} \cong \mathrm{L_q}(\mathcal{K})$ for some quasivariety $\mathcal{K}$.

By a *representation of* $(\mathbf{L}, \gamma)$ *as* $(\mathrm{S_p}(\mathbf{S}, H), \Gamma)$ we mean an isomorphism ξ : $\mathbf{L} \cong \mathrm{S_p}(\mathbf{S}, H)$ such that $\xi(\gamma(x)) = \Gamma(\xi(x))$ for all $x \in L$, where Γ is the natural weak equaclosure operator. When **S** is finite, $\mathrm{S_p}(\mathbf{S}, H)$ becomes $\mathrm{Sub}(\mathbf{S}, \wedge, 1, H)$. The definition for a representation of $(\mathbf{L}, \gamma)$ as $(\mathrm{L_q}(\mathcal{K}), \Gamma)$ is similar.

Recall that for a preclop γ on a lattice **L**, we let $\tau(x)$ denote the least element t such that $\gamma(t) = \gamma(x)$. Let $\tau(\mathbf{L}) = \{\tau(x) : x \in L\}$. It turns out that the problem of finding the algebraic lattice **S** for the representation is much simpler when every join irreducible element is in $\tau(\mathbf{L})$, i.e., $\mathrm{J}(\mathbf{L}) \subseteq \tau(\mathbf{L})$. For in that case, if any such lattice **S** exists, we can just take the dual companion lattice, $\mathbf{S} = \gamma^d(\mathbf{L})$. That case is the subject of Chap. 5. When $\mathrm{J}(\mathbf{L}) \nsubseteq \tau(\mathbf{L})$, then we must take an overlattice of $\gamma^d(\mathbf{L})$, using either *ad hoc* methods or, better, constructions such as that of Theorem 4.30. That is the topic of Chap. 6.

Aside It is tempting to conjecture that if $\mathrm{J}(\mathbf{L}) \subseteq \tau(\mathbf{L})$, then **L** is lower bounded of rank at most 1, i.e., $\mathrm{J}(\mathbf{L}) \subseteq \mathrm{D}_1(\mathbf{L})$. The lattices in Fig. 5.1 and Example 5.1 show that not to be true, but the condition does add restrictions. For example, if $p \neq q \in \mathrm{J}(\mathbf{L})$ and $p\ \mathrm{D}^n\ q$ for some n, then $q \leq \gamma(p)$. If $\mathrm{J}(\mathbf{L}) \subseteq \tau(\mathbf{L})$, then $p \nleq q$, and that limits the types of minimal nontrivial join covers that can occur in **L**. Compare the lattice **J** in Fig. 4.4, with $p\ \mathrm{D}\ r\ \mathrm{D}\ t$ and $p \leq t$, which admits no preclop at all.

K. Adaricheva et al., *A Primer of Subquasivariety Lattices*, CMS/CAIMS Books in Mathematics 3, https://doi.org/10.1007/978-3-030-98088-7_5

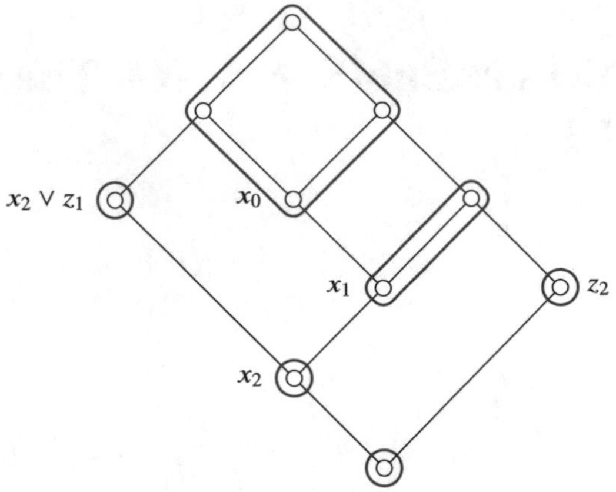

Fig. 5.1 The construction of Example 5.1 yields pairs $(\mathbf{L}_n, \gamma_n)$ with $\mathrm{J}(\mathbf{L}_n) \subseteq \tau(\mathbf{L}_n)$ and $\mathbf{L}_n$ lower bounded of any rank $n \geq 1$. The lattice pictured above is a sublattice of $\mathbf{L}_2$

Example 5.1 For $n \geq 1$, let us construct a sequence of pairs $(\mathbf{L}_n, \gamma_n)$ such that $\mathbf{L}_n$ is an (upper and lower) bounded lattice of rank n, i.e., $\mathrm{J}(\mathbf{L}_n) \subseteq \mathrm{D}_n(\mathbf{L}_n)$, and γ_n is a preclop on $\mathbf{L}_n$ with $\mathrm{J}(\mathbf{L}_n) \subseteq \tau(\mathbf{L}_n)$. We use the familiar technique of representing $\mathbf{L}_n$ as a join semilattice with 0, presented by a generating set X_n of join irreducibles and a set of join dependencies, as in [15, 49]. This is equivalent to representing the lattice as a closure operator on X_n and is particularly useful when the lattices get hard to draw.

Let $X_n = \{x_0, x_1, \ldots, x_n, z_1, \ldots, z_n\}$. The relations defining $\mathbf{L}_n$ as a join semilattice with 0 are

$$x_0 > x_1 > \cdots > x_n$$
$$x_0 \leq x_1 \vee z_1$$
$$\cdots$$
$$x_{n-1} \leq x_n \vee z_n.$$

A sublattice of L_2 is illustrated in Fig. 5.1. The preclop γ_n has, for $w \in L_n$ and $0 < k < n$,

$$\gamma_n(w) = 1 = x_0 \vee z_1 \vee \ldots \vee z_n \text{ if } x_0 \leq w,$$
$$\gamma_n(w) = w \vee z_{k+1} \vee \ldots \vee z_n \text{ if } x_k \leq w \text{ and } x_{k-1} \nleq w,$$
$$\gamma_n(w) = w \text{ if } x_n \nleq w.$$

There are many details to be checked, none of which is very hard. Since the example is not crucial to our further progress, we leave this to the reader. It is straightforward that $\mathbf{L}_n$ is lower bounded, because it has only the indicated minimal join covers of join irreducibles. To see that it is semidistributive, use Theorem 4 of [49]. For boundedness, use Theorem 2.64 of [46].

5.1 Companion Lattices Rise Again

We are given a pair $(\mathbf{L}, \gamma)$ with $\mathbf{L}$ a finite, lower bounded lattice and γ a preclop, satisfying properties (I1)–(I7), though (I7) always holds for $\mathbf{L}$ finite. Assume that the pair has a representation as $(\mathrm{S_p}(\mathbf{S}, H), \Gamma)$. We want to see how this is reflected in properties of the lattice $\gamma(\mathbf{L})$.

The classes of the equapartition on $\mathbf{L}$ are denoted by $[a] = \{x \in L : \gamma(x) = \gamma(a)\}$. Similarly, for H-closed algebraic subsets of $\mathbf{S}$, let $[A] = \{X \in \mathrm{S_p}(\mathbf{S}, H) : \Gamma(X) = \Gamma(A)\}$. As in Sect. 3.3, we represent the lattice $\gamma(\mathbf{L})$ using the classes of the equapartition. That is, $\gamma(\mathbf{L}) \cong \mathbf{T}$ where

- $T = \{[x] : x \in \tau(\mathbf{L})\}$,
- $[x] \leq [y]$ iff $\gamma x \leq \gamma y$,
- the least element of $\mathbf{T}$ is $[z]$ where $z = \tau(1_L)$,
- $[x] \vee [y] = [x \vee y]$,
- $[x] \wedge [y] = [\tau(\gamma x \wedge \gamma y)]$.

Note that $[b] = [c]$ implies $\tau b = \tau c$, which is the least element of that class of the equapartition. Hence we can write $\tau[b]$ unambiguously.

It turns out that we need to work, not just with H, but with the extended set of maps $\mathbb{DMO}$; see Sect. 3.2. If we assume that H is closed under composition and contains the identity and the constant map 1, then $\mathbb{DMO}$ consists of all maps $k : S \to S$ such that k is the join of a nonempty up-directed set of meets of maps from H. The maps in $\mathbb{DMO}$ need not be operators; all we are guaranteed is that they preserve finite meets in $\mathbf{S}$, by Theorem 3.11.

For any $x \in S$, the H-closed algebraic subset generated by x is $\mathrm{Ag}(x) = \mathbb{DMO}(x)$. The least element of $\mathrm{Ag}(x)$ is $x_0 = \bigwedge_{k \in \mathbb{DMO}} kx$. But you do not need directed joins to get the least element, so in fact $x_0 = \bigwedge_{g \in H} gx$. This rather trivial observation will prove useful.

Now for $h \in H$, or more generally $h \in \mathbb{DMO}$, we define h^*, first on $\mathrm{S_p}(\mathbf{S}, H)$ and then on $\Gamma(\mathrm{S_p}(\mathbf{S}, H))$. Each $h \in H$ operates on $\mathbf{S}$, but that induces a map on $\mathrm{S_p}(\mathbf{S}, H)$:

$$h^*(A) = \mathrm{Ag}(\{h(a) : a \in A\}).$$

Now $h(A) = \{h(a) : a \in A\}$ is a subset of A. For a fixed h, the set $h(A)$ need not be H-closed nor an algebraic subset, but $h^*(A) = \mathrm{Ag}(h(A))$ still satisfies $h^*(A) \leq A$ since A is H-closed. Moreover, the least element of $h^*(A)$ is $\bigwedge_{g \in H} gh(a_0)$, where a_0 is the least element of A. So if $\Gamma(A) = \Gamma(B)$, then $a_0 = b_0$, and hence $\Gamma(h^*(A)) = \Gamma(h^*(B))$. Thus we may regard the induced map h^* as acting on $\Gamma(\mathrm{S_p}(\mathbf{S}, H))$, with $h^*[A] = [h^*(A)]$. Explicitly,

$$h^*[A] = [\Gamma(h^*(A))] = [\uparrow \bigwedge_{g \in H} gh(a_0)].$$

Lemma 5.2 *If $k \in \mathbb{DMO}$, then k^* preserves finite joins: if $A_1, \ldots, A_m \in S_p(\mathbf{S}, H)$, then*

$$k^*([A_1] \vee \ldots \vee [A_m]) = k^*[A_1] \vee \ldots \vee k^*[A_m].$$

If $h \in H$, then h^* preserves arbitrary joins in $\Gamma(S_p(\mathbf{S}, H))$.

Proof If A_i $(i \in I)$ are H-closed algebraic subsets, each with least element a_{i0}, then the least element of $\bigvee_i A_i$ is $\bigwedge_i a_{i0}$. We calculate, using the fact that k preserves finite meets in $\mathbf{S}$,

$$\begin{aligned} k^*([A_1] \vee \ldots \vee [A_m]) &= [\uparrow \bigwedge_{g \in H} gk(a_{10} \wedge \ldots \wedge a_{m0})] \\ &= [\uparrow \bigwedge_{g \in H} (gk(a_{10}) \wedge \ldots \wedge gk(a_{m0}))] \\ &= [\uparrow (\bigwedge_{g \in H} gk(a_{10}) \wedge \ldots \wedge \bigwedge_{g \in H} gk(a_{m0}))] \\ &= [\uparrow \bigwedge_{g \in H} gk(a_{10})] \vee \ldots \vee [\uparrow \bigwedge_{g \in H} gk(a_{m0})] \\ &= k^*[A_1] \vee \ldots \vee k^*[A_m]. \end{aligned}$$

When $h \in H$, it preserves arbitrary meets, and the same calculation shows that $h^*[\bigvee A_i] = \bigvee h^*[A_i]$ for any index set. □

A similar argument yields the following observation.

Lemma 5.3 *If $\{k_j : j \in J\} \subseteq \mathbb{DMO}$, then $(\bigwedge_j k_j)^* = \bigvee_j k_j^*$.*

Now let us record some properties of the set of maps $H^* = \{h^* : h \in H\}$ on $\Gamma(S_p(\mathbf{S}, H))$.

Lemma 5.4 *If $(\mathbf{L}, \gamma)$ is representable as $(S_p(\mathbf{S}, H), \Gamma)$, then*

(1) $\tau[a] \leq \tau[b]$ *implies there exists $k \in \mathbb{DMO}$ such that* $k^*[b] = [a]$.

Moreover, the following hold for each h^* ($h \in \mathbb{DMO}$) acting on $\gamma(\mathbf{L})$.

(2) $h^*[x] \leq [x]$,
(3) $\tau h^*[x] \leq \tau[x]$,
(4) $h^*[0] = [0]$,
(5) $[x] \leq [y]$ *implies* $h^*[x] \leq h^*[y]$,
(6) $h^*([x] \vee [y]) = h^*[x] \vee h^*[y]$.

Proof Item (1) uses Lemma 3.19 (4): *If $x, y \in S$, then $x \in \mathrm{Ag}(y)$ if and only if $x = k(y)$ for some $k \in \mathbb{DMO}$.* The hypothesis $\tau[a] \leq \tau[b]$ means that the corresponding least elements are such that $a_0 = k(b_0)$ for some $k \in \mathbb{DMO}$. The least element in $k^*[b]$ is $\bigwedge_{g \in H} gk(b_0) = \bigwedge_{g \in H} g(a_0) = a_0$, whence $k^*[b] = [a]$.

For notational convenience, let the isomorphism $\sigma : \mathbf{L} \cong \mathrm{S_p}(\mathbf{S}, H)$ be such that $\sigma(x) = X$, $\sigma(y) = Y$, etc. Items (2) and (3) follow from the discussion before the statement of the lemma, while (4) holds since $h(1_\mathbf{S}) = 1_\mathbf{S}$ for $h \in \mathbb{DMO}$.

For (5), $[X] \leq [Y]$ means $x_0 \geq y_0$, whence $f(x_0) \geq f(y_0)$ for all $f \in \mathbb{DMO}$. Therefore, $\bigwedge gh(x_0) \geq \bigwedge gh(y_0)$.

For (6), note that the least element of $[X] \vee [Y]$ is $x_0 \wedge y_0$, while the least element of $h^*([X] \vee [Y])$ is the least possible value of $gh(x_0 \wedge y_0) = gh(x_0) \wedge gh(y_0)$. The least element of $h^*[X]$ is some minimum $g_1 h(x_0)$, and likewise the least element of $h^*[Y]$ is some $g_2 h(y_0)$. Remembering that $\mathbb{DMO}$ is closed under meets of maps, we can take $g = g_1 \wedge g_2$ to obtain the least possible $gh(x_0 \wedge y_0)$, which yields the conclusion. □

Example 5.5 Let $\mathbf{L}$ be a 3-element chain, $0 < A < B$, and γ just the identity map. A perfectly good representation of $(\mathbf{L}, \gamma)$ as $(\mathrm{S_p}(\mathbf{S}, H), \Gamma)$ can be obtained by letting $\mathbf{S}$ be the $\omega + 2$-chain $b_0 < b_1 < b_2 < \cdots < a_0 < 1$. The operators in H are h_j $(j \geq 0)$ and g, given by

$$h_j(x) = \begin{cases} b_j & \text{if } x \leq b_j, \\ x & \text{otherwise;} \end{cases}$$

$$g(x) = \begin{cases} b_{j-1} & \text{if } x = b_j \text{ with } j > 0, \\ x & \text{otherwise.} \end{cases}$$

The H-closed algebraic subsets form a 3-element chain $\{1\} < \uparrow a_0 < \uparrow b_0$. We have $A = \tau(A) \leq \tau(B) = B$ in $\mathbf{L}$, and the map $k = \bigvee_{j \geq 0} h_j$ in $\mathbb{DMO}$ has $k(b_0) = a_0$, but no map in H does. In fact, $h_j^*[b_0] = g^*[b_0] = b_0$. So we really do need to go to $\mathbb{DMO}$ for the representation.

The end product of the construction is the chain $[1] < [a_0] < [b_0]$ with the operator $k^*[b_0] = k^*[a_0] = [a_0]$, $k^*[1] = [1]$.

The existence of a complete set of h's satisfying Lemma 5.4 is necessary for representability of a finite pair $(\mathbf{L}, \gamma)$. In general, it need not be sufficient. However, when $\mathrm{J}(\mathbf{L}) \subseteq \tau(\mathbf{L})$, that is, $p = \tau[p]$ for every join irreducible $p \in \mathrm{J}(\mathbf{L})$, it is sufficient.

Theorem 5.6 *Let $(\mathbf{L}, \gamma)$ be a finite, lower bounded lattice with a preclop. If $(\mathbf{L}, \gamma)$ satisfies $\mathrm{J}(\mathbf{L}) \subseteq \tau(\mathbf{L})$, then $(\mathbf{L}, \gamma)$ has a representation as $\mathrm{S_p}(\mathbf{S}, H)$ with the natural weak equaclosure operator if and only if there exists a set of (join, 0-preserving) operators K on $\gamma(\mathbf{L})$ satisfying the conditions:*

(1) *$0 < \tau[a] \leq \tau[b]$ implies there exists $k \in K$ such that $k[b] = [a]$,*

and for each $k \in K$ and elements $[x], [y] \in \gamma(\mathbf{L})$:

(2) *$k[x] \leq [x]$,*
(3) *$\tau k[x] \leq \tau[x]$,*

(4) $k[0] = [0]$,
(5) $[x] \leq [y]$ *implies* $k[x] \leq k[y]$,
(6) $k([x] \vee [y]) = k[x] \vee k[y]$.

If such a set of operators exists, then $\mathbf{L} \cong \mathrm{Sub}(\gamma(\mathbf{L}), 0, \vee, K)$.

Proof It remains to show that there is a representation when the conditions of the theorem hold. Define $\varphi : \mathbf{L} \to \mathrm{Sub}(\gamma(\mathbf{L}), 0, \vee, H^*)$ by

$$\varphi(x) = \{[b] : \tau b \leq x\}.$$

This is dual to the representation on $\gamma^d\mathbf{L}$.

First we note that $\varphi(x)$ is a subalgebra. Surely $[0] \in \varphi(x)$. If τb, $\tau c \leq x$, then $\tau[b \vee c] \leq \tau b \vee \tau c \leq x$. If $\tau b \leq x$, then $\tau h^*[b] \leq x$ by property (3).

We need to show that every subalgebra is in the range of φ. For a subalgebra U, let

$$x = \bigvee \{\tau[u] : [u] \in U\}.$$

Clearly $U \subseteq \varphi(x)$. For the reverse inclusion, suppose $[b] \in \varphi(x)$, i.e., $\tau b \leq \bigvee\{\tau[u] : [u] \in U\}$. Then

$$\tau b = \tau b \wedge \gamma b \leq \bigvee (\tau[u] \wedge \gamma b) \leq \gamma b$$

by the meet distributivity of γb. Thus $[b] = \bigvee[\tau[u] \wedge \gamma b]$. But each $[d] = [\tau[u] \wedge \gamma b]$ has $\tau[d] \leq \tau[u]$, whence $[d] = k^*[u]$ for some $k \in K$ by property (1), and thus $[d] \in U$ since it is a subalgebra. Then $[b]$, the join of such $[d]$, is also in U. Thus $\varphi(x) \subseteq U$.

Clearly φ is order preserving: $x \leq y$ implies $\varphi(x) \leq \varphi(y)$. On the other hand, if $x \nleq y$, then there is a join irreducible $p = \tau p$ with $p \leq x$, $p \nleq y$. This gives $[p] \in \varphi(x)$ while $[p] \notin \varphi(y)$, so $\varphi(x) \nleq \varphi(y)$. □

Example 3.27 uses Theorem 5.6 to show that a particular pair $(\mathbf{K}, \gamma)$ cannot be represented as $(\mathrm{S_p}(\mathbf{S}, H), \Gamma)$ since the representation with $\mathbf{S} = \gamma^d(\mathbf{K})$ fails. There will be more such examples in the next section.

To summarize, whenever $(\mathbf{L}, \gamma)$ can be represented as $(\mathrm{S_p}(\mathbf{S}, H), \Gamma)$, we always have $\gamma^d(\mathbf{L}) \leq \mathbf{S}$ as a complete sublattice by Corollary 3.21. When $\mathbf{L}$ is finite and $J(\mathbf{L}) \subseteq \tau(\mathbf{L})$, we can take $\mathbf{S} = \gamma^d(\mathbf{L})$. When $J(\mathbf{L}) \nsubseteq \tau(\mathbf{L})$, additional elements are generally needed, and we have a limited understanding of how to fit them in, though it is not hard in some examples. Moreover, property (2) of Lemma 5.4, $h^*[a] \leq [a]$, applies only to the natural representation of pairs with $J(\mathbf{L}) \subseteq \tau(\mathbf{L})$ on $\mathrm{Sub}(\gamma(\mathbf{L}), 0, \vee, H^*)$. It need not hold when $J(\mathbf{L}) \nsubseteq \tau(\mathbf{L})$, while pairs with $J(\mathbf{L}) \subseteq \tau(\mathbf{L})$ may have other representations not satisfying it.

The preceding comments assume that a representation exists. Lest we forget, there are necessary conditions for the representability of a pair $(\mathbf{L}, \gamma)$ as presented in Sect. 3.1, *viz.*, (I1)–(I7), (K9)–(K11), and probably more.

5.2 An Algorithm to Test for Representability When $\mathbf{J(L)} \subseteq \tau(\mathbf{L})$

The previous section was in terms of $\gamma(\mathbf{L})$ because our motivation comes from the induced action of an operator on $\Gamma(\mathbf{S})$. This section is in terms of $\gamma^d(\mathbf{L})$ because that is how we apply it in Chap. 7. Our apologies to the reader!

Given is a pair $(\mathbf{L}, \gamma)$, with $\mathrm{J}(\mathbf{L}) \subseteq \tau(\mathbf{L})$, which we want to represent as $(\mathrm{S_p}(\mathbf{S}, H), \Gamma)$. By Theorem 5.6, we can take $\mathbf{S}$ to be the finite semilattice $\hat{\mathbf{S}} = \gamma^d(\mathbf{L})$, in which case the representation is on $\mathrm{Sub}(\hat{\mathbf{S}}, \wedge, 1, H)$. Recall that

- $\hat{S} = \{\hat{x} : x \in \tau(\mathbf{L})\}$,
- $\hat{x} \leq \hat{y}$ iff $\gamma x \geq \gamma y$,
- the largest element of $\hat{\mathbf{S}}$ is $\hat{0}$,
- $\hat{x} \wedge \hat{y} = \hat{\tau}(x \vee y)$,
- $\hat{x} \vee \hat{y} = \hat{\tau}(\gamma x \wedge \gamma y)$.

For convenience and future reference, here is the dual of Theorem 5.6.

Theorem 5.7 *Let $(\mathbf{L}, \gamma)$ be a finite, lower bounded lattice with a preclop. If $(\mathbf{L}, \gamma)$ satisfies $\mathrm{J}(\mathbf{L}) \subseteq \tau(\mathbf{L})$, then $(\mathbf{L}, \gamma)$ has a representation as $\mathrm{S_p}(\mathbf{S}, H)$ with the natural weak equaclosure operator if and only if there exists a set of (meet, 1-preserving) operators K on $\gamma^d(\mathbf{L})$ satisfying the conditions:*

(1) $0 < \tau a \leq \tau b$ *implies there exists* $k \in K$ *such that* $k(\hat{b}) = \hat{a}$,

and for each $k \in K$ and all $\hat{x}, \hat{y} \in \gamma^d(\mathbf{L})$,

(2) $k(\hat{x}) \geq \hat{x}$,
(3) $k(\hat{x}) = \hat{u}$ *implies* $u \leq x$,
(4) $k(\hat{0}) = \hat{0}$,
(5) $\hat{x} \leq \hat{y}$ *implies* $k(\hat{x}) \leq k(\hat{y})$,
(6) $k(\hat{x} \wedge \hat{y}) = k(\hat{x}) \wedge k(\hat{y})$.

If such a set of operators exists, then $\mathbf{L} \cong \mathrm{Sub}(\gamma^d(\mathbf{L}), \hat{0}, \wedge, K)$.

Let $\mathbf{P} = \tau(\mathbf{L})$, with the order inherited from $\mathbf{L}$. Consider $b > a$ in $\mathbf{P}$, which implies $\gamma b \geq \gamma a$ and hence $\hat{b} < \hat{a}$ in $\hat{\mathbf{S}}$. In the present finite case, it suffices to consider pairs (a, b) with $b > a$ in $\mathbf{P}$. We want to determine whether there exists an operator k on $\hat{\mathbf{S}}$ satisfying

(1) $k(\hat{b}) = \hat{a}$

and conditions (2)–(6) from Theorem 5.7. If we can find an operator for each such pair $b > a$ in $\mathbf{P}$, then we will have a representation. If there is no operator for some pair $b > a$, then we will know why $(\mathbf{L}, \gamma)$ cannot be represented as $\mathrm{S_p}(\mathbf{S}, H)$, much less as $\mathrm{L_q}(\mathcal{K})$.

Fixing such a pair (a, b) with $b > a$ in $\mathbf{P}$, we recursive define a sequence of maps k_n which will either terminate in a k_n satisfying (1)–(6) or show that no such k exists. As usual, let $\hat{z}$ denote the least element of $\hat{\mathbf{S}}$, where $z = \tau(1_L)$.

Map k_0

We define $k_0 : \hat{\mathbf{S}} \to \hat{\mathbf{S}}$ recursively, starting at the bottom with $\hat{z}$. Let

$$k_0(\hat{z}) = \begin{cases} \hat{a} & \text{if } z = b, \\ \hat{z} & \text{otherwise.} \end{cases}$$

Now suppose we have $k_0(\hat{y})$ for all $\hat{y} < \hat{x}$. Let $\hat{w} = \bigvee_{\hat{y}<\hat{x}} k_0(\hat{y})$, where the join is taken in $\hat{\mathbf{S}}$. Define

$$k_0(\hat{x}) = \begin{cases} \hat{a} & \text{if } x = b, \\ \hat{\tau}(x \wedge \gamma w) & \text{otherwise.} \end{cases}$$

Lemma 5.8 *The map k_0 satisfies* (1)–(5)*; indeed, it is the least such map.*

Proof Note that $k_0(\hat{x}) \geq \hat{x}$, and $k_0(\hat{x}) = \hat{x}$ if $\hat{x} \ngeq \hat{b}$. Properties (1)–(4) are immediate, as is (5) except when $\hat{x} \geq \hat{b}$. So consider $\hat{x} > \hat{b}$. Then $\gamma(x \wedge \gamma w) \leq \gamma w$, whence $k_0(\hat{x}) \geq \hat{w} = \bigvee_{\hat{y}<\hat{x}} k_0(\hat{y})$, as desired. □

Now k_0 need not preserve meets, so in general we will need to continue with a sequence of maps $k_0 \leq k_1 \leq \ldots$ as indicated below. But sometimes k_0 does preserve meets, in which case we are done for this pair.

Corollary 5.9 *If* **L** *is a finite lower bounded lattice with* J(**L**) ⊆ τ(**L**) *and* $\hat{\mathbf{S}} = \gamma^d(\mathbf{L})$ *is a finite chain, then* $(\mathbf{L}, \gamma)$ *is representable as* $\mathrm{Sub}(\hat{\mathbf{S}}, \wedge, 1, H)$.

For another example, the pentagon in Fig. 7.4 has one pair (a, z) in **P** to consider, and the map k_0 with $k_0(\hat{z}) = \hat{a}$ preserves meets, yielding a representation. It is impossible to get a contradiction in the first step, so if meets are not preserved, we continue.

Map k_1

Now assume there is a pair $(\hat{r}, \hat{s})$ such that

$$\hat{p} = k_0(\hat{r} \wedge \hat{s}) < k_0(\hat{r}) \wedge k_0(\hat{s}) = \hat{q}.$$

Let us construct $k_1 \geq k_0$ such that $k_1(\hat{r} \wedge \hat{s}) \geq \hat{q}$. Let

$$k_1(\hat{z}) = \begin{cases} \hat{\tau}(\tau z \wedge \gamma q) & \text{if } \hat{z} = \hat{r} \wedge \hat{s}, \\ k_0(\hat{z}) & \text{otherwise.} \end{cases}$$

Now suppose we have $k_1(\hat{y})$ for all $\hat{y} < \hat{x}$. Let $\hat{w} = k_0(\hat{x}) \vee \bigvee_{\hat{y}<\hat{x}} k_1(\hat{y})$. Define

$$k_1(\hat{x}) = \begin{cases} \hat{\tau}(\tau x \wedge \gamma q) & \text{if } \hat{x} = \hat{r} \wedge \hat{s}, \\ \hat{\tau}(\tau x \wedge \gamma w) & \text{otherwise.} \end{cases}$$

Thus we recursively define k_1 on all of $\hat{S}$.

Let $\hat{r} \wedge \hat{s} = \hat{u}$, so that in fact $u = \tau(r \vee s)$, though that is not used explicitly. One can now prove, by induction on the order $\leq$ of $\hat{\mathbf{S}}$, that:

- if $\hat{x} \ngeq \hat{u}$, then $k_1(\hat{x}) = k_0(\hat{x})$,
- $k_1(\hat{u}) \geq \hat{q} > \hat{p} = k_0(\hat{u})$,
- if $\hat{x} > \hat{u}$, then $k_1(\hat{x}) \geq \hat{q} \vee k_0(\hat{x})$,
- k_1 satisfies properties (2)–(5),
- k_1 is the least map satisfying all these conditions.

The (many) details are left to the reader. It is an exercise in repeatedly applying the principle $\hat{x} \leq \hat{y}$ iff $\gamma x \geq \gamma y$, induction on the order in $\hat{\mathbf{S}}$, and the fact that k_0 has properties (2)–(5). In particular, for $m \in L$ and $v \in \tau(\mathbf{L})$, we have $\hat{\tau}(m) \geq \hat{v}$ iff $\gamma\tau(m) \leq \gamma v$ iff $\gamma m \leq \gamma v$ iff $m \leq \gamma v$. For (3), remember that our notation convention is that $\hat{x} \in \hat{S}$ implies $x = \tau x$; the latter is written in the formulas for k_0 and k_1 only for emphasis.

Thus, when k_0 does not preserve meets, we have found a map $k_1 \geq k_0$ that satisfies (2)–(5) and addresses one failure of (6). To obtain a map k that satisfies (2)–(6), the algorithm is this:

Given $(\mathbf{L}, \gamma)$, for all $b > a$ in $\mathbf{P}$, recursively construct $k_0 \leq k_1 \leq \cdots \leq k_n$ until it stabilizes satisfying (2)–(6). If $k_n(\hat{b}) = \hat{a}$ then you have the map for the pair (a, b); if not, no map works for (a, b), and $(\mathbf{L}, \gamma)$ is not representable as $(\text{Sub}(\mathbf{S}, \wedge, 1, H), \Gamma)$.

To see that the algorithm works, fix $b > a$. The map k_n satisfies (2)–(6). If $k_n(\hat{b}) = \hat{a}$, so that k_n satisfies (1)–(6), we are done. On the other hand, suppose there is a map h satisfying (1)–(6). It is easy to see that $k_0 \leq h$, and inductively, if $k_j \leq h$ then $k_{j+1} \leq h$. Now

$$\hat{a} = k_0(\hat{b}) \leq k_n(\hat{b}) \leq h(\hat{b}) = \hat{a}$$

so that k_n satisfies (1) also. Thus, for each $b > a$ in $\mathbf{P}$, the map k_n (for sufficiently large n) satisfies (1)–(6) if and only if any such map exists, in which case k_n is the least such map.

Example 5.10 Consider the pair $(\mathbf{K}, \gamma)$ in Fig. 5.2. Clearly J(**K**) $\subseteq \tau(\mathbf{K})$, and we need operators h, ℓ such that $h(\hat{c}) = \hat{a}$ and $\ell(\hat{c}) = \hat{b}$. Following the algorithm for the pair (a, c), we get $k_0(\hat{z}) = \hat{z}$ and $k_0(\hat{c}) = \hat{a}$. In the calculations for $\hat{a}$ and $\hat{b}$, the element $w = a$ since $\hat{c} \leq \hat{a}, \hat{b}$ and $k_0(\hat{c}) = \hat{a}$. Thus $k_0(\hat{a}) = \hat{\tau}(a \wedge \gamma a) = \hat{a}$ but $k_0(\hat{b}) = \hat{\tau}(b \wedge \gamma a) = \hat{0}$. Finally, $k_0(\hat{0}) = \hat{0}$. Now k_0 preserves meets, so it is the desired map.

The map for the pair (b, c), say ℓ_0, is symmetric, and thus we obtain the representation of $(\mathbf{K}, \gamma)$ as $\text{Sub}(\gamma^d(\mathbf{K}), \wedge, \hat{0}, k_0, \ell_0)$. Of course, this one was not hard, but it illustrates how the process works.

Example 5.11 Now suppose we try to represent the pair $(\mathbf{L}, \gamma)$ in Fig. 5.3. Again J(**L**) $\subseteq \tau(\mathbf{L})$, and we need operators to satisfy property (1), including an operator with $k(\hat{d}) = \hat{b}$. Straightforward calculations for k_0 yield

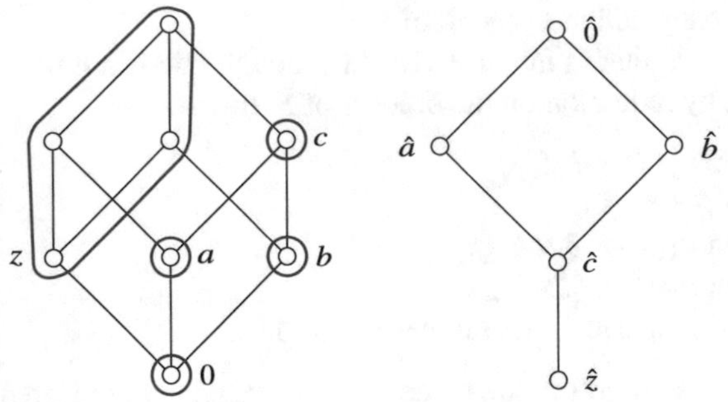

Fig. 5.2 Lattice **K** and $\gamma^d(\mathbf{K})$ for Example 5.10

$$k_0(\hat{z}) = \hat{z} \qquad k_0(\hat{d}) = \hat{b}$$
$$k_0(\hat{e}) = \hat{e} \qquad k_0(\hat{a}) = \hat{\tau}(a \wedge \gamma b) = \hat{0}$$
$$k_0(\hat{b}) = \hat{b} \qquad k_0(\hat{c}) = \hat{c} \qquad k_0(\hat{0}) = \hat{0}$$

But now k_0 does not preserve meets:

$$\hat{z} = k_0(\hat{z}) = k_0(\hat{a} \wedge \hat{c}) < k_0(\hat{a}) \wedge k_0(\hat{c}) = \hat{0} \wedge \hat{c} = \hat{c}.$$

To fix that, we will need to have $k_1(\hat{z}) \geq \hat{c}$, which in turn implies $k_1(\hat{d}) \geq k_0(\hat{d}) \vee \hat{c} = \hat{b} \vee \hat{c} = \hat{0}$. But that defeats the original purpose, which was to have $k(\hat{d}) = \hat{b}$. We conclude that no representation exists.

It was observed elsewhere (see Fig. 3.3) that this pair fails (‡), so we knew this was going to happen. But again, it is interesting to see how the algorithm shows that the desired map does not exist.

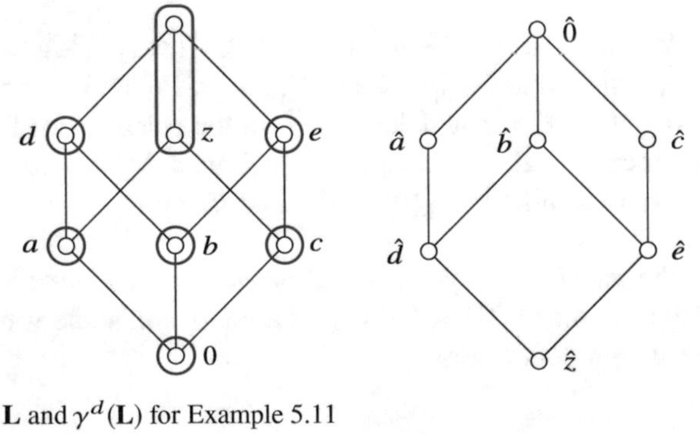

Fig. 5.3 Lattice **L** and $\gamma^d(\mathbf{L})$ for Example 5.11

The pairs in Figs. 10.1, 10.2, and 10.3 are more challenging, but they can all be represented using the algorithm in this section.

5.3 Extension to a Class of Infinite Lattices

We would like to extend the results of Sect. 5.1 to infinite lattices, but there are severe technical difficulties. The source of the problem is that we need to use maps k^* with $k \in \mathbb{DMO}$, as shown by Example 5.5. But maps in $\mathbb{DMO}$ may preserve only finite meets, and not necessarily arbitrary meets and directed joins; see Theorem 3.11. In the end, all that survives is a version of the second half of Theorem 5.6, giving a representation of $(\mathbf{L}, \gamma)$ as $(\mathrm{S}^{\mathrm{d}}_{\mathrm{p}}(\gamma(\mathbf{L})), \Gamma')$ when $\mathbf{L}$ is both algebraic and dually algebraic, and the completely join irreducible elements $\mathrm{CJI}(\mathbf{L})$ are contained in $\tau(\mathbf{L})$. But that remnant is nice enough to give some interesting examples, so let us proceed.

The next equivalence is a standard exercise.

Theorem 5.12 *The following are equivalent for an algebraic lattice* $\mathbf{L}$.

(1) $\mathbf{L}$ *is also dually algebraic.*
(2) *Every completely meet irreducible element of* $\mathbf{L}$ *is dually compact.*
(3) $\mathbf{L}$ *is lower continuous.*

Analogously to Sect. 5.1, we will use dually algebraic subsets. A subset X of a complete lattice $\mathbf{T}$ is *dually algebraic* if it is closed under arbitrary joins (including $\bigvee \varnothing = 0$) and down-directed meets. A *dual operator* on $\mathbf{T}$ is a map $k : T \to T$ that preserves arbitrary joins and down-directed meets. Let $\mathrm{S}^{\mathrm{d}}_{\mathrm{p}}(\mathbf{T}, K)$ denote the lattice of all K-closed dually algebraic subsets of $\mathbf{T}$.

Every dually algebraic subset X has a largest element x_1. The natural (weak) dual equaclosure operator Γ' on $\mathrm{S}^{\mathrm{d}}_{\mathrm{p}}(\mathbf{T}, K)$ is given by $\Gamma'(X) = {\downarrow} x_1$.

Theorem 5.13 *Let* $(\mathbf{L}, \gamma)$ *be a pair consisting of a dually algebraic lattice with a weak equaclosure operator. Assume that*

(i) *each* γx *is completely meet distributive, that is,* $\gamma x \wedge \bigvee y_i = \bigvee(\gamma x \wedge y_i)$ *for any index set,*
(ii) $\mathrm{CJI}(\mathbf{L}) \subseteq \tau(\mathbf{L})$.

Suppose there exists a set K *of dual operators on* $\gamma(\mathbf{L})$ *such that*

(iii) $\tau(k([x])) \leq \tau[x]$ *for all* $k \in K$, $x \in L$,
(iv) $\tau[a] \leq \tau[b]$ *implies there exists* $k \in K$ *such that* $k[b] = [a]$.

Then the pair $(\mathrm{S}^{\mathrm{d}}_{\mathrm{p}}(\gamma(\mathbf{L}), K), \Gamma')$ *represents* $(\mathbf{L}, \gamma)$.

Observe that when $\mathbf{L}$ is also upper continuous (in particular, algebraic), the meet distributivity of γx implies (i). Note that (iii) implies $k[x] \leq [x]$.

Proof By Lemma 3.4, which is a consequence of property (I7) for equaclosure operators, $\gamma(\mathbf{L})$ is a dually algebraic lattice, and we are assuming that the maps in K are dual operators. Again define $\varphi : \mathbf{L} \to \mathrm{S}_\mathrm{p}^\mathrm{d}(\gamma(\mathbf{L}), K)$ by

$$\varphi(x) = \{[b] : \tau b \leq x\}.$$

First we show that $\varphi(x)$ is a K-closed dually algebraic subset of $\gamma(\mathbf{L})$. Surely $[0] \in \varphi(x)$. For join closure, assume $[b_i] \in \varphi(x)$ for all $i \in I$. Then each $\tau b_i \leq x$, whence $\tau \bigvee [b_i] \leq \bigvee \tau[b_i] \leq x$, so $\bigvee [b_i] \in \varphi(x)$. Next consider a down-directed subset $[b_i]$ ($i \in I$) of $\varphi(x)$. Let $c_i = \gamma b_i$; these also form a down-directed set. By property (K11) of equaclosure operators, $\tau(\bigwedge_i c_i) \leq x$, so that $[\bigwedge_i c_i] = \bigwedge_i [b_i] \leq x$ and thus $\bigwedge_i [b_i] \in \varphi(x)$. Also, if $\tau b \leq x$ and $k \in K$, then $\tau k[b] \leq x$ by (iii). Therefore, each $\varphi(x)$ is a K-closed dually algebraic subset of $\gamma(\mathbf{L})$.

We need to show that every such subset is in the range of φ. For a K-closed dually algebraic subset U, let

$$x = \bigvee \{\tau[u] : [u] \in U\}.$$

Clearly $U \subseteq \varphi(x)$. For the reverse inclusion, suppose $[b] \in \varphi(x)$, i.e., $\tau b \leq \bigvee \{\tau[u] : [u] \in U\}$. Then using the complete meet distributivity of assumption (i),

$$\tau b = \tau b \wedge \gamma b \leq (\bigvee (\tau[u]) \wedge \gamma b = \bigvee (\tau[u] \wedge \gamma b) \leq \gamma b\,.$$

Thus $[b] = \bigvee [\tau[u] \wedge \gamma b]$. But each $[d] = [\tau[u] \wedge \gamma b]$ has $\tau[d] \leq \tau[u]$, whence $[d] = k[u]$ for some $k \in K$ by (iv). Since U is K-closed that implies $[d] \in U$. Then $[b]$, the join of such $[d]$, is also in U. Thus $\varphi(x) \subseteq U$.

Clearly φ is order preserving: $x \leq y$ implies $\varphi(x) \leq \varphi(y)$. On the other hand, if $x \nleq y$, then since $\mathbf{L}$ is dually algebraic there is a completely join irreducible p with $p \leq x$, $p \nleq y$. By assumption (ii), $p = \tau[p]$. This gives $[p] \in \varphi(x)$ while $[p] \notin \varphi(y)$, so $\varphi(x) \nleq \varphi(y)$.

Finally, we want to show that $\Gamma'(\varphi(x)) = \downarrow[\gamma x]$. Clearly $[\gamma x] \in \varphi(x)$ as $\tau(\gamma x) \leq x$. Moreover, if $[b] \in \varphi(x)$ then $\tau b \leq x$, whence $\gamma b \leq \gamma x$, that is, $[b] \leq [\gamma x]$. Thus $[\gamma x]$ is the largest element of $\varphi(x)$, and $\Gamma'(\varphi(x)) = \{[b] : \gamma b \leq \gamma x\}$. It follows that $\varphi(x)$ is Γ'-closed iff $x = \gamma x$. □

Example 5.14 Let $\mathbf{B}_\omega$ be the Boolean lattice of all subsets of the ordinal ω, which is both algebraic and dually algebraic. There is a familiar equaclosure operator on $\mathbf{B}_\omega$: for $\varnothing \neq S \subseteq \omega$,

$$\gamma(S) = \uparrow S = \uparrow \min(S)\,.$$

Thus, including $\varnothing$, $\gamma(\mathbf{B}_\omega) \cong (\omega + 1)^d$. The subset $\tau(\mathbf{B}_\omega)$ consists of its atoms (singletons), so no operators are needed. Theorem 5.13 applies, and after dualizing, we obtain the representation of $(\mathbf{B}_\omega, \gamma)$ as $\mathrm{S}_\mathrm{p}(\omega + 1)$. By Theorem 2.65, $\mathbf{B}_\omega$ is isomorphic to $\mathrm{L}_\mathrm{q}(\mathcal{U})$ for a quasivariety of 1-element structures.

This example raises the hope that one might be able to use Theorem 5.13 to provide a simplified proof of Theorem 2.65, the representation of algebraic, dually algebraic, atomistic lattices. But the hard step of that theorem is (4) implies (3), and we would need to prove that if such a lattice admits an equaclosure operator γ, then it also admits one γ' with the property that $\tau'(\mathbf{L})$ consists of its atoms. That should be true, given Theorem 2.65, but the proof may require the original arguments from [14].

Example 5.15 Figure 5.4 illustrates a couple of easy applications of Theorem 5.13. The lattices pictured there, with the indicated closure operators, satisfy the conditions of the theorem. Finding the necessary operators on $\gamma^d(\mathbf{L})$ is straightforward and is left as an exercise for the reader. Thus both pairs can be represented as $(\mathrm{S_p}(\mathbf{S}, H), \Gamma)$. We do not know whether they are subquasivariety lattices, but in both cases, if you remove the atom on the left side, the resulting lattice is $\mathrm{L_q}(\mathcal{K})$ for a quasivariety with equality by using Theorem 7.7.

One can make more examples similar to these. The pentagons in the figure are stacked over ω as an index set, but one could also stack lattices over ω^d or $\mathbb{Z}$, taking care to be sure that the result is algebraic and dually algebraic.

Yet another exercise is to consider the various equaclosure operators on $(\omega+1)\times\mathbf{2}$, or more generally $\alpha \times \beta$ for ordinals α, β, that satisfy $\mathrm{CJI}(\mathbf{L}) \subseteq \tau(\mathbf{L})$.

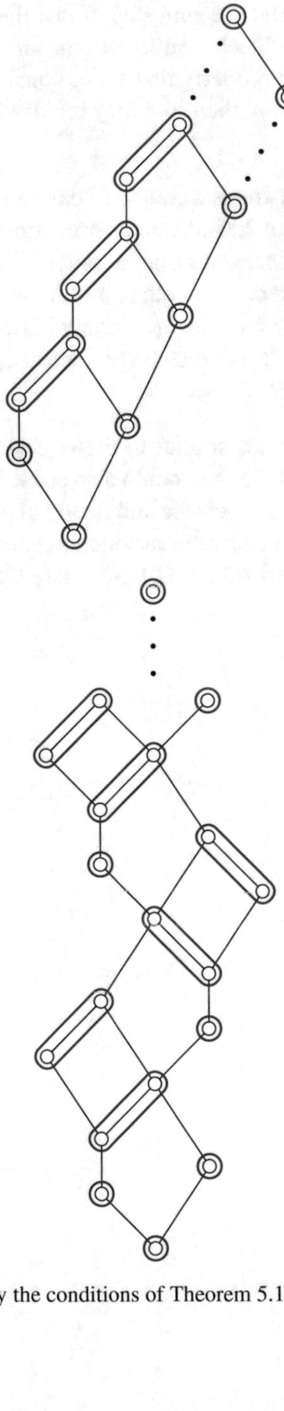

Fig. 5.4 Pairs $(\mathbf{L}, \gamma)$ that satisfy the conditions of Theorem 5.13, and which can be represented as $(\mathrm{S_p}(\mathbf{S}, H), \Gamma)$

Chapter 6
Finite Lattices as Sub(S, $\wedge$, 1, H): The Case J(L) $\nsubseteq$ τ(L)

Men will live like billy goats if they are left alone. – Mattie Ross in *True Grit* (Charles Purvis)

Все смешалось в доме Облонских. – Лев Толстой, Анна Каренина

In this chapter, we turn to the case of representing a finite lattice by subsemilattices with operators when $\mathrm{J}(\mathbf{L}) \nsubseteq \tau(\mathbf{L})$. To iterate the problem, we are given a pair $(\mathbf{L}, \gamma)$ where $\mathbf{L}$ is a finite lower bounded lattice and γ is an equaclosure operator on it. We seek a representation of $(\mathbf{L}, \gamma)$ in the form $(\mathrm{Sub}(\mathbf{S}, \wedge, 1, H), \Gamma)$ where $\mathbf{S}$ is a finite $\wedge$-semilattice with 1 and H is a set of operators on $\mathbf{S}$ preserving $\wedge$ and 1, and Γ is the natural equaclosure operator on $\mathrm{Sub}(\mathbf{S}, \wedge, 1, H)$, so that for $\mathbf{A} \leq \mathbf{S}$, $\Gamma(\mathbf{A}) = \uparrow a_0$ where a_0 is the least element of $\mathbf{A}$.

Section 6.1 discusses how we might go about constructing such a representation and provides some examples. Such *ad hoc* methods work well for simple cases, but for more complicated cases, we turn to a systematic approach based on the construction of Theorem 4.30. This is the topic of Sect. 6.2.

6.1 *Ad hoc* Representations

We are given a pair $(\mathbf{L}, \gamma)$ consisting of a finite lower bounded lattice $\mathbf{L}$ and an equaclosure operator γ on it. The objective is to construct a finite semilattice with operators $\widehat{\mathbf{S}} = (\hat{S}, \wedge, 1, H)$ such that $\mathrm{Sub}(\hat{\mathbf{S}}, \wedge, 1, H)$ along with the natural equaclosure operator Γ represents $(\mathbf{L}, \gamma)$.

Our plan is to apply the method of Sect. 4.4. To do so, we need a map $\lambda : \hat{S} \to L$ satisfying the properties (a)–(e) required by Theorems 4.37 to 4.39 with the corresponding embedding $\xi : \mathbf{L} \leq \mathrm{Sub}\,\widehat{\mathbf{S}}$ to be given by $\xi(a) = \{s \in \hat{S} : \lambda s \leq a\}$. Those conditions translate to the current situation thusly: for all $h \in H$ and $x, y \in \hat{S}$ and $Y \subseteq \hat{S}$,

K. Adaricheva et al., *A Primer of Subquasivariety Lattices*, CMS/CAIMS Books in Mathematics 3, https://doi.org/10.1007/978-3-030-98088-7_6

(a1) $x = \bigwedge Y$ implies $\lambda x \le \bigvee \lambda Y$;
(a2) $x = h(y)$ implies $\lambda x \le \lambda y$;
(b) $\lambda x = 0_{\mathbf{L}}$ iff $x = 1$, and $h(1) = 1$;
(c) if $\lambda x \le \bigvee \lambda Y$ is a minimal nontrivial join cover, then $x \in \mathrm{Sg}(Y)$;
(d) $\mathrm{J}(\mathbf{L}) \subseteq \lambda(\hat{S})$;
(e) $0_{\mathbf{L}} < \lambda x \le \lambda y$ implies there exists $h \in H$ such that $x = h(y)$,

where $\lambda Y = \{\lambda y : y \in Y\}$.

We will expand upon the conditions (a1)–(e) for embedding momentarily, but we also want to ensure that the equaclosure operators correspond. Recall that whenever $(\mathbf{L}, \gamma)$ is represented as $(\mathrm{Sub}\ \widehat{\mathbf{S}}, \Gamma)$, then $\gamma^d(\mathbf{L}) \le \widehat{\mathbf{S}}$ by Corollary 3.21. The elements of $\tau(\mathbf{L})$ are in 1-1 correspondence with $\gamma^d(\mathbf{L})$. So we can assume there is a (proper) subset $T \subseteq \hat{S}$ given by

- $T = \{\hat{x} : x \in \tau(\mathbf{L})\}$

with the order $\hat{x} \le \hat{y}$ in $\widehat{\mathbf{S}}$ iff $\gamma x \ge \gamma y$ in $\mathbf{L}$. Moreover, as a convention we let $z = \tau(1_{\mathbf{L}})$, so that the least element of $\widehat{\mathbf{S}}$ is $\hat{z}$, while the greatest element 1 of $\widehat{\mathbf{S}}$ is $\hat{0}_{\mathbf{L}}$, which we just denote as $\hat{0}$.

The subalgebras in our putative representation are of the form $\xi(a) = \{s \in \hat{S} : \lambda s \le a\}$ for $a \in L$. The elements of T are to be the least elements of subalgebras of $\widehat{\mathbf{S}}$. For that, we need two more conditions on λ: for $s \in \hat{S}$,

(f) $\lambda s \le a$ implies $s \ge \hat{\tau}(a)$;
(g) $\lambda \hat{a} = a$ for $a \in \tau(\mathbf{L})$,

where $\hat{\tau}(a) = \widehat{\tau(a)}$.

Returning to the embedding conditions, (d) requires that $\mathrm{J}(\mathbf{L}) \subseteq \lambda(\hat{S})$. In Chap. 5 we assumed that $\mathrm{J}(\mathbf{L}) \subseteq \tau(\mathbf{L})$ and showed that the representation could use $\widehat{\mathbf{S}} = \gamma^d(\mathbf{L})$, i.e., $\hat{S} = T$. In this chapter, we are considering the case when $\mathrm{J}(\mathbf{L}) \nsubseteq \tau(\mathbf{L})$, and will need more elements in $\hat{S}$.

Now whenever $\xi : \mathbf{L} \le \mathrm{Sub}\,\mathbf{S}$ and $p \in \mathrm{J}(\mathbf{L})$, then there is at least one element $p_1 \in \xi(p) \setminus \xi(p_*)$. Thus set

- $J = \{p_1 : p \in \mathrm{J}(\mathbf{L}) \setminus \tau(\mathbf{L})\}$

and $\lambda p_1 = p$ for $p \in J$. By (f), this implies $p_1 \ge \hat{\tau}(p)$.

So far we have some necessary elements of $\hat{S}$ and some necessary order relations, which can be summarized as follows:

- $J \cup T \subseteq \hat{S}$,
- $\lambda\hat{x} = x$ for $\hat{x} \in T$,
- $\lambda p_1 = p$ for $p_1 \in J$,
- $\hat{0}$ is the largest element of $\widehat{\mathbf{S}}$,
- $\hat{z}$ is the least element, where $z = \tau(1_{\mathbf{L}})$.

The situation is sketched in Fig. 6.1.

However, the information listed thus far is generally not enough to determine $\widehat{\mathbf{S}}$. Because $J \cup T \subseteq \hat{S}$, we *do* have $\mathrm{J}(\mathbf{L}) \subseteq \lambda(\hat{S})$. Hence the set $\xi(L)$ of all $\xi(a)$ for $a \in L$, ordered by set containment, is isomorphic to $\mathbf{L}$. We want to specify an order

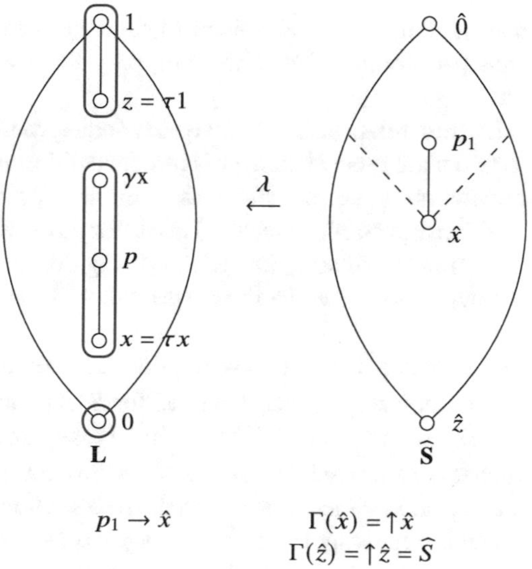

Fig. 6.1 Schematic diagram of $(\mathbf{L}, \gamma)$ and $\widehat{\mathbf{S}}$. Here $x, z \in \tau(\mathbf{L})$ so in $\widehat{\mathbf{S}}$ we have $\hat{x}, \hat{z} \in T$, while $p \notin \tau(\mathbf{L})$ thus $p \in J$. Also $\lambda\hat{x} = x$, $\lambda\hat{z} = z$, $\lambda p_1 = p$

and operations H on $\hat{S}$ to make it a semilattice with operators, so that the conditions (a1)–(g) are satisfied. The process may require additional order relations or meets of elements in $J \cup T$. It may be necessary to duplicate some of the elements of $J \cup T$, so that say $\hat{x}$ is replaced by $\{\hat{x}, x_1, \ldots, x_n\}$ with $\hat{x} \leq x_j$ for all j, as in some of the examples below. Sometimes it is easy enough to see how to do this, while other cases may require the more complicated but systematic approach of Sect. 6.2. The end result should be $\widehat{\mathbf{S}} = (\hat{S}, \wedge, \hat{0}, H)$ and $\lambda : \hat{S} \longrightarrow L$ satisfying the conditions.

We use *arrow relations* to guide the process of building $\widehat{\mathbf{S}}$, where $Y \longrightarrow x$ means that $x \in \mathrm{Sg}(Y)$ should hold. If $x, y \in \hat{S}$ and $\lambda x \leq \lambda y$, then we write $y \longrightarrow x$ to indicate that by condition (e), there should be an $h \in H$ with $x = h(y)$. If $x \in \hat{S}$ and $\lambda x \leq \bigvee \lambda Y$ minimally, then by condition (c) it should hold that $x \in \mathrm{Sg}(Y)$. This could be $x = \bigwedge Y$ in $\widehat{\mathbf{S}}$, but more generally there could exist $x' \in \hat{S}$ and $Y' \subseteq \hat{S}$ such that $\lambda x' = \lambda x$, $\lambda Y' = \lambda Y$, and $x' = \bigwedge Y'$ (since $\lambda x = \lambda x'$ implies $\mathrm{Sg}(x) = \mathrm{Sg}(x')$ by (e), and likewise for $y \in Y$).

It is useful to list the arrow relations as part of the construction process. We will include them in the figures for our examples. When $|Y| = 2$, the arrow may be written as $y_1 \,\&\, y_2 \longrightarrow x$.

If s and t are elements of $\hat{S}$ such that $\lambda s \leq \lambda t$, then we want to devise a map h such that $h(t) = s$ to be an operator in H. The catch is that h must be meet-preserving and satisfy $\lambda h(u) \leq \lambda u$ for all $u \in \hat{S}$ (condition (a2)). Let $\mathbf{P} = \{\lambda x : x \in \hat{S}\}$, with the order inherited from $\mathbf{L}$. By composing operators, it suffices to find operators for pairs with $\lambda s \leq \lambda t$ in $\mathbf{P}$. For pairs with $\lambda s = \lambda t$, we need maps in both directions, which is indicated by $s \leftrightarrow t$.

Note that the greatest element $\hat{0}$ of $\widehat{\mathbf{S}}$ is a constant of the type, and thus in every subalgebra. Condition (b) requires that $\lambda(s) = 0_{\mathbf{L}}$ only for $s = \hat{0}$. So the arrows $s \to \hat{0}$ may be omitted.

On the other hand, arrow relations $Y \to x$, corresponding to minimal join covers $\lambda x \le \bigvee \lambda Y$, come in two main types: when λx is join reducible and $\lambda x = \bigvee \lambda Y$ is the canonical join representation of λx, and when λx is join irreducible and $\lambda x < \bigvee \lambda Y$ is a minimal nontrivial join cover. If property (c) holds for those minimal join covers, then it holds for non-minimal (refinable) join covers, using the fact that $a \le b$ implies $\xi(a) \le \xi(b)$. That allows us to cover the third case, when λx is join reducible and $\lambda x < \bigvee \lambda Y$.

To realize $x \in \mathrm{Sg}(Y)$ when $Y \to x$ may require that we add elements and/or comparabilities to $\widehat{\mathbf{S}}$. As indicated earlier, we need for $\hat{S}$ to contain x' and Y' such that $\lambda x' = \lambda x$, $\lambda Y' = \lambda Y$, and $x' = \bigwedge Y'$. There will be many examples, but exactly *how* to build $\widehat{\mathbf{S}}$, or when it is even possible, in the end remains more art than science. In Examples 6.2–6.4 we use *ad hoc* methods, and in the next section a more systematic method, but neither one is a real algorithm that always works.

One type of arrow relation automatically holds in our constructions. When $\{a\} \cup B \subseteq \tau(\mathbf{L})$ and $a \le \bigvee B$, then we have the arrow $\hat{B} \to \hat{a}$. In this case $\hat{a} \in \mathrm{Sg}(\hat{B})$ follows from the other properties. First recall Lemma 3.18, where $\hat{\tau}(x)$ denotes $\widehat{\tau(x)}$.

Lemma 6.1 *The subset $T = \{\hat{x} : x \in \tau(\mathbf{L})\}$ is a 0-1-sublattice of $\widehat{\mathbf{S}}$, such that*

(1) *the greatest element is $\hat{0}$,*
(2) *the least element is $\hat{z}$, where $z = \tau(1_{\mathbf{L}})$,*
(3) $\hat{x} \wedge \hat{y} = \hat{\tau}(x \vee y)$,
(4) $\hat{x} \vee \hat{y} = \hat{\tau}(\gamma x \wedge \gamma y)$.

Consider an arrow of the form say $\hat{b} \,\&\, \hat{c} \to \hat{a}$, corresponding to the inclusion $a \le b \vee c$ in $\mathbf{L}$, where $a, b, c \in \tau(\mathbf{L})$. From $a \le b \vee c$ we obtain $\gamma a \le \gamma(b \vee c)$, whence $\hat{a} \ge \hat{\tau}(b \vee c) = \hat{b} \wedge \hat{c}$ in $\widehat{\mathbf{S}}$. If property (e) holds, there is an $h \in H$ such that $\hat{a} = h(\hat{b} \wedge \hat{c}) = h(\hat{b}) \wedge h(\hat{c})$. Thus $\hat{a} \in \mathrm{Sg}(\hat{b}) \vee \mathrm{Sg}(\hat{c})$, as desired.

However, for arrow relations $Y \to x$ with $Y \cup \{x\} \not\subseteq T$, the construction of $\widehat{\mathbf{S}}$ should include meets to witness that $x \in \mathrm{Sg}(Y)$.

Now let us actually do some representations!

Example 6.2 Our first example is in Fig. 6.2. For this pair, $T = \{\hat{0}, \hat{a}, \hat{b}, \hat{z}\}$ and $J = \{p_1\}$ as $J(L) \setminus \tau(L) = \{p\}$. With the order of $\gamma^d(\mathbf{L})$ on T and $p_1 \ge \hat{b}$, the set $\hat{S} = T \cup J$ is a lattice, as shown in the figure. Set $\lambda\hat{x} = x$ for $x = 0, a, b, z$ and $\lambda p_1 = p$. Write down the $\to$ relations as given in the figure, and find operators h, k to realize them, remembering to ensure that $\lambda h(s) \le \lambda s$ and $\lambda k(s) \le \lambda s$ for each $s \in \hat{S}$. For example, $h(\hat{z}) = \hat{a}$ enforces $\hat{z} \to \hat{a}$. Note that $\hat{a} \,\&\, \hat{b} \to \hat{z}$ is realized by $\hat{a} \wedge \hat{b} = \hat{z}$, as observed above.

Then let H be the monoid generated by h and k, which in this case is $H = \{h, k, id, \hat{0}\}$ where $\hat{0}$ is a constant map. Thus $\hat{S} = \{\hat{0}, \hat{a}, \hat{b}, \hat{z}, p_1\}$ and the monoid H give a representation of $(\mathbf{L}, \gamma)$ as $(\mathrm{Sub}(\widehat{\mathbf{S}}, \wedge, \hat{0}, H), \Gamma)$. Of course, we really only need h and k as operations.

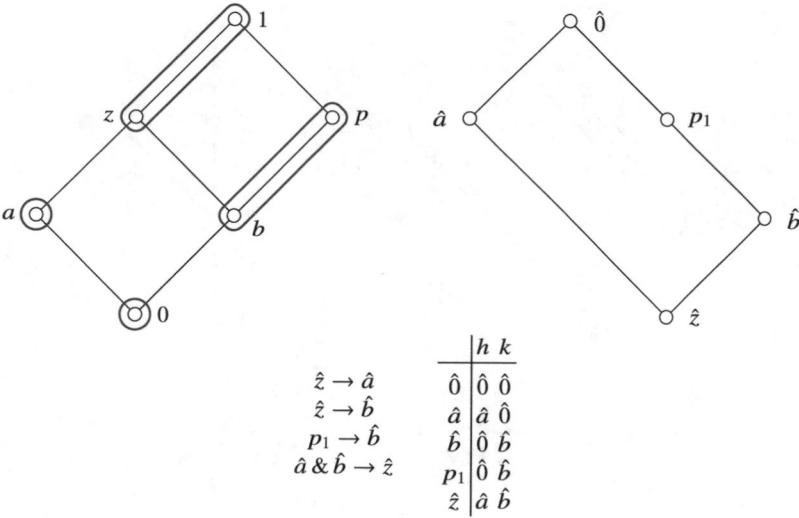

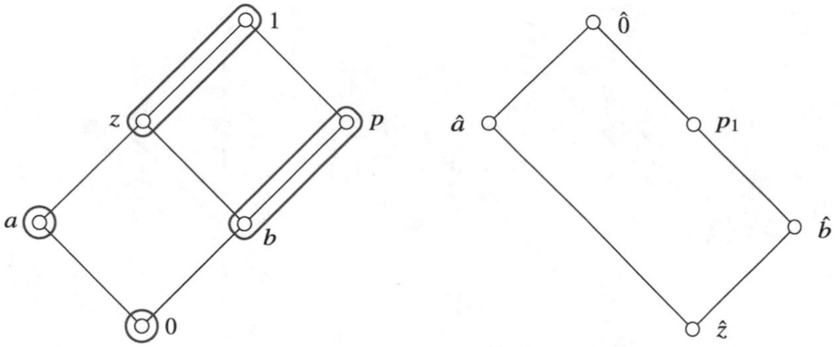

$\hat{z} \to \hat{a}$
$\hat{z} \to \hat{b}$
$p_1 \to \hat{b}$
$\hat{a}\,\&\,\hat{b} \to \hat{z}$

	h	k
$\hat{0}$	$\hat{0}$	$\hat{0}$
$\hat{a}$	$\hat{a}$	$\hat{0}$
$\hat{b}$	$\hat{0}$	$\hat{b}$
p_1	$\hat{0}$	$\hat{b}$
$\hat{z}$	$\hat{a}$	$\hat{b}$

Fig. 6.2 Example 6.2 with $J \nsubseteq T$

Example 6.3 Now we turn to the pentagon with the equaclosure operator shown in Fig. 6.3; this will be Running Example 5 in Chap. 7. From the inclusion $p \geq z$ in $\mathbf{L}$ we get the arrow $p_1 \to \hat{z}$, and the minimal nontrivial join cover $p \leq z \vee b$ in $\mathbf{L}$ gives $\hat{z}\,\&\,\hat{b} \to p_1$. Now $J \cup T = \{\hat{0}, \hat{b}, \hat{z}, p_1\}$, but we cannot realize $\hat{z}\,\&\,\hat{b} \to p_1$ by $\hat{z} \wedge \hat{b} = p_1$, which would be the first option, because $\hat{z}$ is the least element of $\widehat{\mathbf{S}}$. Thus we need to add another element, say z_1, such that $z_1 \wedge \hat{b} = p_1$, and make z_1 equivalent to z in the sense that $\lambda z_1 = \lambda z$. Since $\hat{z}$ is the least element of $\mathrm{Sg}(\hat{z})$, we make $z_1 > \hat{z}$, and this gives the order on $\widehat{\mathbf{S}}$ shown in Fig. 6.3. Add the double arrow $\hat{z} \leftrightarrow z_1$, to indicate that we need operators with $h(\hat{z}) = z_1$ and $k(z_1) = \hat{z}$. The operators given in the figure do that, along with $k(p_1) = \hat{z}$ to realize $p_1 \to \hat{z}$.

Thus with $\hat{S} = \{\hat{0}, \hat{b}, \hat{z}, p_1, z_1\}$ and the monoid H on $\hat{S}$ generated by h and k, we achieve a representation of $(\mathbf{L}, \gamma)$ as $(\mathrm{Sub}(\hat{\mathbf{S}}, \wedge, \hat{0}, H), \Gamma)$. In Sect. 7.1, we will convert this representation to one with a subquasivariety lattice $\mathrm{L_q}(\mathcal{K})$ and its natural equaclosure operator.

Example 6.4 The pair $(\mathbf{N}, \gamma_3)$ illustrated in Fig. 6.4 is similar; it also appears in the discussion in Example A.1 in the Appendix. Exactly as in the preceding example with the pentagon, we add an element z_1 such that $\lambda z_1 = \lambda \hat{z}$ and $z_1 \wedge \hat{b} = p_1$, due to the join cover $p \leq z \vee b$ in $\mathbf{L}$. However, we cannot have $z_1 \wedge \hat{a} = p_1$, because $p \nleq z \vee a$ in $\mathbf{L}$. This problem is avoided by putting $z_1 \leq \hat{a}$. With that admittedly *ad hoc* adjustment, operators are easily found to achieve the representation, as given in the figure.

More examples of such representations are given in Appendix A.1.

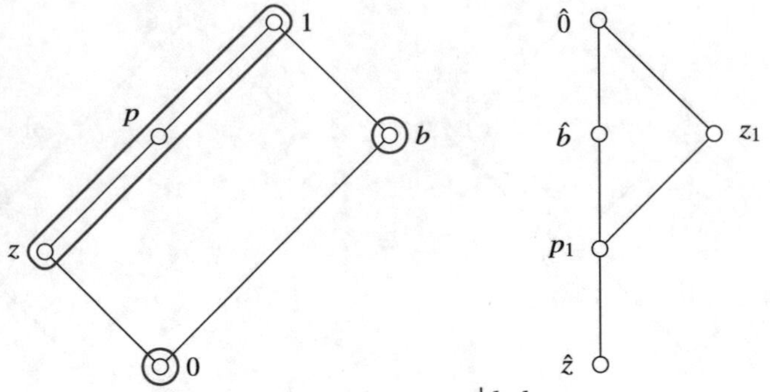

$p_1 \to \hat{z}$
$\hat{z}\,\&\,\hat{b} \to p_1$
$\hat{z} \leftrightarrow z_1$

	h	k
$\hat{0}$	$\hat{0}$	$\hat{0}$
$\hat{z}$	z_1	$\hat{z}$
z_1	z_1	$\hat{z}$
$\hat{b}$	$\hat{0}$	$\hat{0}$
p_1	z_1	$\hat{z}$

Fig. 6.3 Example 6.3, which is Running Example 5 in Chap. 7

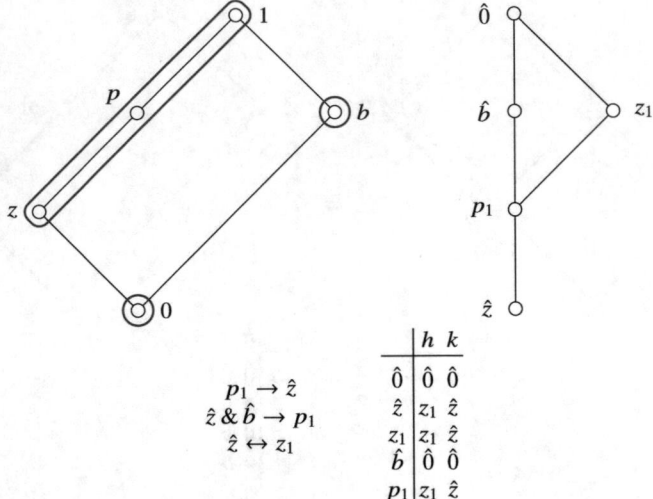

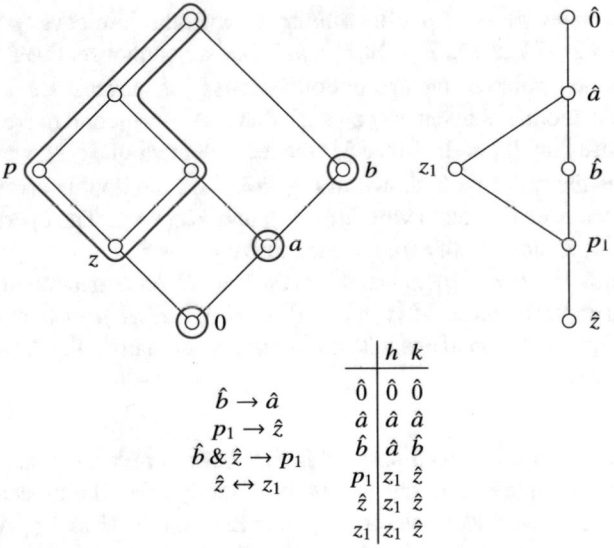

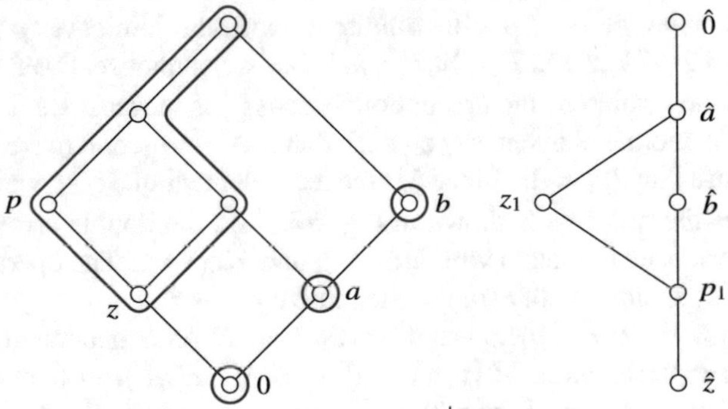

$\hat{b} \to \hat{a}$
$p_1 \to \hat{z}$
$\hat{b}\,\&\,\hat{z} \to p_1$
$\hat{z} \leftrightarrow z_1$

	h	k
$\hat{0}$	$\hat{0}$	$\hat{0}$
$\hat{a}$	$\hat{a}$	$\hat{a}$
$\hat{b}$	$\hat{a}$	$\hat{b}$
p_1	z_1	$\hat{z}$
$\hat{z}$	z_1	$\hat{z}$
z_1	z_1	$\hat{z}$

Fig. 6.4 Example 6.4 with $J \nsubseteq T$: $(\mathbf{N}, \gamma_3)$

6.2 Representations Based on Embeddings into Sub S

The technique of this section extends the embedding of a finite lower bounded lattice **L** into Sub **P** for a meet semilattice **P** constructed in Theorem 4.30.

At this point it behooves us to review the entire setup, as a guide to the reader (and the authors). We are given a pair $(\mathbf{L}, \gamma)$ consisting of a finite, lower bounded lattice and an equaclosure operator on it. To "represent" this pair we want to find a (finite) semilattice with operators, $\mathbf{S} = (S, \wedge, 1, H)$, and an isomorphism $\xi : \mathbf{L} \cong \mathrm{Sub}(\mathbf{S}, \wedge, 1, H)$ in such a way that $\xi(\gamma(x)) = \Gamma(\xi(x))$ for all $x \in L$. Here Γ is the natural equaclosure operator on Sub **S**, so that for a subalgebra $X \leq \mathbf{S}$, we have $\Gamma(X) = \uparrow x_0$ where x_0 is the least element of X.

We begin the construction by embedding **L** into $\mathrm{Sub}(\mathbf{P}, \wedge, 1)$ following the recipe of Theorem 4.30. Then two types of modifications are usually required. First, we may have to adjust **P** by adding relations, so that the images of elements in the same γ-class can be subalgebras with the same least element. Our **S** will be the modified semilattice. Then we need to add operators so that the embedding of **L** into $\mathrm{Sub}(\mathbf{S}, \wedge, 1)$ becomes an isomorphism onto $\mathrm{Sub}(\mathbf{S}, \wedge, 1, H)$. Sometimes this also requires adding extra elements or relations.

To that end, we briefly recall the construction of Theorem 4.30. Find $\mathrm{J}(\mathbf{L})$ and all minimal nontrivial join covers $x \leq \bigvee U$ in $\mathrm{J}(\mathbf{L})$. This enables us to draw the ordered set $(\mathrm{J}(\mathbf{L}), \overline{\mathrm{D}})$. Form the set G of all sequences $d_1 \ldots d_k$ $(k \geq 1)$ such that

(1) d_1 is minimal in $(\mathrm{J}(\mathbf{L}), \overline{\mathrm{D}})$, and
(2) $d_i \mathrel{\mathrm{D}} d_{i+1}$ for $1 \leq i < k$.

Then let **P** be the meet semilattice with 1 freely generated by G subject to the relations that whenever $d_k \leq u_1 \vee \ldots \vee u_m$ is a minimal nontrivial join cover in **L**, then

$$d_1 \ldots d_k = d_1 \ldots d_k u_1 \sqcap \cdots \sqcap d_1 \ldots d_k u_m$$

in **P**. We use $\sqcap$ for the meet in **P**, and later **S**, to avoid confusion with the meet in **L**.

In this section, we are going to represent the lattice **W** with all its preclops μ, γ_j $(1 \leq j \leq 6)$ from Fig. 4.1. There is one minimal nontrivial join cover in **W**, which is $b \leq a \vee d$. The lattice and $(\mathrm{J}(\mathbf{W}), \overline{\mathrm{D}})$ are reproduced in Fig. 6.5. The construction gives us the generating set $G = \{b, ba, bd, c\}$, and **P** is the semilattice with 1 generated by G subject to the relations

[1] $b = ba \sqcap bd$,
[2] $b \leq ba$, $b \leq bd$,

where [2] is of course a consequence of [1]. The result is the semilattice **P** drawn in Fig. 6.6. The embedding $\xi : \mathbf{W} \to \mathrm{Sub}(\mathbf{P}, \sqcap, 1)$ is defined via $\xi(u)$ is the $\sqcap$-subsemilattice generated by $\{d_1 \ldots d_k \in \mathbf{P} : d_k \leq u\}$. Specifically,

$$a \longmapsto \{ba, 1\}$$
$$c \longmapsto \{c, 1\}$$
$$e \longmapsto \{ba, c, ba \sqcap c, 1\}$$
$$d \longmapsto \{bd, c, bd \sqcap c, 1\}$$
$$b \longmapsto \{b, ba, c, ba \sqcap c, b \sqcap c, 1\}$$

and of course $\xi(1) = P$, $\xi(0) = \{1\}$.

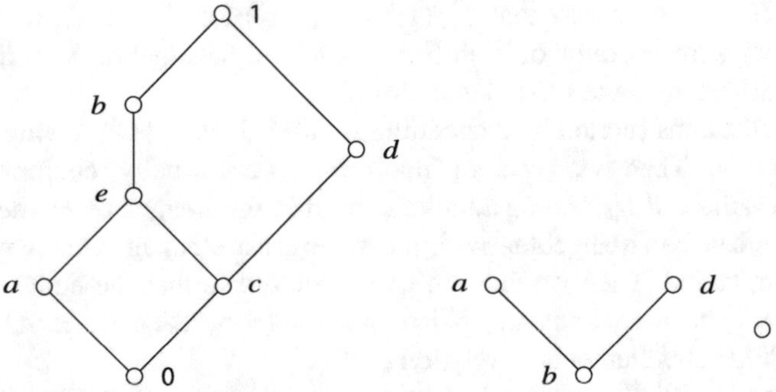

Fig. 6.5 The lattice **W** and (J(**W**), $\overline{\mathrm{D}}$)

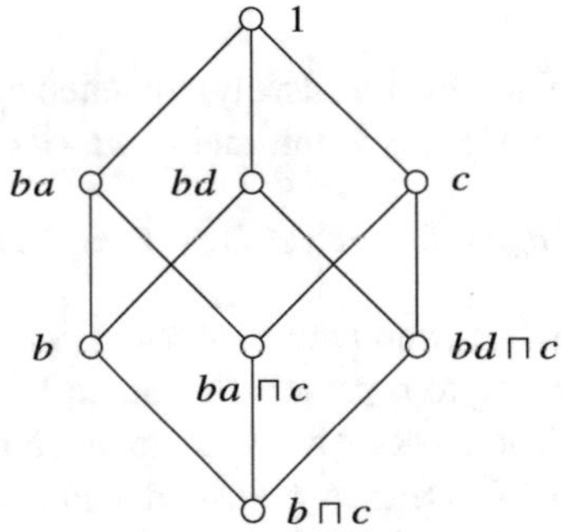

Fig. 6.6 The semilattice **P** such that **W** $\leq$ Sub(**P**, $\sqcap$, 1)

Now let us implement the program of Sect. 4.4, and especially Theorem 4.39 for our situation. That is, we want a map $\lambda : S \to L$ such that $\xi(\lambda(x)) = \mathrm{Sg}(x)$ for each $x \in S$. As in the proof of Theorem 4.30, let $\lambda(d_1 \dots d_k) = d_k$, the last term in the sequence $\overline{d}$; there it was denoted $l(\overline{d})$. For meets in **S**, while the requirement of Theorem 4.37 is only that $\lambda(\overline{u} \sqcap \overline{v}) \leq \lambda\overline{u} \vee \lambda\overline{v}$, we take $\lambda(\overline{u} \sqcap \overline{v}) = \lambda\overline{u} \vee \lambda\overline{v}$ unless otherwise specified.

For our particular lattice **W**, we have $\lambda(b) = b$, $\lambda(c) = c$, $\lambda(ba) = a$, and $\lambda(bd) = d$. Thus $\lambda(ba \sqcap c) = e$, $\lambda(bd \sqcap c) = d$, and $\lambda(b \sqcap c) = b$.

We use *arrow relations*, $y \to x$ or $U \to x$, to indicate the properties that must hold in order for λ to yield the right subalgebra inclusions.

- For $x, y \in S$, the arrow $y \to x$ means that $\lambda x \le \lambda y$, whence it should be that $x \in \mathrm{Sg}(y)$. For meet semilattices with operators, this usually means $x = h(y)$ for some $h \in H$.
- For $x \in S$ and $U \subseteq S$, the arrow $U \to x$ means that $\lambda x \le \bigvee \lambda(U)$, so that we want $x \in \mathrm{Sg}(U)$ to hold. For meet semilattices with operators, if $U = \{u_1, \ldots, u_m\}$, this means that there exist $s, t_1, \ldots, t_m$ such that $\lambda s = \lambda x$, $\lambda t_j = \lambda u_j$ for all j, and $s = t_1 \sqcap \ldots \sqcap t_m$. It suffices to do this when $\lambda x \le \bigvee \lambda(U)$ is a minimal nontrivial join cover in **L**.

The first type of arrow is just part of ensuring that $\xi : \mathbf{L} \to \mathrm{Sub}\,\mathbf{S}$ is a homomorphism; cf. Theorem 4.37. But the second type of arrow is essential to guarantee that ξ is surjective; see Theorem 4.39. Moreover, for this general method, sometimes **S** will be a larger semilattice than required, meaning that there may be more arrows than we are used to with the *ad hoc* constructions.

Thus in **P** we have the arrow relations

$$\begin{aligned} bd &\to c \\ b &\to ba \\ b &\to c \\ \{ba, bd\} &\to b \end{aligned}$$

plus those obtained from the law $\lambda(\overline{u} \sqcap \overline{v}) = \lambda\overline{u} \vee \lambda\overline{v}$.

This concludes our summary of the method. The lattice **W** has 7 preclops: $\mu < \gamma_1 < \gamma_2 < \gamma_5$, $\mu < \gamma_6 < \gamma_3 < \gamma_4$ and $\gamma_5 \wedge \gamma_4 = \gamma_2$, $\gamma_2 \wedge \gamma_3 = \gamma_1$, $\gamma_1 \wedge \gamma_6 = \mu$, as illustrated in Fig. 6.7. In the following sections we will represent each of the pairs $(\mathbf{W}, \mu)$ and $(\mathbf{W}, \gamma_j)$ $(1 \le j \le 6)$ as $\mathrm{Sub}(\mathbf{S}, \sqcap, 1, H)$ for an appropriate **S** and H.

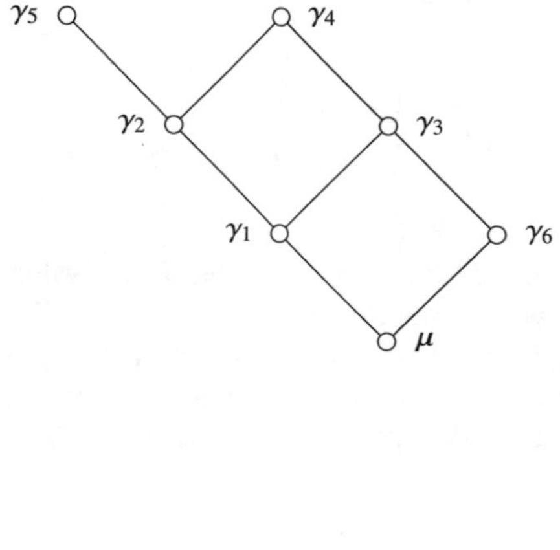

Fig. 6.7 Pre(**W**)

6.2.1 $(\mathbf{W}, \mu)$

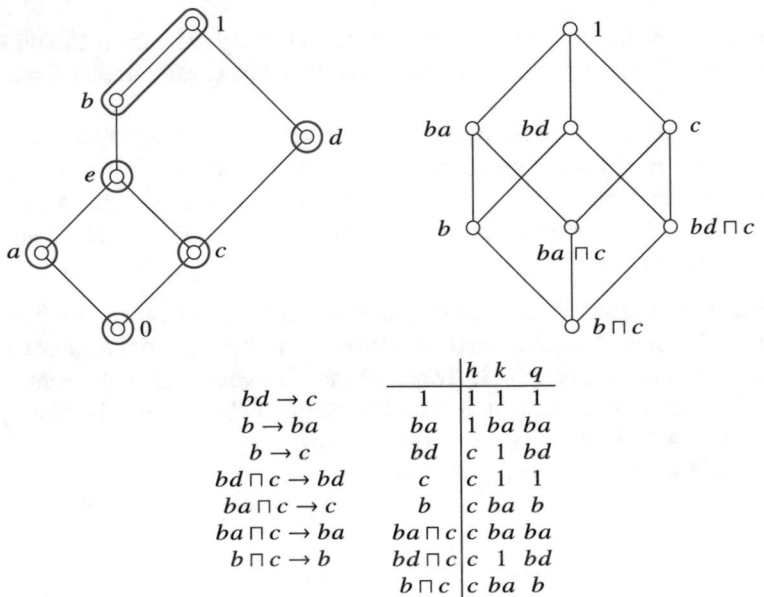

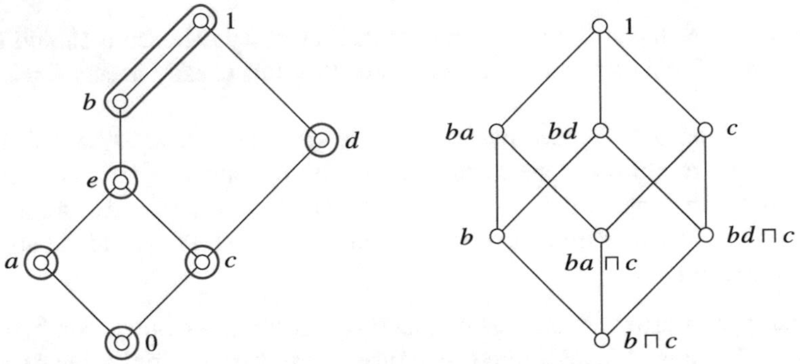

$$
\begin{aligned}
bd &\to c\\
b &\to ba\\
b &\to c\\
bd \sqcap c &\to bd\\
ba \sqcap c &\to c\\
ba \sqcap c &\to ba\\
b \sqcap c &\to b
\end{aligned}
$$

	h	k	q
1	1	1	1
ba	1	ba	ba
bd	c	1	bd
c	c	1	1
b	c	ba	b
$ba \sqcap c$	c	ba	ba
$bd \sqcap c$	c	1	bd
$b \sqcap c$	c	ba	b

Fig. 6.8 Calculations for $(\mathbf{W}, \mu)$

The equapartition for μ has one nontrivial class $[b, 1]$. For representation we take $\mathbf{S}_\mu = \mathbf{P}$ with just the relations [1], [2] from page 145. Thus, $S_\mu = \{b, ba, bd, c, ba \sqcap c, bd \sqcap c, b \sqcap c, 1\}$ as shown in Fig. 6.8. The embedding $\xi : \mathbf{W} \longrightarrow \mathrm{Sub}(\mathbf{S}_\mu, \sqcap, 1)$ is given by

$$
\begin{aligned}
a &\longmapsto \{ba, 1\}\\
c &\longmapsto \{c, 1\}\\
e &\longmapsto \{ba, c, ba \sqcap c, 1\}\\
d &\longmapsto \{bd, c, bd \sqcap c, 1\}\\
b &\longmapsto \{b, ba, c, ba \sqcap c, b \sqcap c, 1\}
\end{aligned}
$$

and of course $\xi(1) = S_\mu$, $\xi(0) = \{1\}$.

The next step is to identify the range of potential operators on $\mathbf{S}$ so that only the subsemilattices above will be subalgebras.

For every element x of the semilattice $\mathbf{S}_\mu$, we identify the need to include an element y, which becomes $f(x)$ for a potential operator f on S_μ, if y belongs to the minimal subsemilattice $W_x \in \xi(\mathbf{W})$ that contains x. One may interpret W as the range of a closure operator on Sub($\mathbf{S}_\mu, \sqcap, 1$), so that $\{x \longrightarrow y : x \in S_\mu,\ y \in W_x\}$ would be the set of implications defining it.

In Table 6.1 we list each $x \in S_\mu$ together with W_x and the implications $x \longrightarrow y$ that follow. Only implications with $y \in \{ba, bd, b, c\}$ are needed, because the meets will be forced by the virtue of $\sqcap$-closedness of elements in Sub($\mathbf{S}_\mu, \sqcap, 1$).

Table 6.1 S_μ and $x \to y$

x	W_x	$x \to y$
bd	$\{bd, c, bd \sqcap c, 1\}$	$bd \to c$ (1)
b	$\{b, ba, c, ba \sqcap c, b \sqcap c, 1\}$	$b \to ba$ (2)
		$b \to c$ (3)
$bd \sqcap c$	$\{bd, c, bd \sqcap c, 1\}$	$bd \sqcap c \to bd$ (4)
$ba \sqcap c$	$\{ba, c, ba \sqcap c, 1\}$	$ba \sqcap c \to c$ (5)
		$ba \sqcap c \to ba$ (6)
$b \sqcap c$	$\{b, ba, c, ba \sqcap c, b \sqcap c, 1\}$	$b \sqcap c \to b$ (7)

Note that we do not include $bd \sqcap c \longrightarrow c$ for $x = bd \sqcap c$, for example, because we included $bd \longrightarrow c$ for $x = bd$, and the operators will be closed under composition. Similarly, for $x = b \sqcap c$ we included only $b \sqcap c \longrightarrow b$, because other implications for this x will follow under the composition from those included for $x = b$.

Now define a set of operators H_μ on ($\mathbf{S}_\mu, \sqcap, 1$) so that $x \longrightarrow y$ in (1)–(7) are realized as $p(x) = y$ for some $p \in H_\mu$. These operators are given in Table 6.2; the numbers in parentheses indicate the implication from the previous table that is realized by the corresponding operator.

Table 6.2 Operators on $\mathbf{S}_\mu$

Operator p	$p(ba)$	$p(bd)$	$p(c)$	$p(b)$	$p(ba \sqcap c)$	$p(bd \sqcap c)$	$p(b \sqcap c)$
h	1	c (1)	c	c (3)	c (5)	c	c
k	ba	1	1	ba (2)	ba (6)	1	ba
q	ba	bd	1	b	ba	bd (4)	b (7)

We postpone ($\mathbf{W}, \gamma_1$) until last.

6.2.2 ($\mathbf{W}, \gamma_2$)

The equapartition for γ_2 has nontrivial classes $[e, 1]$ and $[c, d]$. To reflect the fact that $d = \gamma_2(c)$ in $\mathbf{W}$, we want bd to be in the filter $\uparrow c$ in $\mathbf{S}_2$. To achieve this take an additional relation on $\mathbf{P}$ from Theorem 4.30:

[3] $c \leq bd$.

Thus we let $\mathbf{S}_2$ be the semilattice generated by $\{b, ba, bd, c\}$ subject to [1], [2], [3] as shown in Fig. 6.9, so that $S_2 = \{b, ba, bd, c, b \sqcap c, 1\}$.

Note that [3] implies $ba \sqcap c = b \sqcap c$, the least element of $\mathbf{S}_2$, which fits the requirement that the minimal element in $\mathbf{S}_2$, i.e., $\hat{\tau}(1)$, is generated by elements from $e = a \vee c$.

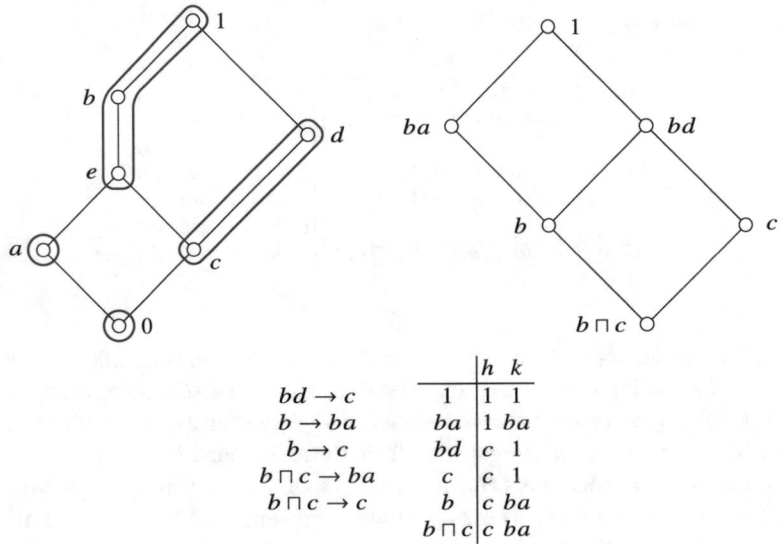

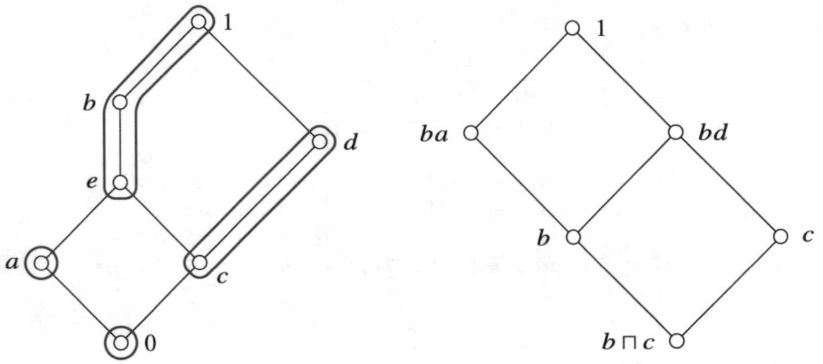

$bd \to c$
$b \to ba$
$b \to c$
$b \sqcap c \to ba$
$b \sqcap c \to c$

	h	k
1	1	1
ba	1	ba
bd	c	1
c	c	1
b	c	ba
$b \sqcap c$	c	ba

Fig. 6.9 Calculations for $(\mathbf{W}, \gamma_2)$

The embedding $\xi : \mathbf{W} \to \mathrm{Sub}(\mathbf{S}_2, \sqcap, 1)$ is

$$
\begin{aligned}
a &\longmapsto \{ba, 1\} \\
c &\longmapsto \{c, 1\} \\
e &\longmapsto \{ba, c, b \sqcap c, 1\} \\
d &\longmapsto \{bd, c, 1\} \\
b &\longmapsto \{b, ba, c, b \sqcap c, 1\}
\end{aligned}
$$

along with $\xi(1) = S_2$, $\xi(0) = \{1\}$.

Table 6.3 gives the implications needed to represent $\mathbf{W}$ as $\mathrm{Sub}(\mathbf{S}_2, \sqcap, 1, H_2)$ with a potential set of operators H_2.

Now define a set of operators H_2 on $(S_2, \sqcap, 1)$ so that $x \longrightarrow y$ in (1)–(5) are realized as $p(x) = y$ for some $p \in H_2$. These operators are given in Table 6.4; the numbers in parentheses indicate the implication from the previous table that is realized by the corresponding operator.

Table 6.3 $\mathbf{S}_2$ and $x \to y$

x	W_x	$x \to y$
bd	$\{bd, c, 1\}$	$bd \to c$ (1)
b	$\{b, ba, c, b \sqcap c, 1\}$	$b \to ba$ (2)
		$b \to c$ (3)
$b \sqcap c$	$\{ba, c, b \sqcap c, 1\}$	$b \sqcap c \to ba$ (4)
		$b \sqcap c \to c$ (5)

Table 6.4 Operators on $\mathbf{S}_2$

Operator p	$p(ba)$	$p(bd)$	$p(c)$	$p(b)$	$p(b \sqcap c)$
h	1	c (1)	c	c (3)	c (5)
k	ba	1	1	ba (2)	ba (4)

6.2.3 $(\mathbf{W}, \gamma_3)$

The corresponding equapartition has only one nontrivial class $[a, 1]$. In particular, $\tau(1) = a$ for γ_3 on $\mathbf{W}$. For representation we take an additional element a for a generator and additional relations on $\mathbf{P}$ from Theorem 4.30:

[3] $a \leq b, a \leq c$.

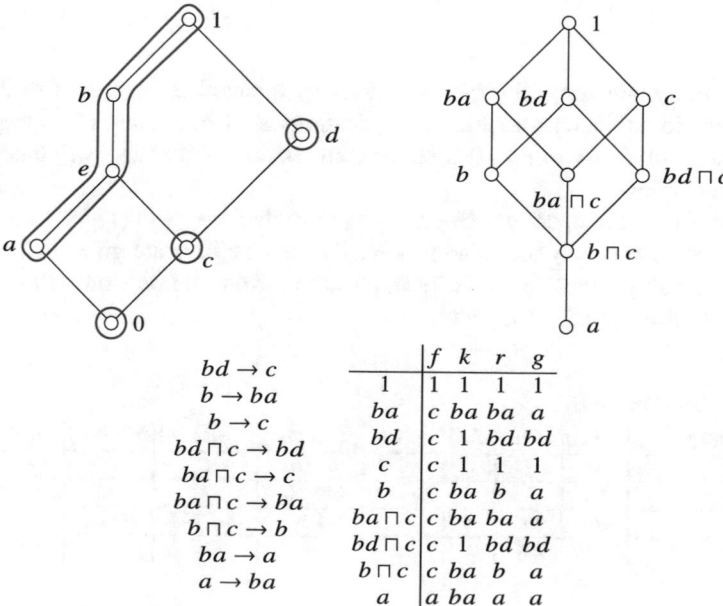

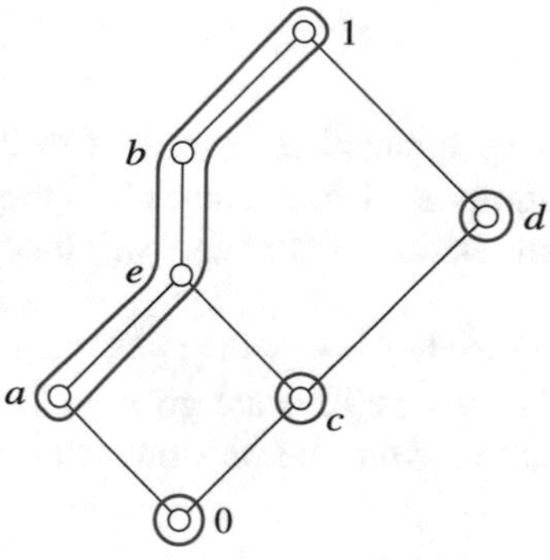

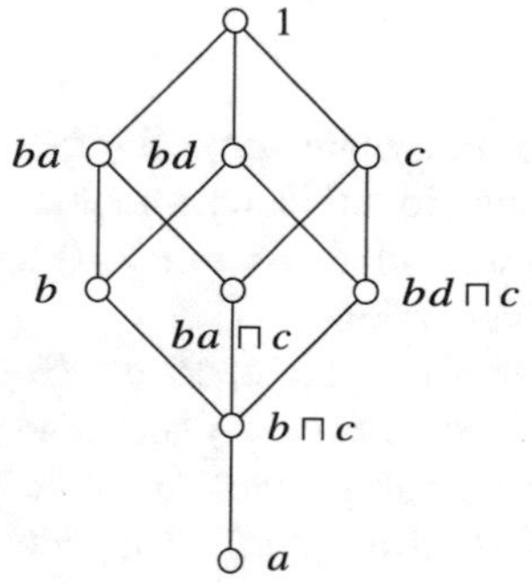

$bd \to c$
$b \to ba$
$b \to c$
$bd \sqcap c \to bd$
$ba \sqcap c \to c$
$ba \sqcap c \to ba$
$b \sqcap c \to b$
$ba \to a$
$a \to ba$

	f	k	r	g
1	1	1	1	1
ba	c	ba	ba	a
bd	c	1	bd	bd
c	c	1	1	1
b	c	ba	b	a
$ba \sqcap c$	c	ba	ba	a
$bd \sqcap c$	c	1	bd	bd
$b \sqcap c$	c	ba	b	a
a	a	ba	a	a

Fig. 6.10 Calculations for $(\mathbf{W}, \gamma_3)$

Now generate a free semilattice $\mathbf{S}_3$ defined by the extended set of generators $\{a, ba, bd, b, c\}$ and relations [1], [2] and [3]. Thus, we get $\mathbf{S}_3 = \{a, b, ba, bd, c, c \sqcap ba, c \sqcap bd, c \sqcap b\}$ as shown in Fig. 6.10. The embedding $\xi : \mathbf{W} \longrightarrow \text{Sub}(\mathbf{S}_3, \sqcap, 1)$ is

$$a \longmapsto \{a, ba, 1\}$$
$$c \longmapsto \{c, 1\}$$
$$e \longmapsto \{a, ba, c, ba \sqcap c, 1\}$$
$$d \longmapsto \{bd, c, bd \sqcap c, 1\}$$
$$b \longmapsto \{a, b, ba, c, ba \sqcap c, b \sqcap c, 1\}$$

along with $\xi(1) = S_3$, $\xi(0) = \{1\}$.

Proceed to the implications needed to represent $\mathbf{W}$ as $\text{Sub}(\mathbf{S}_3, \sqcap, 1, H_3)$ with a potential set of operators H_3.

Table 6.5 $\mathbf{S}_3$ and $x \to y$

x	W_x	$x \to y$
bd	$\{bd, c, bd \sqcap c, 1\}$	$bd \to c$ (1)
b	$\{a, b, ba, c, ba \sqcap c, b \sqcap c, 1\}$	$b \to ba$ (2)
		$b \to c$ (3)
$bd \sqcap c$	$\{bd, c, bd \sqcap c, 1\}$	$bd \sqcap c \to bd$ (4)
$ba \sqcap c$	$\{a, ba, c, ba \sqcap c, 1\}$	$ba \sqcap c \to c$ (5)
		$ba \sqcap c \to ba$ (6)
$b \sqcap c$	$\{a, b, ba, c, ba \sqcap c, b \sqcap c, 1\}$	$b \sqcap c \to b$ (7)
ba	$\{ba, a, 1\}$	$ba \to a$ (8)
a	$\{ba, a, 1\}$	$a \to ba$ (9)

Note that while some W_x get extensions by element a, compared to the case $(\mathbf{W}, \mu)$, we do not include additional implications $x \to a$, because all of them have ba, and we include $ba \to a$ as (8). So the composition of operators will take care of those implications.

Now define set of operators H_3 on $(\mathbf{S}_3, \sqcap, 1)$ so that $x \to y$ in (1)–(9) of Table 6.5 are realized as $p(x) = y$ for some $p \in H_3$. These operators are given in Table 6.6; the numbers in parentheses indicate implications from the previous table that are realized by corresponding operator.

Table 6.6 Operators on $\mathbf{S}_3$

Operator p	$p(ba)$	$p(bd)$	$p(c)$	$p(b)$	$p(ba \sqcap c)$	$p(bd \sqcap c)$	$p(b \sqcap c)$	$p(a)$
f	c	c (1)	c	c (3)	c (5)	c	c	a
k	ba	1	1	ba (2)	ba (6)	1	ba	ba (9)
r	ba	bd	1	b	ba	bd (4)	b (7)	a
g	a (8)	bd	1	a	a	bd	a	a

6.2.4 $(\mathbf{W}, \gamma_4)$

The equapartition for γ_4 has two nontrivial classes: $[a, 1]$ and $[c, d]$. As in the previous case of γ_3, we take an additional element a for a generator and additional relations on **P** from Theorem 4.30:

[3] $a \leq b, a \leq c$;
[4] $c \leq bd$.

These relations correspond to $\gamma_4(a) = 1$ and $\gamma_4(c) = d$ in **W**.

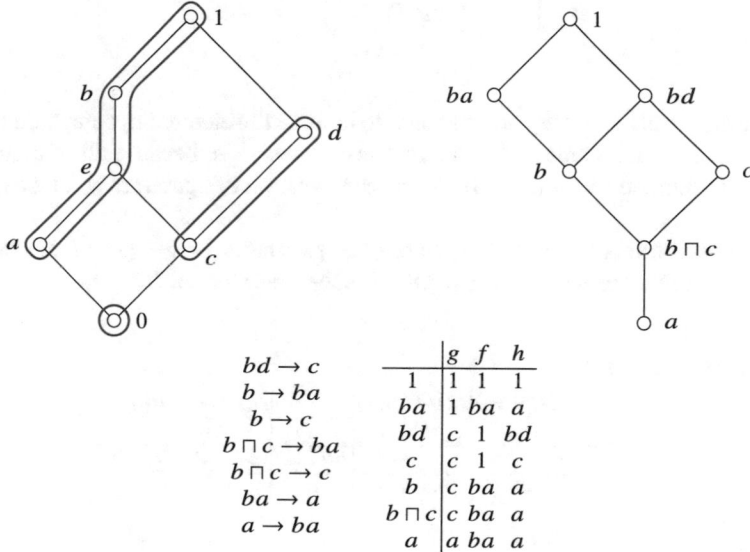

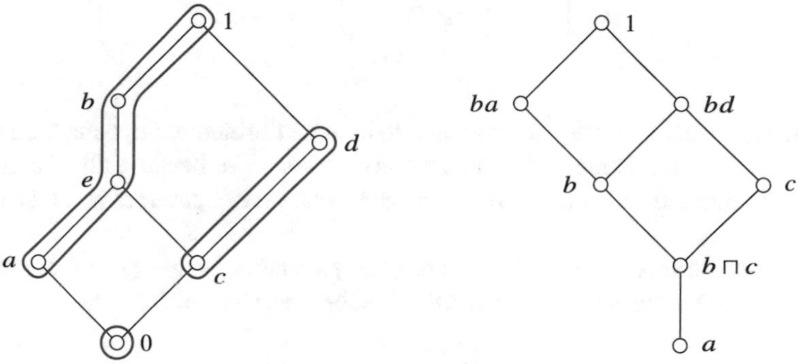

$bd \to c$
$b \to ba$
$b \to c$
$b \sqcap c \to ba$
$b \sqcap c \to c$
$ba \to a$
$a \to ba$

	g	f	h
1	1	1	1
ba	1	ba	a
bd	c	1	bd
c	c	1	c
b	c	ba	a
$b \sqcap c$	c	ba	a
a	a	ba	a

Fig. 6.11 Calculations for $(\mathbf{W}, \gamma_4)$

Now generate the free semilattice $\mathbf{S}_4$ defined by the extended set of generators $\{a, ba, bd, b, c\}$ and relations [1], [2], [3], and [4]. Note that [4] forces $b \sqcap c = ba \sqcap c$. Thus, $S_4 = \{a, b, ba, bd, c, c \sqcap b\}$ as shown in Fig. 6.11. The embedding $\xi : \mathbf{W} \to \mathrm{Sub}(\mathbf{S}_4, \sqcap, 1)$:

$$a \longmapsto \{a, ba, 1\}$$
$$c \longmapsto \{c, 1\}$$
$$e \longmapsto \{a, ba, c, b \sqcap c, 1\}$$
$$d \longmapsto \{bd, c, 1\}$$
$$b \longmapsto \{a, b, ba, c, b \sqcap c, 1\}$$

along with $\xi(1) = S_4$, $\xi(0) = \{1\}$.

The implications needed to represent **W** as Sub($\mathbf{S}_4, \sqcap, 1, H_4$) with a potential set of operators H_4 are in Table 6.7.

Table 6.7 $\mathbf{S}_4$ and implications $x \to y$

x	W_x	$x \to y$
bd	$\{bd, c, 1\}$	$bd \to c$ (1)
b	$\{a, b, ba, c, b \sqcap c, 1\}$	$b \to ba$ (2)
		$b \to c$ (3)
$b \sqcap c$	$\{a, ba, c, b \sqcap c, 1\}$	$b \sqcap c \to ba$ (4)
		$b \sqcap c \to c$ (5)
ba	$\{ba, a, 1\}$	$ba \to a$ (6)
a	$\{ba, a, 1\}$	$a \to ba$ (7)

Note that while some W_x are extended to include the element a, compared to case $(\mathbf{W}, \mu)$, we do not include additional implications $x \to a$, because all of them have ba, and we include $ba \to a$ as (8). So the composition of operators will take care of those.

Now define a set of operators H_4 on $(\mathbf{S}_4, \sqcap, 1)$ so that $x \to y$ in (1)–(7) are realized as $p(x) = y$ for some $p \in H_4$. Table 6.8 gives the operators in H_4.

Table 6.8 Operators on $\mathbf{S}_4$

Operator p	$p(ba)$	$p(bd)$	$p(c)$	$p(b)$	$p(b \sqcap c)$	$p(a)$
g	1	c (1)	c	c (3)	c (5)	a
f	ba	1	1	ba (2)	ba (4)	ba (7)
h	a (6)	bd	c	a	a	a

6.2.5 $(\mathbf{W}, \gamma_5)$

The equapartition for γ_5 has one nontrivial class $[c, 1]$. For representation we take an additional relation on **P** from Theorem 4.30:

[3] $c \leq b$,

reflecting the fact that $\gamma_5(c) = 1$ in **W**.

Now generate the free semilattice $\mathbf{S}_5$ defined by [1], [2], and [3]. Thus, $S_5 = \{b, ba, bd, c\}$ as shown in Fig. 6.12.

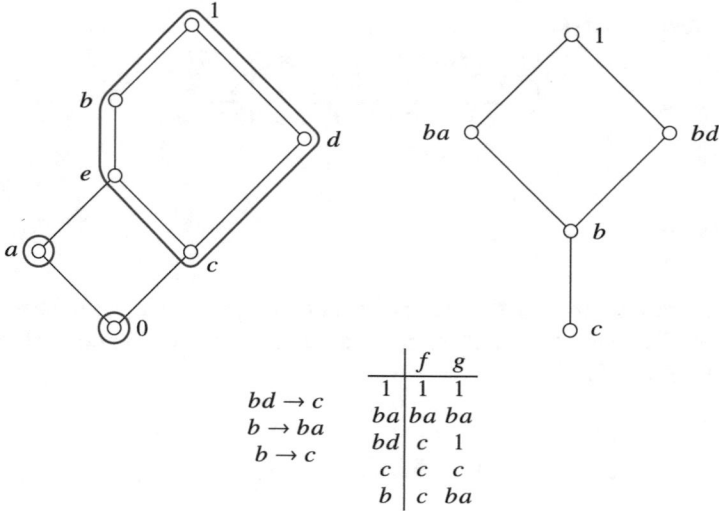

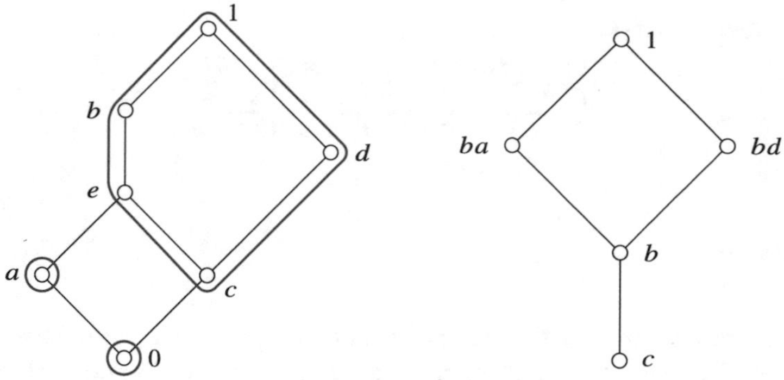

$bd \to c$
$b \to ba$
$b \to c$

	f	g
1	1	1
ba	ba	ba
bd	c	1
c	c	c
b	c	ba

Fig. 6.12 Calculations for $(\mathbf{W}, \gamma_5)$

The embedding $\xi : \mathbf{W} \to \mathrm{Sub}(\mathbf{S}_5, \sqcap, 1)$ is

$$
\begin{aligned}
a &\longmapsto \{ba, 1\} \\
c &\longmapsto \{c, 1\} \\
e &\longmapsto \{ba, c, 1\} \\
d &\longmapsto \{bd, c, 1\} \\
b &\longmapsto \{b, ba, c, 1\}
\end{aligned}
$$

along with $\xi(1) = S_5$, $\xi(0) = \{1\}$.

Table 6.9 gives the implications needed to represent **W** as $\mathrm{Sub}(\mathbf{S}_5, \sqcap, 1, H_5)$ with a potential set of operators H_5.

Table 6.9 $\mathbf{S}_5$ and $x \to y$

x	W_x	$x \to y$
bd	$\{bd, c, 1\}$	$bd \to c$ (1)
b	$\{b, ba, c, 1\}$	$b \to ba$ (2)
		$b \to c$ (3)

Now define a set of operators H_5 on $(\mathbf{S}_5, \sqcap, 1)$ so that $x \to y$ in (1)–(3) are realized as $p(x) = y$ for some $p \in H_5$. The operators in H_5 are given in Table 6.10.

Table 6.10 Operators on $\mathbf{S}_5$

Operator p	$p(ba)$	$p(bd)$	$p(c)$	$p(b)$
f	1	c (1)	c	c (3)
g	ba	1	c	ba (2)

6.2.6 $(\mathbf{W}, \gamma_6)$

The equapartition for γ_6 has two nontrivial classes $[b, 1]$ and $[a, e]$. Note that $c \leq \gamma_6(a)$ in $\mathbf{W}$. To realize this in the representation, we take an additional relation on $\mathbf{P}$ from Theorem 4.30:

[3] $ba \leq c$.

Note that [3] implies that $ba \sqcap c = ba$, and $b \leq c \sqcap bd$.

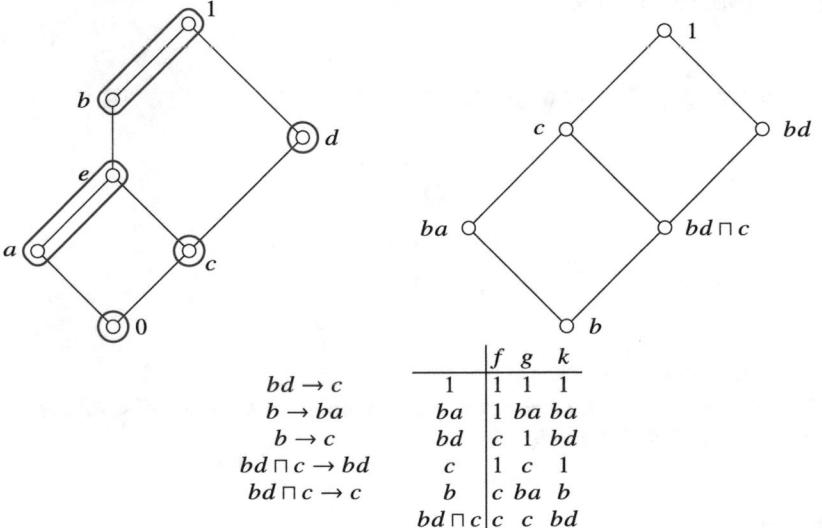

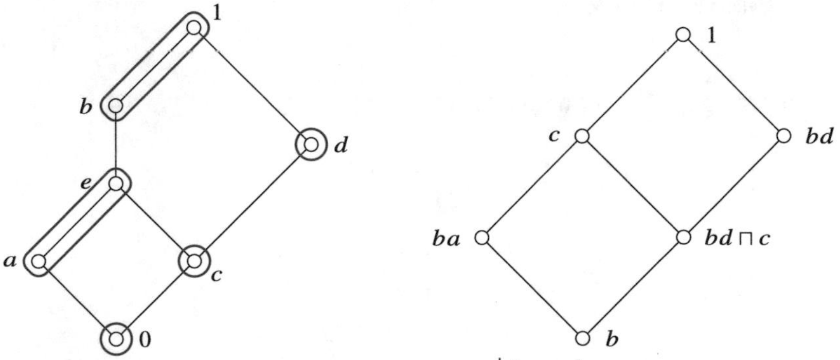

$bd \to c$
$b \to ba$
$b \to c$
$bd \sqcap c \to bd$
$bd \sqcap c \to c$

	f	g	k
1	1	1	1
ba	1	ba	ba
bd	c	1	bd
c	1	c	1
b	c	ba	b
$bd \sqcap c$	c	c	bd

Fig. 6.13 Calculations for $(\mathbf{W}, \gamma_6)$

Now generate the free semilattice $\mathbf{S}_6$ defined by [1], [2], and [3]. Thus, $S_6 = \{b, ba, bd, c, bd \sqcap c\}$ as in Fig. 6.13. The embedding $\xi : \mathbf{W} \to \mathrm{Sub}(\mathbf{S}_6, \sqcap, 1)$ is

$$
\begin{aligned}
a &\longmapsto \{ba, 1\}\\
c &\longmapsto \{c, 1\}\\
e &\longmapsto \{ba, c, 1\}\\
d &\longmapsto \{bd, c, bd \sqcap c, 1\}\\
b &\longmapsto \{b, ba, c, 1\}
\end{aligned}
$$

along with $\xi(1) = S_6$, $\xi(0) = \{1\}$.

The implications needed to represent **W** as Sub$(\mathbf{S}_6, \sqcap, 1, H_6)$ with a potential set of operators H_6 are in Table 6.11.

Table 6.11 $\mathbf{S}_6$ and $x \to y$

x	W_x	$x \to y$
bd	$\{bd, c, bd \sqcap c, 1\}$	$bd \to c$ (1)
b	$\{b, ba, c, 1\}$	$b \to ba$ (2)
		$b \to c$ (3)
$bd \sqcap c$	$\{bd, c, bd \sqcap c, 1\}$	$bd \sqcap c \to bd$ (4)
		$bd \sqcap c \to c$ (5)

Now define a set of operators H_6 on $(\mathbf{S}_6, \sqcap, 1)$ so that $x \longrightarrow y$ in (1)–(5) are realized as $p(x) = y$ for some $p \in H_6$. Table 6.12 gives the operators in H_6.

Table 6.12 Operators on $\mathbf{S}_6$

Operator p	$p(ba)$	$p(bd)$	$p(c)$	$p(b)$	$p(bd \sqcap c)$
f	1	c (1)	1	c (3)	c (5)
g	ba	1	c	ba (2)	c
k	ba	bd	1	b	bd (4)

6.2.7 $(\mathbf{W}, \gamma_1)$

The equapartition for γ_1 has one nontrivial class $[e, 1]$. For the representation we add an additional generator and take an additional relation on **P** from Theorem 4.30:

[3] Since $\gamma_1 c < \gamma_1 d$, we add the relation $bd \leq c$;
[4] Add an element a and relations $a < ba$ and $a \sqcap c \leq b$.

Compare relation [3] with $c \leq bd$ in the representation for $(\mathbf{W}, \gamma_2)$. There we had $\gamma_2 c = \gamma_2 d$, which forced the element c to be below bd in $\mathbf{S}_2$.

The relation $bd \leq c$ implies $b \leq ba \sqcap c$. We cannot force $b = ba \sqcap c$, because this would imply $b \leq a \vee c$ in **W**. Therefore, it will stay $b < ba \sqcap c$ in $\mathbf{S}_1$. On the other hand, $\gamma_1(e) = 1$ forces us to have a minimal element in $\mathbf{S}_1$ that is generated from the images of a and c in $\mathbf{W}$. So we are adding the element $a < ba$ and a relation $a \sqcap c < b$ that makes $a \sqcap c$ the smallest element of $\mathbf{S}_1$.

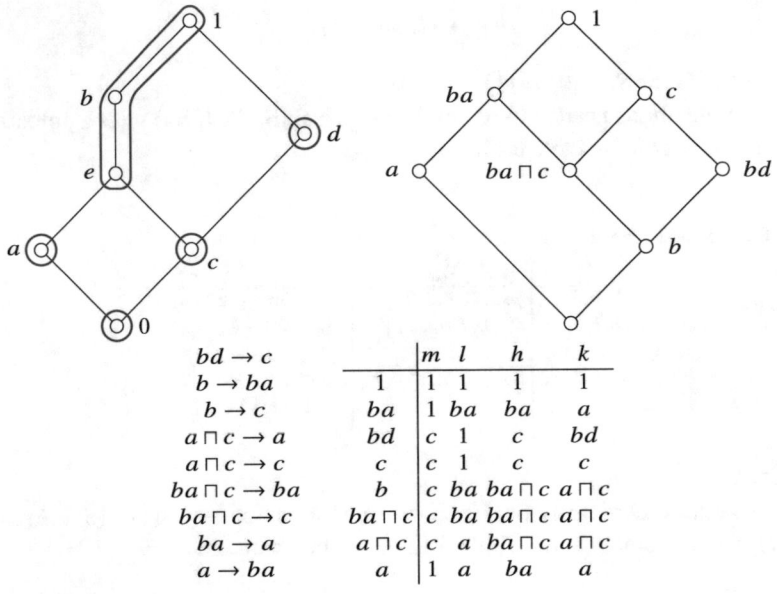

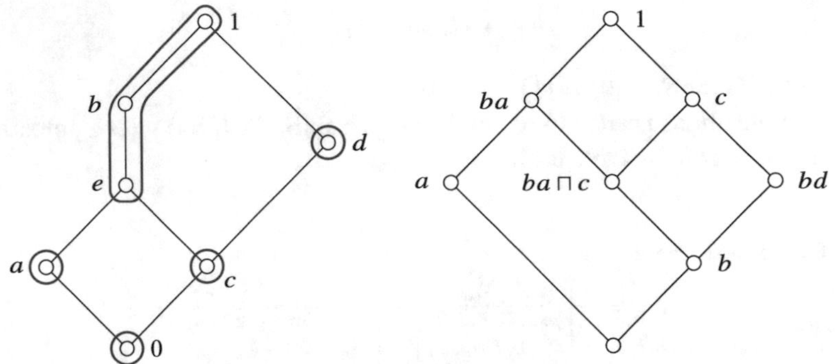

$$bd \to c$$
$$b \to ba$$
$$b \to c$$
$$a \sqcap c \to a$$
$$a \sqcap c \to c$$
$$ba \sqcap c \to ba$$
$$ba \sqcap c \to c$$
$$ba \to a$$
$$a \to ba$$

	m	l	h	k
1	1	1	1	1
ba	1	ba	ba	a
bd	c	1	c	bd
c	c	1	c	c
b	c	ba	$ba \sqcap c$	$a \sqcap c$
$ba \sqcap c$	c	ba	$ba \sqcap c$	$a \sqcap c$
$a \sqcap c$	c	a	$ba \sqcap c$	$a \sqcap c$
a	1	a	ba	a

Fig. 6.14 Calculations for $(\mathbf{W}, \gamma_1)$

Now generate the free semilattice $\mathbf{S}_1$ defined by [1], [2], [3], and [4]. Thus, $S_1 = \{a, b, ba, bd, c, ba \sqcap c, a \sqcap c\}$ as in Fig. 6.14. The embedding $\xi : \mathbf{W} \longrightarrow \mathrm{Sub}(\mathbf{S}_1, \sqcap, 1)$ is

$$a \longmapsto \{a, ba, 1\}$$
$$c \longmapsto \{c, 1\}$$
$$e \longmapsto \{a, ba, c, ba \sqcap c, a \sqcap c, 1\}$$
$$d \longmapsto \{bd, c, 1\}$$
$$b \longmapsto \{a, b, ba, c, ba \sqcap c, a \sqcap c, 1\}$$

along with $\xi(1) = S_1$, $\xi(0) = \{1\}$.

Table 6.13 gives the implications needed to represent $\mathbf{W}$ as $\mathrm{Sub}(\mathbf{S}_1, \sqcap, 1, H_1)$ with a potential set of operators H_1.

Operators on $(\mathbf{S}_1, \sqcap, 1)$ that realize all the implications (1)–(9) are given in Table 6.14.

Table 6.13 S_1 and $x \to y$

x	W_x	$x \to y$
bd	$\{bd, c, 1\}$	$bd \to c$ (1)
b	$\{a, b, ba, c, ba \sqcap c, a \sqcap c, 1\}$	$b \to ba$ (2)
		$b \to c$ (3)
$a \sqcap c$	$\{a, ba, c, ba \sqcap c, a \sqcap c, 1\}$	$a \sqcap c \to a$ (4)
		$a \sqcap c \to c$ (5)
$ba \sqcap c$	$\{a, ba, c, ba \sqcap c, a \sqcap c, 1\}$	$ba \sqcap c \to ba$ (6)
		$ba \sqcap c \to c$ (7)
ba	$\{a, ba, 1\}$	$ba \to a$ (8)
a	$\{a, ba, 1\}$	$a \to ba$ (9)

Table 6.14 Operators on S_1

Operator p	$p(ba)$	$p(bd)$	$p(c)$	$p(b)$	$p(ba \sqcap c)$	$p(a \sqcap c)$	$p(a)$
m	1	c (1)	c	c (3)	c (7)	c (5)	1
l	ba	1	1	ba (2)	ba (6)	a (4)	a
h	ba	c	c	$ba \sqcap c$	$ba \sqcap c$	$ba \sqcap c$	ba (9)
k	a (8)	bd	c	$a \sqcap c$	$a \sqcap c$	$a \sqcap c$	a

Chapter 7
The Six-Step Program: From $(\mathbf{L}, \gamma)$ to $(\mathbf{L}_\mathrm{q}(\mathcal{K}), \Gamma)$

There is one almost infallible way to find honest food at just prices in blue highways America: count the wall calendars in a cafe. ... I once found a six-calendar cafe in the Ozarks, which served fried chicken, peach pie, and chocolate malts, that left me searching for another ever since. I've never seen a seven-calendar place. But old-time travelers – road men in a day when cars had running boards and lunchroom windows said AIR COOLED *in blue letters with icicles dripping from the tops – those travelers have told me the golden legends of seven-calendar cafes.* – William Least Heat-Moon in *Blue Highways*

In this chapter, we concentrate on a method for representing pairs $(\mathbf{L}, \gamma)$, where $\mathbf{L}$ is a finite lower bounded lattice and γ an equaclosure operator on it, as $(\mathrm{L}_\mathrm{q}(\mathcal{K}), \Gamma)$ for some quasivariety $\mathcal{K}$ and its natural equaclosure operator Γ. The whole process is rather roundabout, and there are not one but two places where it may fail. On the other hand, it often succeeds, and it is the only general method we have for addressing the problem.

Before going into the details, let us outline the process.

(0) Given is a pair $(\mathbf{L}, \gamma)$ consisting of a finite lower bounded lattice and an equaclosure operator.

(1) Construct a finite meet semilattice $\widehat{\mathbf{S}}$ with operators h_i and greatest element $\hat{0}$ such that $\mathrm{Sub}(\widehat{\mathbf{S}}, \wedge, \hat{0}, h_1, h_2, \ldots) \cong \mathbf{L}$ with the natural equaclosure operator Γ on $\widehat{\mathbf{S}}$ corresponding to γ. Recall that $\mathrm{S}_\mathrm{p}(\mathbf{S}, H) = \mathrm{Sub}(\mathbf{S}, \wedge, 1, H)$ when $\mathbf{S}$ is finite.

(2) Dualize: $\mathrm{Sub}(\widehat{\mathbf{S}}, \wedge, \hat{0}, h_1, h_2, \ldots) \cong^d \mathrm{Con}(\widehat{\mathbf{S}}, +, \hat{z}, \eta_1, \eta_2, \ldots)$ by using adjoint maps.

(3) Do scratch work: find the congruence lattice of $\widehat{\mathbf{S}}$ explicitly, assign predicates to the elements of $\widehat{\mathbf{S}}$ as in [91], and find the laws of a quasivariety $\mathcal{K}_0$ in a language *without* equality that will represent $(\mathbf{L}, \gamma)$.

(4) Draw the lattice of implicational theories $\mathrm{ITh}(\mathcal{K}_0)$, which is isomorphic to $\mathrm{Con}(\widehat{\mathbf{S}}, +, \hat{z}, \eta_1, \eta_2, \ldots)$.

K. Adaricheva et al., *A Primer of Subquasivariety Lattices*, CMS/CAIMS Books in Mathematics 3, https://doi.org/10.1007/978-3-030-98088-7_7

(5) Redualize: $\mathrm{L_q}(\mathcal{K}_0) \cong^d \mathrm{ITh}(\mathcal{K}_0)$.
(6) Convert to a quasivariety $\mathcal{K}_1$ *with* equality by interpreting the least quasivariety as $\langle x \approx e\rangle$ for a constant e.

We keep five running examples in Sect. 7.1, and there are more examples throughout the remainder of the text.

Step 1, finding a representation of $(\mathbf{L}, \gamma)$ as $(\mathrm{Sub}(\widehat{\mathbf{S}}, \wedge, \hat{0}, H), \Gamma)$, was the topic of Chaps. 4–6, and it is fair to say that we still do not know exactly when or how this can be achieved. As shown in Theorem 5.6, if $\mathrm{J}(\mathbf{L}) \subseteq \tau(\mathbf{L})$, then we should take $\widehat{\mathbf{S}} = \gamma^d(\mathbf{L})$ and try to find operators that work, if possible, or why no such operators exist, if not. When $\mathrm{J}(\mathbf{L}) \not\subseteq \tau(\mathbf{L})$ we have the methods of Chap. 6, which are admittedly open-ended when it comes to how to add extra elements. We should add that when $\mathbf{L}$ is not lower bounded, there are infinite analogues of Theorem 4.30 that may work; these will be used in Chap. 8.

If step 1 succeeds, though, then steps 2–5 are quite routine, taking us from $\mathrm{Sub}(\widehat{\mathbf{S}}, \wedge, 1, H)$ to $\mathrm{L_q}(\mathcal{K}_0) \cong \mathbf{L}$ for a quasivariety in a language *without* equality and preserving γ, as in [91]. That paper is hard to read, but the process is straightforward enough and will be explained here and illustrated by the running examples.

Step 6, converting $\mathcal{K}_0$ to a quasivariety $\mathcal{K}_1$ in a language with equality, is more art than science. Sections 7.2 and 7.4 give two distinct sufficient conditions for a successful translation, denoted (ϖ) and (ß), while Sect. 7.5 discusses the serious difficulties that arise when neither of those conditions holds. Section 7.6 resolves those problems for the examples in Sect. 7.5, though not in general.

7.1 The Six-Step Program

Each step of the six-step program will be illustrated in the five Running Examples.

7.1.1 Step 1

Given $(\mathbf{L}, \gamma)$, construct a semilattice with operators such that $\mathrm{Sub}(\widehat{\mathbf{S}}, \wedge, \hat{0}, h_1, h_2, \ldots)$ is isomorphic to $\mathbf{L}$ with the natural equaclosure operator Γ corresponding to γ.

Running Example 1 The lattice $\mathbf{L}_1$ is $\mathbf{2}{\times}\mathbf{2}$ with $0 < z, b < 1$ and $\gamma(0) = 0$, $\gamma(b) = b$, and $\gamma(z) = \gamma(1) = 1$. See Fig. 7.1. Then $\mathrm{J}(\mathbf{L}_1) \subseteq \tau(\mathbf{L}_1)$, so we follow the prescription of Theorem 5.7. Thus we take $\hat{S} = \{\hat{0}, \hat{z}, \hat{b}\}$ ordered as a chain, $\hat{z} < \hat{b} < \hat{0}$. There are no $\rightarrow$ relations, so no operators are required, and indeed $\mathrm{Sub}(\hat{S}, \wedge, \hat{0}) \cong \mathbf{L}_1$. Recall that the natural equaclosure operator on Sub $\widehat{\mathbf{S}}$ has $\Gamma(A) = {\uparrow} a_0$, where a_0 is the least element of a subalgebra A. In this case, $\Gamma(\{\hat{0}, \hat{z}\}) = \hat{S}$, which mimics the action of γ on $\mathbf{L}_1$, where $\gamma(z) = 1_{\mathbf{L}_1}$.

Running Example 2 The lattice $\mathbf{L}_2$ is the chain $0 < z < p$ with $\gamma(0) = 0$ and $\gamma(z) = \gamma(p) = p$. See Fig. 7.2. In this case $\mathrm{J}(\mathbf{L}_2) \not\subseteq \tau(\mathbf{L}_2)$, since $p \in \mathrm{J}(\mathbf{L}_2) \setminus \tau(\mathbf{L}_2)$, and we follow the prescription of Sect. 6.1. Here $T = \{\hat{0}, \hat{z}\}$ and $J = \{\hat{0}, \hat{z}, p_1\}$. Arrows

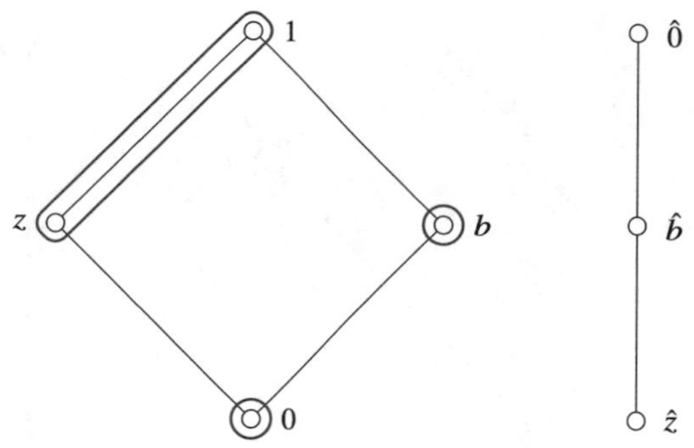

Fig. 7.1 Running Example 1

are introduced whenever join irreducible elements are comparable in **L**, giving the relation $p_1 \to \hat{z}$. This leads to $\widehat{\mathbf{S}} = \{\hat{0}, \hat{z}, p_1\}$ ordered as a chain $\hat{z} < p_1 < \hat{0}$. The map $h(p_1) = h(\hat{z}) = \hat{z}$, $h(\hat{0}) = \hat{0}$ enforces $p_1 \to \hat{z}$ for subalgebras, as it makes $\hat{z} \in \mathrm{Sg}(p_1)$. Thus we obtain $\mathrm{Sub}(\hat{S}, \wedge, \hat{0}, h) \cong \mathbf{L}_2$, again with the natural equaclosure operator corresponding to γ.

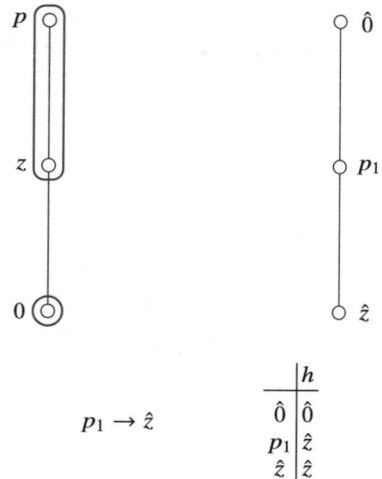

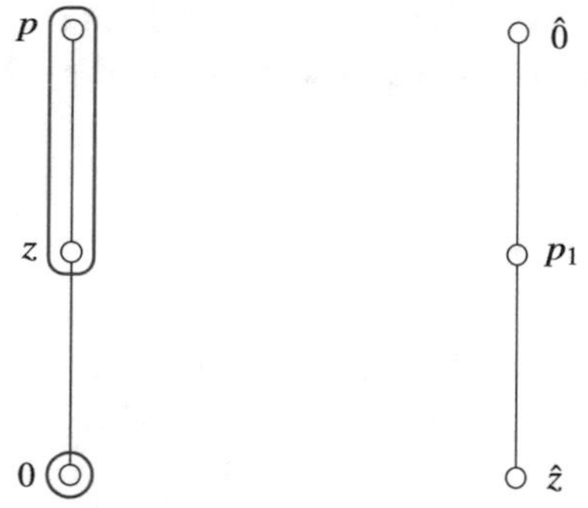

$p_1 \to \hat{z}$

	h
$\hat{0}$	$\hat{0}$
p_1	$\hat{z}$
$\hat{z}$	$\hat{z}$

Fig. 7.2 Running Example 2

The next two examples have $\mathrm{J}(\mathbf{L}) \subseteq \tau(\mathbf{L})$ and step 1 is easy, following the construction in Theorem 5.7.

Running Example 3 See Fig. 7.3. Since $\mathrm{J}(\mathbf{L}_3) \subseteq \tau(\mathbf{L}_3)$ and $\widehat{\mathbf{S}}$ is a chain, Corollary 5.9 applies, meaning that the map k_0 from the algorithm in Sect. 5.2 is the operator we are looking for.

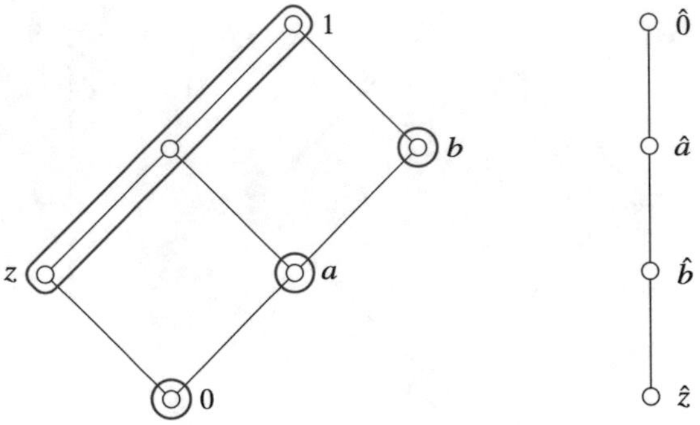

$\hat{b} \to \hat{a}$

	h
$\hat{0}$	$\hat{0}$
$\hat{a}$	$\hat{a}$
$\hat{b}$	$\hat{a}$
$\hat{z}$	$\hat{z}$

Fig. 7.3 Running Example 3

Running Example 4 See Fig. 7.4. The arrow relation $\hat{a} \,\&\, \hat{b} \to \hat{z}$ reflects the non-trivial join cover $z \leq a \vee b$ in $\mathbf{L}_4$. This is realized by $\hat{a} \wedge \hat{b} = \hat{z}$ in $\widehat{\mathbf{S}}$, so that $\hat{z} \in \mathrm{Sg}(\{\hat{a}, \hat{b}\})$.

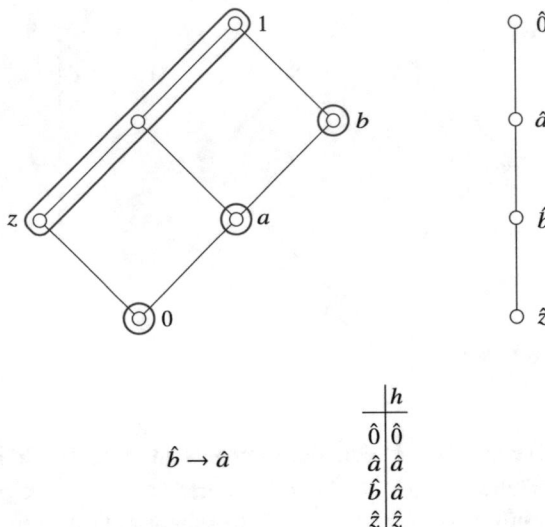

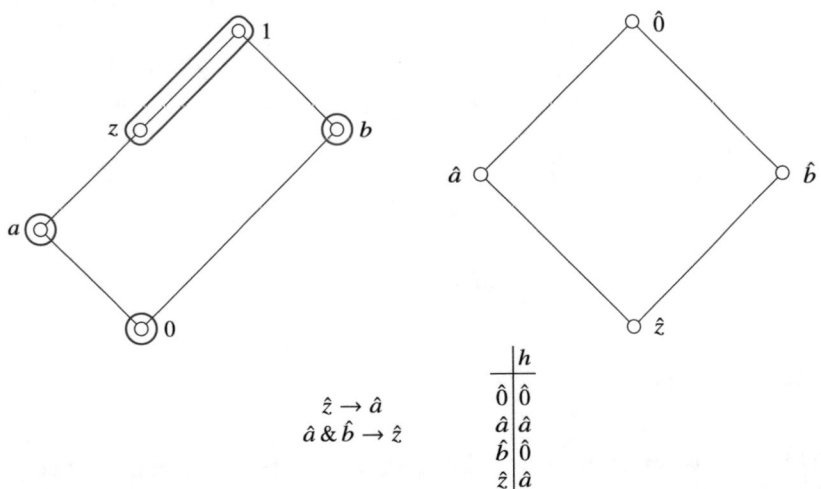

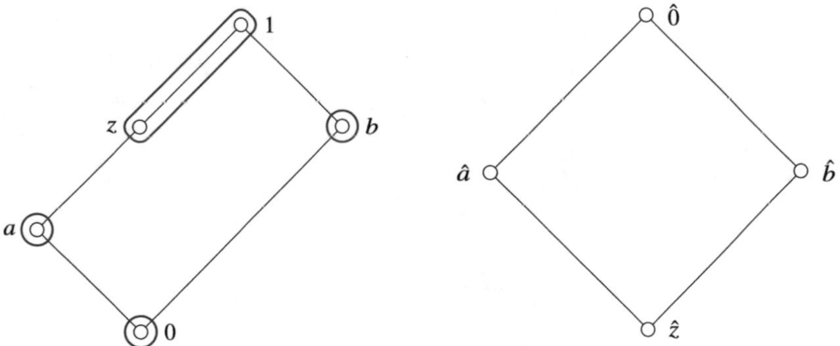

$\hat{z} \to \hat{a}$
$\hat{a} \,\&\, \hat{b} \to \hat{z}$

	h
$\hat{0}$	$\hat{0}$
$\hat{a}$	$\hat{a}$
$\hat{b}$	$\hat{0}$
$\hat{z}$	$\hat{a}$

Fig. 7.4 Running Example 4

Running Example 5 See Fig. 7.5. Now $\mathrm{J}(\mathbf{L}_5) \not\subseteq \tau(\mathbf{L}_5)$ and we must add an extra element z_1 with relations $\hat{z} \leftrightarrow z_1$. Step 1 for this lattice was worked out in Example 6.3; see Fig. 6.3.

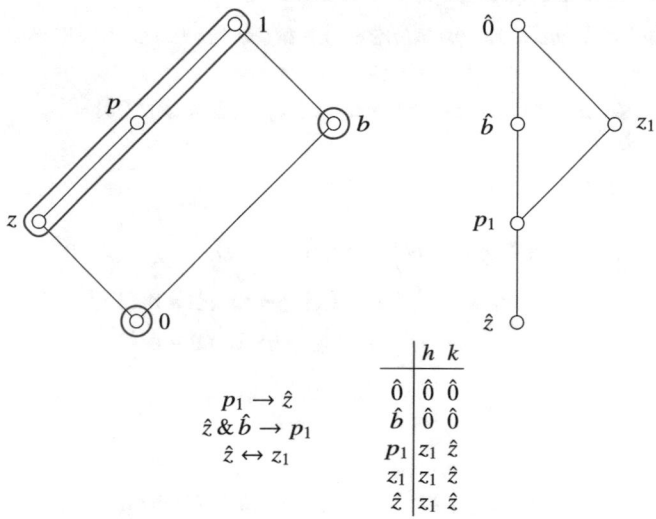

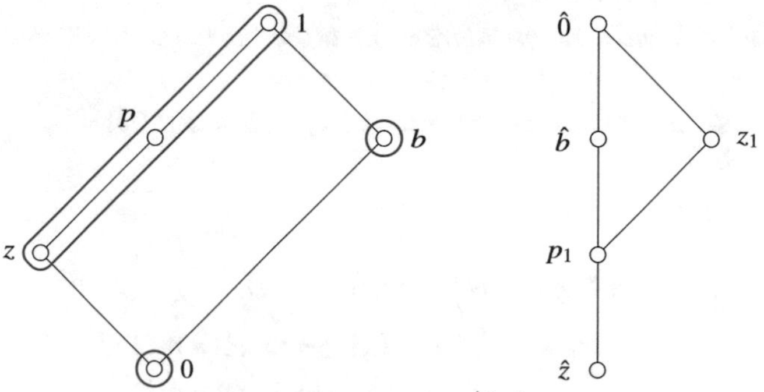

$p_1 \to \hat{z}$
$\hat{z} \,\&\, \hat{b} \to p_1$
$\hat{z} \leftrightarrow z_1$

	h	k
$\hat{0}$	$\hat{0}$	$\hat{0}$
$\hat{b}$	$\hat{0}$	$\hat{0}$
p_1	z_1	$\hat{z}$
z_1	z_1	$\hat{z}$
$\hat{z}$	z_1	$\hat{z}$

Fig. 7.5 Running Example 5

7.1.2 Step 2

Dualize by finding adjoint maps that make

$$\mathrm{Sub}(\widehat{\mathbf{S}}, \wedge, \hat{0}, h_1, h_2, \ldots) \cong^d \mathrm{Con}(\widehat{\mathbf{S}}, +, \hat{z}, \eta_1, \eta_2, \ldots).$$

Since $\widehat{\mathbf{S}}$ is a finite meet semilattice with largest element $\hat{0}$, it is a lattice with a join operation + and least element $\hat{z}$. Corresponding to each operator $h \in H$ there is an adjoint operator h' that preserves joins (Lemma 7.1). For the intermediate steps, we are going to pass from considering the meet semilattice $(\widehat{\mathbf{S}}, \wedge, \hat{0}, H)$ to the join semilattice $(\widehat{\mathbf{S}}, +, \hat{z}, H')$. The rationale, which can be formulated in the more general context of H-closed algebraic subsets, needs to be explained.

Let h be an operator on an algebraic lattice $\mathbf{S}$. Define the *adjoint map* $\eta = h' : S \to S$ by

$$\eta(s) = \bigwedge\{x \in S : h(x) \geq s\}$$

so that

$$h(x) \geq s \text{ iff } x \geq \eta(s).$$

Hidden in this calculation is the fact that, because h preserves meets, $h\eta(s) \geq s$. It avoids a lot of confusion to use the notation $h' = \eta$, $k' = \kappa$, etc. Now straightforward calculations prove the basic properties of adjoints.

Lemma 7.1 *Let h and k be operators on an algebraic lattice $\mathbf{S}$. Denote $h' = \eta$ and $k' = \kappa$.*

(1) *The adjoint maps preserve arbitrary joins, including $\eta(0_{\mathbf{S}}) = 0_{\mathbf{S}}$.*
(2) $(hk)' = k'h' = \kappa\eta$.

Proof If $U \subseteq S$, then

$$\begin{aligned} x \geq \eta\left(\bigvee U\right) &\text{ iff } h(x) \geq \bigvee U \\ &\text{ iff } h(x) \geq u \text{ for all } u \in U \\ &\text{ iff } x \geq \eta(u) \text{ for all } u \in U \\ &\text{ iff } x \geq \bigvee_{u \in U} \eta(u). \end{aligned}$$

Similarly,

$$hk(x) \geq s \text{ iff } k(x) \geq \eta(s) \text{ iff } x \geq \kappa\eta(s).$$

This brings us to the duality used in Step 2.

Theorem 7.2 *Let $\mathbf{S}$ be an algebraic lattice, with a monoid of operators H. Let H' denote the monoid of adjoints of operators in H. Then $\mathrm{S_p}(\mathbf{S}, H) \cong^d \mathrm{Con}(\mathbf{S}, \vee, 0, H')$.*

The theorem is the main result of Hyndman, Nation, and Nishida [64], growing out of earlier results by Fajtlowicz and J. Schmidt [42], Freese and Nation [48], and E. T. Schmidt [105, 106]. This transition allows us to employ in subsequent steps the connections between lattices of implicational theories and congruence lattices of semilattices with operators, as developed in Adaricheva and Nation [20, 91]. The connections will be explained as we go along.

For now, let us do the conversion from $\mathrm{Sub}(\widehat{\mathbf{S}}, \wedge, \hat{0}, H)$ to $\mathrm{Con}(\widehat{\mathbf{S}}, +, \hat{z}, H')$ for each of the running examples, using the formula

$$\eta(s) = \bigwedge \{x \in S : h(x) \geq s\}$$

for each $h \in H$ and $s \in S$.

Running Example 1 We still have $\widehat{\mathbf{S}}$ ordered as a chain $\hat{z} < \hat{b} < \hat{0}$ with no maps to adjust, only now regarded as a join semilattice with least element $\hat{z}$.

Running Example 2 Now $\widehat{\mathbf{S}}$ is the chain $\hat{z} < p_1 < \hat{0}$ regarded as a join semilattice. The adjoint $h' = \eta$ has $\eta(p_1) = \eta(\hat{0}) = \hat{0}$ and $\eta(\hat{z}) = \hat{z}$.

Running Example 3 The adjoint map for h is this.

	η
$\hat{0}$	$\hat{0}$
$\hat{a}$	$\hat{b}$
$\hat{b}$	$\hat{b}$
$\hat{z}$	$\hat{z}$

Running Example 4 When calculating adjoints it is convenient to note that $\eta(s) = \eta(t)$ if and only if $[s, \hat{0}] \cap \text{range}(h) = [t, \hat{0}] \cap \text{range}(h)$. Here this implies $\eta(\hat{0}) = \eta(\hat{b})$ and $\eta(\hat{a}) = \eta(\hat{z})$.

	η
$\hat{0}$	$\hat{b}$
$\hat{a}$	$\hat{z}$
$\hat{b}$	$\hat{b}$
$\hat{z}$	$\hat{z}$

Running Example 5 Now there are two adjoint maps. Note that $hk = h$ and $kh = k$, so $\kappa\eta = \eta$ and $\eta\kappa = \kappa$ as predicted by Lemma 7.1(2).

	η	κ
$\hat{0}$	$\hat{b}$	$\hat{b}$
$\hat{b}$	$\hat{b}$	$\hat{b}$
p_1	$\hat{z}$	$\hat{b}$
z_1	$\hat{z}$	$\hat{b}$
$\hat{z}$	$\hat{z}$	$\hat{z}$

7.1.3 Step 3

Do the scratch work for step 4: find the congruence lattice of $\widehat{\mathbf{S}}$ explicitly, assign predicates to the elements of $\widehat{\mathbf{S}}$ as in [91], and find the laws of a quasivariety $\mathcal{K}_0$ in a language *without* equality that will represent $(\mathbf{L}, \gamma)$.

The congruence lattice $\text{Con}(\widehat{\mathbf{S}}, +, \hat{z}, \eta_1, \eta_2, \dots)$ encodes the laws of the subquasivarieties of $\mathcal{K}_0$ corresponding to the various elements of the lattice L. The language of $\mathcal{K}_0$ includes a unary predicate P for each $p \in \widehat{\mathbf{S}} \setminus \{\hat{z}\}$. The map from Con $\widehat{\mathbf{S}}$ to $\text{ITh}(\mathcal{K}_0)$ is such that

- A principal congruence $\text{con}(\hat{z}, \hat{a})$ maps to the theory $\langle Ax \rangle$.
- A principal congruence $\text{con}(\hat{b}, \hat{c})$ with $\hat{z} < \hat{b} < \hat{c}$ maps to the theory $\langle Bx \to Cx \rangle$.

The correspondence between congruences and theories is discussed in more detail after Theorem 7.3, when we have the more general assignment of predicates (such as $P(\eta x)$) at our disposal. To actually find the congruence lattice of a finite semilattice with operators is straightforward.

Assignment The "assigning of predicates" is done as follows. The language will contain a constant e, a monoid of operations η from step 2, and for each element

$p \in \widehat{\mathbf{S}}$ with $p \neq \hat{z}$ a unary predicate P, where $z = \tau(1)$. The element p could be some $\hat{a} \in \gamma^d(\mathbf{L})$, or some $s_1 \notin \gamma^d(\mathbf{L})$. Assign the predicate $P(x)$ to the element p. Whenever $p \neq \hat{z}$ and $\eta p = q$, assign $P(\eta x)$ to q; here $q = \hat{z}$ is a possibility. Thus the assignment associates with each p the predicate $P(x)$ and a collection of predicates of the form $R(\kappa x)$, for those R, κ such that $\kappa r = p$. We denote such an assignment by $p \rightsquigarrow P(x), R(\kappa x), \ldots$.

Laws The laws of $\mathcal{K}_0$ are as follows.

(1) $X(e)$ and $X(\kappa e)$ for every predicate X and operation κ.
(2) $P(x) \leftrightarrow R(\kappa x)$ whenever $p \rightsquigarrow R(\kappa x)$.
(3) $P(\eta x)$ for all predicates assigned to the least element $\hat{z}$.
(4) $P(x) \rightarrow Q(x)$ whenever $p \geq q > \hat{z}$ in $\widehat{\mathbf{S}}$.
(5) $\& P_i(x) \rightarrow Q(x)$ whenever $\bigvee p_i \geq q > \hat{z}$ in $\widehat{\mathbf{S}}$.
(6) $P(\eta(\kappa x)) \leftrightarrow P(\lambda x)$ whenever $\kappa\eta = \lambda$.

Note that (2) makes $Q(\eta x) \leftrightarrow R(\kappa x)$ for any pair assigned to the same element p.

Because we use this so often, a schematic diagram of assignment is given in Fig. 7.6 as a reference.

	η
$\hat{0}$	a
...	...
p	q
...	...
r	$\hat{z}$

$$a \rightsquigarrow A(x), O(\eta x)$$
...
$$q \rightsquigarrow Q(x), P(\eta x)$$
...
$$z \rightsquigarrow R(\eta x)$$

Fig. 7.6 Scheme of assigning predicates to elements. In the laws of $\mathcal{K}$, the rows become $A(x) \leftrightarrow O(\eta x)$, $Q(x) \leftrightarrow P(\eta x)$, and $R(\eta x)$

We should be explicit about (6), which can be confusing. There are two sets of maps, the original $H = \{h, k, \ldots\}$ from step 1 and the adjoints $H' = \{\eta, \kappa, \ldots\}$ from step 2. The *opposite groupoid* $\mathbf{T}^{\mathrm{opp}} = \langle T, \star\rangle$ of a groupoid $\mathbf{T} = \langle T, \circ\rangle$ has the order of operations reversed: $x \star y = y \circ x$. In our case, we apply this to H', so that $\eta \star \kappa = \kappa \circ \eta = (hk)'$. Thus the order of operations on H and H'^{opp} are the same, and the reverse of that on H'. The laws in (6) use the set of operations H' with the composition $\star$, so that

(6)′ $P(\eta(\kappa x)) \leftrightarrow P(\lambda x)$ whenever $\eta \star \kappa = \lambda$.

Running Example 5 provides a nontrivial example of (6)′.

It is useful to categorize the laws of $\mathcal{K}_0$ into 4 types:

- laws saying that e satisfies every possible relation, type (1) above;
- laws derived from the assignments, types (2) and (3) above;

- laws reflecting the order on $\widehat{\mathbf{S}}$, types (4) and (5) above;
- laws reflecting the operations on H'^{opp}, type (6) above.

Now let us apply step 3 to the running examples.

Running Example 1 The congruence lattice of the chain $\hat{z} < \hat{b} < \hat{0}$ is isomorphic to $\mathbf{2} \times \mathbf{2}$ with atoms $\text{con}(\hat{b}, \hat{0})$ and $\text{con}(\hat{z}, \hat{b})$. Assign predicates O, B to $\widehat{\mathbf{S}}$ *via* $\hat{0} \rightsquigarrow O(x)$, $\hat{b} \rightsquigarrow B(x)$, $\hat{z} \rightsquigarrow \varnothing$ (no assignment in this case).

This assignment corresponds to a quasivariety $\mathcal{K}_0$ with a constant e and predicates O, B satisfying the laws:

$$O(e) \qquad B(e) \qquad O(x) \to B(x).$$

Running Example 2 We have the chain $\hat{z} < \hat{b} < \hat{0}$ but now also an operator with $\eta(p_1) = \eta(\hat{0}) = \hat{0}$ and $\eta(\hat{z}) = \hat{z}$. Note $\eta^2 = \eta$. Thus $\text{con}(\hat{z}, p_1) = \nabla$, and the congruence lattice of $\widehat{\mathbf{S}}$ is a 3-element chain: $\Delta < \text{con}(p_1, \hat{0}) < \nabla$.

The predicates corresponding to $\widehat{\mathbf{S}} \setminus \{\hat{z}\}$ are O and P, which we assign as follows:

$$\begin{aligned} \hat{0} &\rightsquigarrow O(x), O(\eta x), P(\eta x) \\ \hat{p} &\rightsquigarrow P(x) \\ \hat{z} &\rightsquigarrow \varnothing. \end{aligned}$$

Thus $\mathcal{K}_0$ has the operations e, η and predicates O, B with laws

$$\begin{gathered} O(e) \qquad O(\eta e) \qquad P(e) \qquad P(\eta e) \\ O(\eta^2 x) \leftrightarrow O(\eta x) \qquad P(\eta^2 x) \leftrightarrow P(\eta x) \\ O(x) \leftrightarrow O(\eta x) \leftrightarrow P(\eta x) \qquad O(x) \to P(x). \end{gathered}$$

The first two lines are standard setup, while the third reflects the assignments and $\hat{0} > \hat{p}$.

Running Example 3 The congruences of $\widehat{\mathbf{S}}$ are Δ, $\text{con}(\hat{b}, \hat{a}) = [\hat{a}, \hat{b}]$, $\text{con}(\hat{a}, \hat{0}) = \text{con}(\hat{b}, \hat{0}) = [0, \hat{a}, \hat{b}]$, $\text{con}(\hat{z}, \hat{b}) = [\hat{z}, \hat{b}]$, $\text{con}(\hat{z}, \hat{a}) = [\hat{a}, \hat{b}, \hat{z}]$, ∇, as illustrated in Fig. 7.7. The predicates corresponding to $\widehat{\mathbf{S}} \setminus \{\hat{z}\}$ are O, A, and B, which are assigned as follows:

$$\begin{aligned} \hat{0} &\rightsquigarrow O(x), O(\eta x) \\ \hat{a} &\rightsquigarrow A(x) \\ \hat{b} &\rightsquigarrow B(x), B(\eta x), A(\eta x) \\ \hat{z} &\rightsquigarrow \varnothing. \end{aligned}$$

Thus $\mathcal{K}_0$ has operations e, η and predicates O, A, B with laws reflecting $\hat{0} > \hat{a} > \hat{b}$:

$$\begin{gathered} X(e) \qquad X(\eta e) \qquad \text{for } X = O, A, B \\ X(\eta^2 x) \leftrightarrow X(\eta x) \qquad \text{for } X = O, A, B \\ O(x) \leftrightarrow O(\eta x) \to A(x) \to B(x) \leftrightarrow B(\eta x) \leftrightarrow A(\eta x). \end{gathered}$$

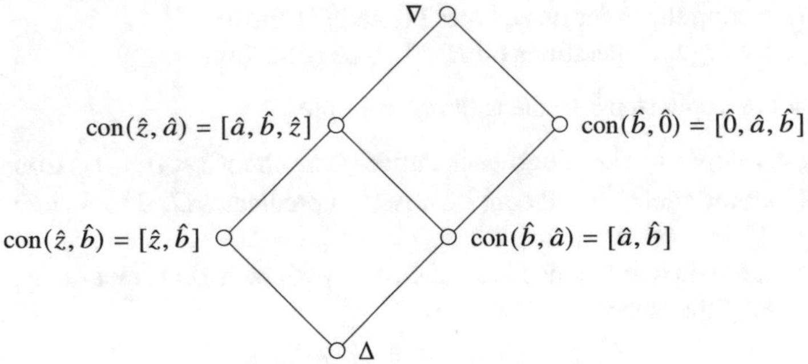

Fig. 7.7 Con $\widehat{\mathbf{S}}$ for Running Example 3. Compare Fig. 7.10

Running Example 4 The congruences of $\widehat{\mathbf{S}}$ are Δ, $\mathrm{con}(\hat{b}, \hat{0}) = [\hat{b}, 0]$, $\mathrm{con}(\hat{z}, \hat{a}) = [\hat{z}, \hat{a}][\hat{b}, 0]$, $\mathrm{con}(\hat{z}, \hat{b}) = [\hat{z}, \hat{b}][\hat{a}, \hat{0}]$, ∇, the pentagon illustrated in Fig. 7.8a. The predicates corresponding to $\widehat{\mathbf{S}} \setminus \{\hat{z}\}$ are O, A, and B, assigned as follows:

$$\begin{aligned}
\hat{0} &\rightsquigarrow O(x)\\
\hat{a} &\rightsquigarrow A(x)\\
\hat{b} &\rightsquigarrow B(x), B(\eta x), O(\eta x)\\
\hat{z} &\rightsquigarrow A(\eta x).
\end{aligned}$$

Thus $\mathcal{K}_0$ has the operations e, η and predicates O, A, B with laws

$$\begin{aligned}
X(e) \qquad X(\eta e) &\qquad \text{for } X = O, A, B\\
X(\eta^2 x) \leftrightarrow X(\eta x) &\qquad \text{for } X = O, A, B\\
O(x) \to A(x) &\qquad O(x) \to B(x) \leftrightarrow B(\eta x) \leftrightarrow O(\eta x)\\
A(x)\ \&\ B(x) \to O(x) &\qquad A(\eta x)
\end{aligned}$$

The last law is due to the assignment of $\hat{z}$.

Running Example 5 The congruences of $\widehat{\mathbf{S}}$ are Δ, $\mathrm{con}(\hat{b}, \hat{0}) = [\hat{b}, \hat{0}]$, $\mathrm{con}(p_1, z_1) = [p_1, z_1][\hat{b}, \hat{0}]$, $\mathrm{con}(\hat{z}, \hat{b}) = [\hat{z}, \hat{b}][z_1, \hat{0}]$, ∇, the pentagon illustrated in Fig. 7.8b. The predicates corresponding to $\widehat{\mathbf{S}} \setminus \{\hat{z}\}$ are O, P, Z_1, and B, assigned as follows:

$$\begin{aligned}
\hat{0} &\rightsquigarrow O(x)\\
\hat{b} &\rightsquigarrow B(x), B(\eta x), B(\kappa x), O(\eta x), O(\kappa x), Z_1(\kappa x), P(\kappa x)\\
\hat{z}_1 &\rightsquigarrow Z_1(x)\\
\hat{p} &\rightsquigarrow P(x)\\
\hat{z} &\rightsquigarrow Z_1(\eta x), P(\eta x).
\end{aligned}$$

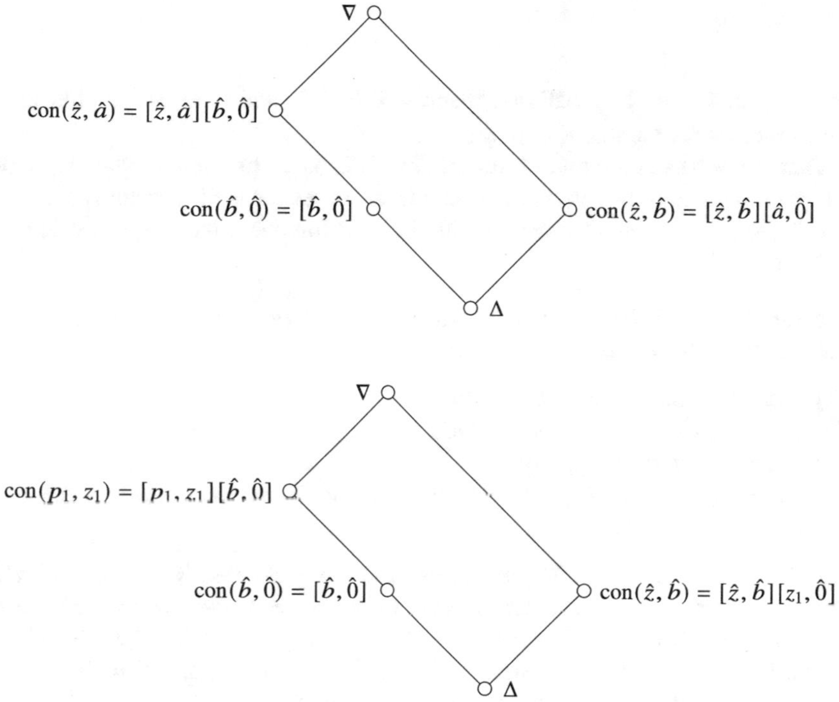

Fig. 7.8 Con $\widehat{\mathbf{S}}$ for Running Examples 4 and 5. Compare Fig. 7.11

Thus $\mathcal{K}_0$ has the operations e, η, κ and predicates O, B, P, Z_1 with the laws

$$
\begin{gathered}
B(x) \leftrightarrow B(\eta x) \leftrightarrow B(\kappa x) \leftrightarrow O(\eta x) \leftrightarrow O(\kappa x) \leftrightarrow Z_1(\kappa x) \leftrightarrow P(\kappa x)\\
X(e) \qquad X(fe) \qquad \text{for } X = O, B, P, Z_1 \text{ and } f = \eta, \kappa\\
X(f^2 x) \leftrightarrow X(fx) \qquad \text{for } X = O, B, P, Z_1 \text{ and } f = \eta, \kappa\\
O(x) \to B(x) \to P(x) \qquad O(x) \to Z_1(x) \to P(x)\\
B(x) \ \& \ Z_1(x) \to O(x)\\
Z_1(\eta x) \qquad P(\eta x).
\end{gathered}
$$

The last laws are due to the assignment of $\hat{z}$. But actually they are not the last laws, because we must deal with composition of operators. Note that $\eta\kappa = \kappa$ and $\kappa\eta = \eta$. As discussed above, the corresponding laws use the *opposite* composition:

$$
\begin{aligned}
X(\eta(\kappa(y))) &\leftrightarrow X((\eta \star \kappa)(y) = X((\kappa\eta)(y) = X(\eta(y))\\
X(\kappa(\eta(y))) &\leftrightarrow X((\kappa \star \eta)(y) = X((\eta\kappa)(y) = X(\kappa(y))
\end{aligned}
$$

for $X = O, B, P, Z_1$.

7.1.4 Step 4

Draw the lattice of implicational theories $\mathrm{ITh}(\mathcal{K}_0)$, which is isomorphic to the congruence lattice $\mathrm{Con}(\widehat{\mathbf{S}}, +, \hat{z}, \eta_1, \eta_2, \ldots)$.

That this works is the main result of [91]. It is based on the following reduction to quasi-identities in one variable, which in turn uses ideas of Gorbunov [52], also used in [20]. A modified version, adding information about the top predicate $O(x)$, is Theorem 7.8 below.

Theorem 7.3 *Let $\mathcal{B}$ be an implicational theory in a language $\mathcal{L}$ without equality, and with the following restrictions and laws.*

(1) *$\mathcal{L}$ has only unary predicate symbols.*
(2) *$\mathcal{L}$ has only unary function symbols.*
(3) *$\mathcal{L}$ has one constant symbol e.*
(4) *$\mathcal{B}$ contains the laws $P(f(e))$ for every predicate P and every formal composition f of functions of $\mathcal{L}$.*

Then every implication holding in a theory extending the theory of $\mathcal{B}$ is equivalent (modulo the laws of $\mathcal{B}$) to a set of implications in only one variable. Hence the lattice of theories of $\mathcal{B}$ is isomorphic to Con $\mathbf{S}$ *where* $\mathbf{S} = \langle \mathbf{T}, \vee, 0, \widehat{\mathcal{E}} \rangle$ *with* $\mathbf{T}$ *the semilattice of compact congruences of* $\mathrm{Con}_{\mathcal{B}}(\mathbf{F})$, $\mathcal{E} = \mathrm{End}\ \mathbf{F}$, *and* $\mathbf{F} = \mathbf{F}_{\mathcal{B}}(1)$.

Moreover, modulo the quasi-equations of $\mathcal{B}$, every quasi-equation in $\mathcal{L}$ is equivalent to a quasi-equation of the form

$$\mathop{\&}_{1 \le i \le k} P_i(f_i(x)) \to Q(g(x))$$

with P_i, Q predicate symbols, f_i, g operation symbols, and possibly $f_i, g = id$.

For a quasi-equation β, we write $\langle \beta \rangle$ to denote the subquasivariety of all structures satisfying β, within the quasivariety under discussion. Examples would be $\langle x \approx e \rangle$ or $\langle Ax \to Bx \rangle$. It should cause no confusion to use the same notation $\langle \mathbf{T} \rangle$ to denote the subquasivariety generated by a structure $\mathbf{T}$. In figures, the brackets may be omitted.

Recall that the language of $\mathcal{K}_0$ includes a predicate $P(x)$ for each $p \in \widehat{S} \setminus \{\hat{z}\}$. The laws of $\mathcal{K}_0$ relate these to each other and, *via* the assignment, to compound predicates such as $Q(\eta x)$. The isomorphism from Con $\widehat{\mathbf{S}}$ to $\mathrm{ITh}(\mathcal{K}_0)$ is determined by its values on the principal congruences, as follows.

- A principal congruence $\mathrm{con}(\hat{z}, \hat{a})$ maps to the theory $\langle Ax \rangle$.
- A principal congruence $\mathrm{con}(\hat{b}, \hat{c})$ with $\hat{z} < \hat{b} < \hat{c}$ maps to the theory $\langle Bx \to Cx \rangle$. Note that $Cx \to Bx$ is a law of $\mathcal{K}_0$.

The top and bottom of the lattices correspond naturally:

- ∇ maps to $\langle Ox \rangle$,
- Δ maps to $\mathcal{K}_0$.

The method will be illustrated in the Running Examples, e.g., as in Figs. 7.7 and 7.10, or Figs. 7.8 and 7.11.

In this subsection we just describe the lattices of theories verbally, as we will be drawing their duals, the lattices of subquasivarieties $\mathrm{L_q}(\mathcal{K}_0)$, in step 5.

Running Example 1 The lattice of theories $\mathrm{ITh}(\mathcal{K}_0)$ is isomorphic to $\mathbf{2} \times \mathbf{2}$ with

(1) the theory of $\mathcal{K}_0$ (as given in step 3) on the bottom,
(2) $\langle B(x) \rangle$ on the one side (corresponding to $\mathrm{con}(\hat{z}, \hat{b})$),
(3) $\langle B(x) \to O(x) \rangle$ on the other side (corresponding to $\mathrm{con}(\hat{b}, \hat{0})$),
(4) $\langle O(x) \rangle$ (which implies everything else) on the top.

Running Example 2 The lattice of theories $\mathrm{ITh}(\mathcal{K}_0)$ is a 3-element chain with

(1) the theory of $\mathcal{K}_0$ (as given in step 3) on the bottom,
(2) $\langle P(x) \to O(x) \rangle$ in the middle (corresponding to $\mathrm{con}(p_1, \hat{0})$),
(3) $\langle O(x) \rangle$ (which implies everything else) on the top.

Running Example 3 The lattice of theories $\mathrm{ITh}(\mathcal{K}_0)$ is $\mathbf{2} \times \mathbf{3}$ with join irreducible theories $\langle B(x) \to A(x) \rangle$, $\langle A(x) \to O(x) \rangle = \langle B(x) \to O(x) \rangle$, and $\langle B(x) \rangle$. Note $\langle A(x) \rangle = \langle B(x) \to A(x) \rangle \vee \langle B(x) \rangle$.

Running Example 4 The lattice of theories $\mathrm{ITh}(\mathcal{K}_0)$ is a pentagon with join irreducible theories $\langle A(x) \rangle$, $\langle B(x) \rangle$, and $\langle B(x) \to O(x) \rangle$.

Running Example 5 The lattice of theories $\mathrm{ITh}(\mathcal{K}_0)$ is a pentagon with join irreducible theories $\langle B(x) \rangle$, $\langle B(x) \to O(x) \rangle$, and $\langle P(x) \to Z_1(x) \rangle$.

7.1.5 Step 5

Redualize $\mathrm{L_q}(\mathcal{K}_0) \cong^d \mathrm{ITh}(\mathcal{K}_0)$.

At this point, if we were successful in finding a semilattice with operators for step 1, the construction guarantees that $(\mathrm{L_q}(\mathcal{K}_0), \Gamma)$ represents the original pair $(\mathbf{L}, \gamma)$, i.e., that $\mathrm{L_q}(\mathcal{K}_0) \cong \mathbf{L}$ and that the natural equaclosure operator Γ mimics the action of γ. (Converting to a quasivariety $\mathcal{K}_1$ with equality in step 6 is where we encounter problems.) Step 5 also provides an opportunity to check that the calculations in the previous steps have been done correctly; it is possible to make a mistake.

Recall that for a subquasivariety $Q \leq \mathcal{K}$, the subquasivariety $\Gamma(Q) = \mathcal{K} \cap \mathbb{HSP}(Q)$ is the least relative subvariety of $\mathcal{K}$ above Q. Thus $\Gamma(Q)$ is determined, within $\mathcal{K}$, by atomic formulas rather than proper implications.

Running Example 1 The lattice of quasivarieties (models) is just dual to the theories:

(1) $\mathcal{K}_0$ (as given in step 3) on the top,
(2) $\langle B(x) \rangle$ on the one side,
(3) $\langle B(x) \to O(x) \rangle$ on the other side,
(4) $\langle O(x) \rangle$ (the trivial quasivariety) on the bottom.

See Fig. 7.9a. Notice that the natural equaclosure operator has $\Gamma\langle B(x) \to O(x)\rangle = \mathcal{K}_0$, as desired!

Running Example 2 Dually the lattice of quasivarieties has

(1) $\mathcal{K}_0$ on the top,
(2) $\langle P(x) \to O(x)\rangle$ in the middle,
(3) $\langle O(x)\rangle$ on the bottom.

See Fig. 7.9b. Again, $\Gamma\langle P(x) \to O(x)\rangle = \mathcal{K}_0$, as desired!

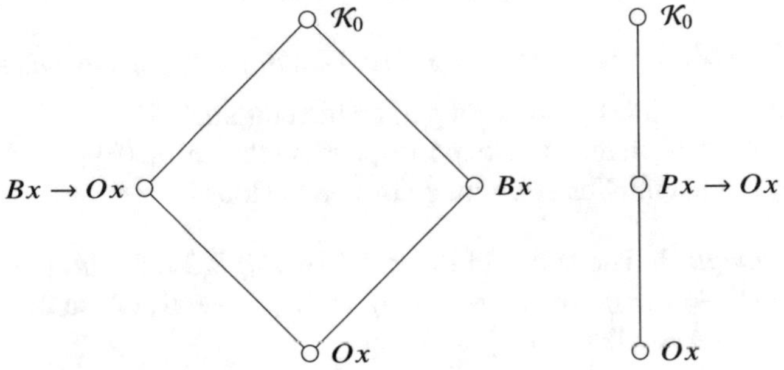

Fig. 7.9 $L_q(\mathcal{K}_0)$ for Running Examples 1 and 2

Running Example 3 The lattice of quasivarieties is dual to the theories in step 4. See Fig. 7.10. Note $\Gamma\langle B(x) \to O(x)\rangle = \mathcal{K}_0$, as desired!

Running Example 4 The lattice of quasivarieties is dual to the theories in step 4. See Fig. 7.11a. Note $\Gamma\langle B(x) \to O(x)\rangle = \mathcal{K}_0$, as desired! (Incidentally, $\langle B(x) \to O(x)\rangle = \langle B(x) \to A(x)\rangle$.)

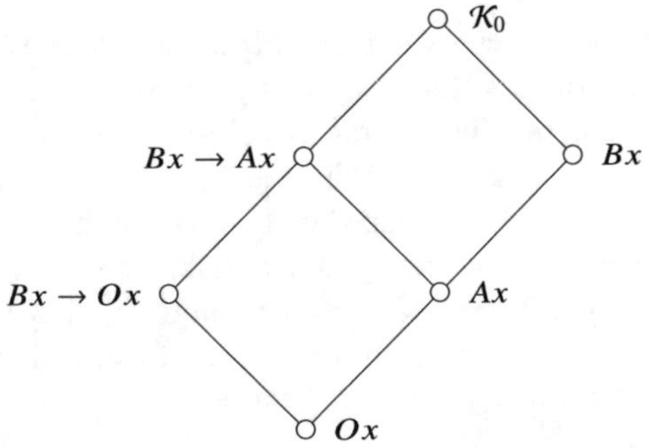

Fig. 7.10 $L_q(\mathcal{K}_0)$ for Running Example 3

Running Example 5 The lattice of quasivarieties is again dual to the theories in step 4. See Fig. 7.11b. Note $\Gamma\langle P(x) \to Z_1(x)\rangle = \mathcal{K}_0$, as desired!

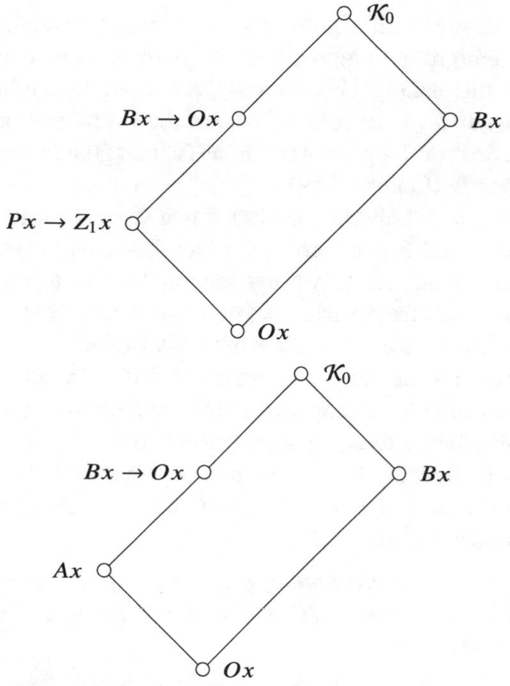

Fig. 7.11 $\mathrm{L_q}(\mathcal{K}_0)$ for Running Examples 4 and 5

7.1.6 Step 6

Convert $\mathcal{K}_0$ to a quasivariety $\mathcal{K}_1$ *with* equality by interpreting the least quasivariety $\langle Ox \rangle$ as $\langle x \approx e \rangle$ for a constant e. If $\langle x \approx e \rangle$ is to be the least subquasivariety, then the laws

$$\eta e \approx e \qquad\qquad P(e)$$

should hold in $\mathcal{K}_1$ for all functions η and predicates P in the language, reflecting the laws $Ox \to O(\eta x)$ and $Ox \to Px$ of $\mathcal{K}_0$.

We use three distinct ways to do the conversion (which, of course, does not exhaust all possible representations). Informally, we call these *longstyle*, *shortstyle*, and *mediumstyle* representations.

The longstyle method is described in detail in Sect. 7.2. It applies whenever $(\widehat{\mathbf{S}}, H)$ satisfies the condition (ϖ) that $h(\hat{x}) = \hat{0}$ implies $\hat{x} = \hat{0}$, in which case it always yields a representation $(\mathrm{L_q}(\mathcal{K}_1), \Gamma)$ with equality; see Theorem 7.7. The structures in a longstyle quasivariety, except the 1-element structure $\{e\}$, are absolutely free as algebras, differing in their unary predicates. Example 7.6 gives a prototypical longstyle quasivariety. We will use the longstyle method in Running Examples 2 and 3, Sect. 7.3, and especially in Chap. 8.

By contrast, if we start with a finite lattice **L**, a shortstyle representation is locally finite. We let the semigroup of adjoints H'^{opp} act on a set and interpret the axioms of $\mathcal{K}_0$ in terms of this action. This always gives us an embedding of $\mathrm{L_q}(\mathcal{K}_0)$ into $\mathrm{L_q}(\mathcal{K}_1)$, which *sometimes* is surjective. We use the shortstyle method for Running Examples 4 and 5. Section 7.4 gives a condition (ß) that guarantees that the shortstyle method works properly (Lemma 7.19).

But that certainly is not always the case: when (ß) fails, it may happen that not every subquasivariety of $\mathcal{K}_1$ is a translation of a subquasivariety of $\mathcal{K}_0$. In Sect. 7.5, we will analyze how things can go wrong. Section 7.6 uses a *mediumstyle* approach to represent the examples from Sect. 7.5. For better or worse, though, we can devise other examples where none of these three methods works.

Recall that a finite structure **S** is *quasicritical* if it is not in the quasivariety generated by its proper substructures. Quasicritical structures play a crucial role in locally finite quasivarieties. Every subquasivariety lattice $\mathrm{L_q}(\mathcal{K})$ is dually algebraic, whence every quasivariety is a join of completely join irreducible subquasivarieties. In the locally finite case, the completely join irreducible subquasivarieties are those generated by a quasicritical structure.

Theorem 7.4 *Let $\mathcal{K}$ be a locally finite quasivariety of finite type. A subquasivariety is completely join irreducible in* $\mathrm{L_q}(\mathcal{K})$ *if and only if it is generated by a single finite quasicritical structure.*

For a proof and discussion of this classical result, see Section 2.1 of [63]. Its extension to non-locally finite quasivarieties is the topic of Appendix A.3.

Running Example 1 From step 3, the laws of $\mathcal{K}_0$ for this example were $O(e)$, $B(e)$, and $O(x) \longrightarrow B(x)$. These laws convert directly to a quasivariety with equality using the translation $O(x) \mapsto x \approx e$. Thus, after removing redundancies, the new quasivariety $\mathcal{K}_1$ has a constant e, a unary predicate B and $\approx$, and the single law $B(e)$. There are two quasicritical structures in $\mathcal{K}_1$:

- **S** has universe $\{e, 0\}$ with $B^{\mathbf{S}} = \{e\}$.
- **T** has universe $\{e, 0\}$ with $B^{\mathbf{T}} = \{e, 0\}$.

The lattice $\mathrm{L_q}(\mathcal{K}_1)$ and the quasicritical structures **S**, **T** are drawn in Fig. 7.12. Compare the subquasivarieties of $\mathcal{K}_0$ in Fig. 7.9. (The longstyle, shortstyle, and mediumstyle methods pertain to how we deal with operators and do not apply to Running Example 1).

An alternate way to represent this pair, using algebras without relational predicates, is to let Q_1 be the quasivariety of all 1-unary algebras satisfying $f^2x \approx x$ and $fx \approx x \rightarrow fy \approx y$. For this version, $B(x)$ becomes $fx \approx x$. The algebras in Q_1 have either no fixed points or all fixed points.

Quasivarieties of 1-unary algebras, which are a useful source of examples, are discussed in Chapter 6 of [63]. This example and the alternate representation Q_2 in Running Example 2 are both principal ideals in Figure 6.4 of [63].

Running Example 2 The longstyle construction of Sect. 7.2 applies whenever the semilattice representation $(\widehat{\mathbf{S}}, H)$ of step 1 has the property (ϖ) that $\hat{x} \neq \hat{0}$ implies $h(\hat{x}) \neq \hat{0}$. That is the case here. Meanwhile, the laws of $\mathcal{K}_0$ for this example include

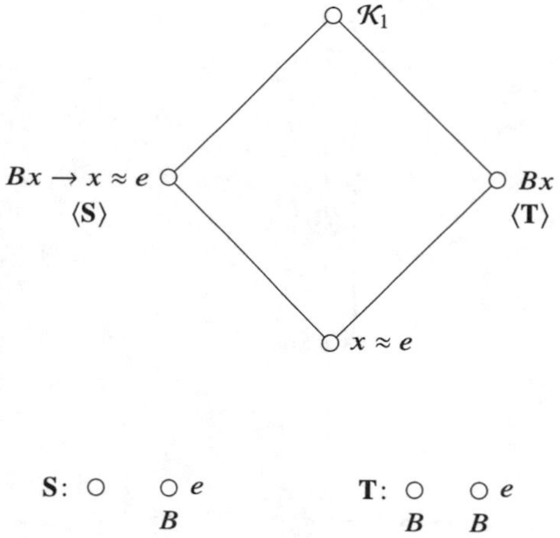

Fig. 7.12 $\mathrm{L_q}(\mathcal{K}_1)$ for Running Example 1

$P(\eta x) \leftrightarrow O(x)$ and a number of other laws that will become redundant after we substitute $O(x) \mapsto x \approx e$. We obtain the laws of $\mathcal{K}_1$ by translating $P(\eta x) \leftrightarrow O(x)$ to $P(\eta x) \leftrightarrow x \approx e$, and adding the longstyle axioms from Theorem 7.9. Thus the new quasivariety $\mathcal{K}_1$ has operations e, η and predicates $P, \approx$ satisfying the laws

$$\begin{array}{cc} P(e) & \eta e \approx e \\ \eta x \approx \eta y \to x \approx y & \eta^k x \approx x \to x \approx e \text{ for all } k \geq 1 \\ & P(\eta x) \to x \approx e. \end{array}$$

Normally we would also have something like $P(\eta^2 x) \leftrightarrow P(\eta x)$ to reflect the corresponding law of $\mathcal{K}_0$, but the last one makes that redundant.

In terms of models, $\mathcal{K}_1$ is the quasivariety $\langle \mathbf{U}_7 \rangle$ from the longstyle quasivariety in Example 7.6; see Fig. 7.17. Figure 7.13 gives the lattice $\mathrm{L_q}(\mathcal{K}_1)$ and its quasicritical structures $\mathbf{U}_1$ and $\mathbf{U}_7$.

The quasivariety $\langle \mathbf{U}_7 \rangle$ is not locally finite. An alternate way to represent the pair of Running Example 2 with a locally finite quasivariety and no relational predicates is to use the quasivariety $\mathcal{Q}_2$ of all 1-unary algebras satisfying $f^2 x \approx x$ and $fx \approx x \,\&\, fy \approx y \to x \approx y$. Again, $P(x)$ becomes $fx \approx x$. The algebras in $\mathcal{Q}_2$ have at most one fixed point, and the middle subquasivariety is $\langle fx \approx x \to x \approx y \rangle$ wherein nontrivial algebras have no fixed point.

Running Example 3 Again the semilattice representation of step 1 satisfies the condition (ϖ), and we can use the longstyle method of Sect. 7.2. The new feature here is that the laws of $\mathcal{K}_0$ include $B(x) \leftrightarrow A(\eta x)$. Thus we can replace the predicate B entirely with the translation $B(x) \mapsto A(\eta x)$. Now translate the laws of $\mathcal{K}_0$ from step 3 using that substitution, $O(x) \mapsto x \approx e$, and the longstyle axioms from Theorem 7.9. The result, after removing redundancies, is that $\mathcal{K}_1$ has operations e, η and predicates

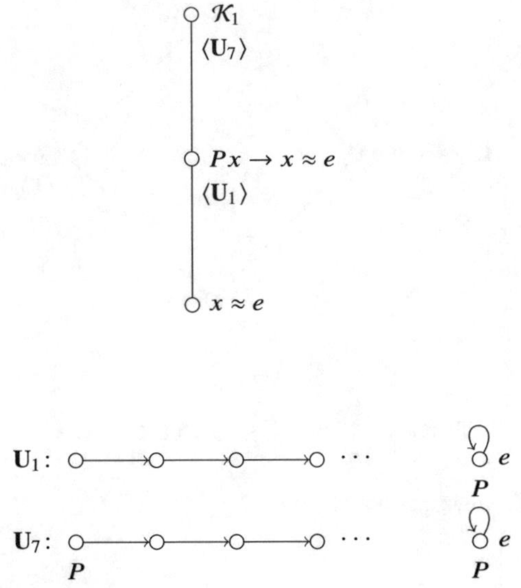

Fig. 7.13 $\mathrm{L}_\mathrm{q}(\mathcal{K}_1)$ for Running Example 2

A, $\approx$ and satisfies the laws

$$\begin{array}{rll} A(e) & \eta e \approx e & \\ \eta x \approx \eta y \to x \approx y & \eta^k x \approx x \to x \approx e & \text{for all } k \geq 1 \\ A(\eta^2 x) \leftrightarrow A(\eta x) & A(x) \to A(\eta x). & \end{array}$$

This is the quasivariety $\langle \mathbf{U}_1, \mathbf{U}_2, \mathbf{U}_6 \rangle$ from Example 7.6; see Fig. 7.17. The lattice $\mathrm{L}_\mathrm{q}(\mathcal{K}_1)$ and its quasicritical structures are drawn in Fig. 7.14.

We do not yet have a locally finite representation for this pair.

Running Example 4 Because $h(\hat{b}) = \hat{0}$ in step 2 for this example, the longstyle method of Sect. 7.2 does not apply, and we must use the shortstyle method here.

The situation here is typical whenever $\eta(\hat{x}) = \hat{0}$ for some $\hat{x} \neq \hat{0}$. To obtain a quasivariety representation with equality, we often need to either interpret the predicates as equations or assume additional laws not given in $\mathcal{K}_0$ from step 5. The models in $\mathcal{K}_1$ become finite algebras, rather than absolutely free algebras, which leads to the moniker *shortstyle*. Sometimes this works, sometimes it doesn't; see Sects. 7.4 and 7.5.

Returning to Running Example 4, as usual, $O(x) \mapsto x \approx e$, and from the equivalence $B(x) \leftrightarrow O(\eta x)$ we let $B(x) \mapsto \eta x \approx e$. (That equivalence comes from $\eta\hat{0} = \hat{b}$.) If we just translate the laws for $\mathcal{K}_0$ directly, we obtain the following laws for $\mathcal{K}_1$:

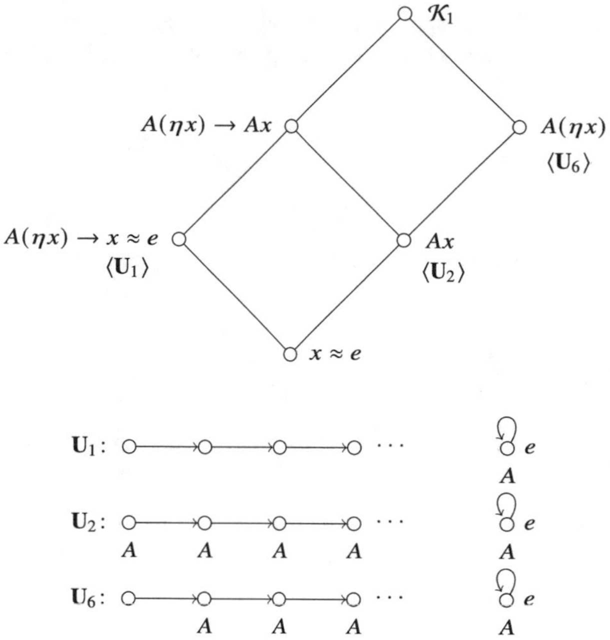

Fig. 7.14 $\mathrm{L_q}(\mathcal{K}_1)$ for Running Example 3

$$A(e) \qquad \eta e \approx e$$
$$\eta^2 x \approx \eta x \qquad A(\eta x)$$
$$A(x) \ \& \ \eta x \approx e \to x \approx e.$$

Unfortunately that leaves us with two problems. First, there are too many models, largely because of the predicate A. Second, the subquasivariety $\langle \eta x \approx x \rangle$ is unaccounted for. We can solve both problems with the translation $A(x) \mapsto \eta x \approx x$, which is consistent with all the previous laws. That gives us the following simple basis for $\mathcal{K}_1$:

$$\eta e \approx e$$
$$\eta^2 x \approx \eta x$$

and the representation given in Fig. 7.15. This is the quasivariety $\mathcal{N}^0_{1,2} = \langle \mathbf{J}^+_{01} \rangle \vee \langle \mathbf{J}_{12} \rangle$ in Figure 6.3 of [63].

At this point, we have an embedding of $\mathrm{L_q}(\mathcal{K}_0)$ into $\mathrm{L_q}(\mathcal{K}_1)$ that is correct for subquasivarieties that are determined by 1-variable quasi-equations relative to $\mathcal{K}_1$. To see that the map is surjective, we need to know that $\mathcal{K}_1$ has no other subquasivarieties, or equivalently, that every quasicritical algebra in $\mathcal{K}_1$ is 1-generated. For $\mathcal{N}^0_{1,2}$, this is an immediate consequence of Corollary 6.13 of [63]. In Sect. 7.4 we give sufficient conditions for every quasicritical structure in a quasivariety to be 1-generated; in particular, Theorem 7.20(1) applies to Running Example 4. The examples in Sect. 7.5 show that step 6 can fail when there are quasicritical structures that are not 1-generated.

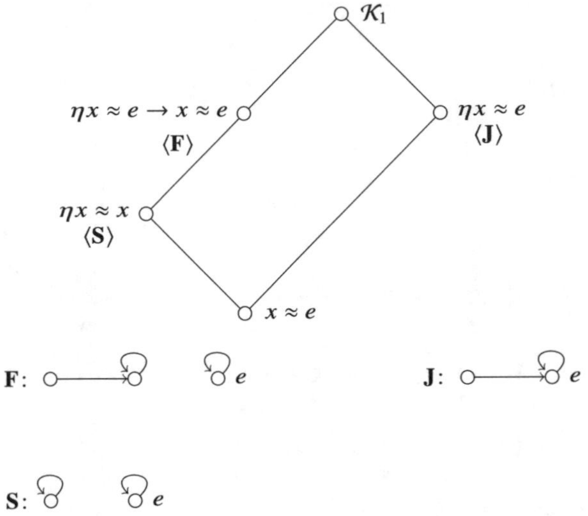

Fig. 7.15 The lattice $L_q(\mathcal{K}_1)$ and quasicritical algebras for Running Example 4. Note $\mathbf{S} \leq \mathbf{F}$

Running Example 5 Again we use the shortstyle method to convert to $\mathcal{K}_1$, with the replacements $O(x) \mapsto x \approx e$ and $B(x) \mapsto \eta x \approx e$. Once more we find ourselves with too many models and the subquasivariety $\langle \eta x \approx x \rangle$ missing, which can be resolved with the interpretation $Z_1(x) \mapsto \eta x \approx x$. This gives us a quasivariety with operations η, κ, e and a unary predicate P, satisfying the laws:

$$P(e) \qquad \eta e \approx e \approx \kappa e$$
$$\eta(\kappa x) \approx \eta x \qquad \kappa(\eta x) \approx \kappa x$$
$$\eta^2 x \approx \eta x \qquad \kappa^2 x \approx \kappa x$$
$$\eta x \approx e \leftrightarrow \kappa x \approx e \leftrightarrow P(\kappa x)$$
$$\eta x \approx e \rightarrow P(x) \qquad P(\eta x).$$

Those laws are translated from $\mathcal{K}_0$. It is useful to observe that the law

$$\eta x \approx \kappa y \longrightarrow \eta x \approx e$$

holds in $\mathcal{K}_1$: for if $\eta x = \kappa y$, then $P(\kappa y)$ holds, whence $\kappa y = e = \eta x$.

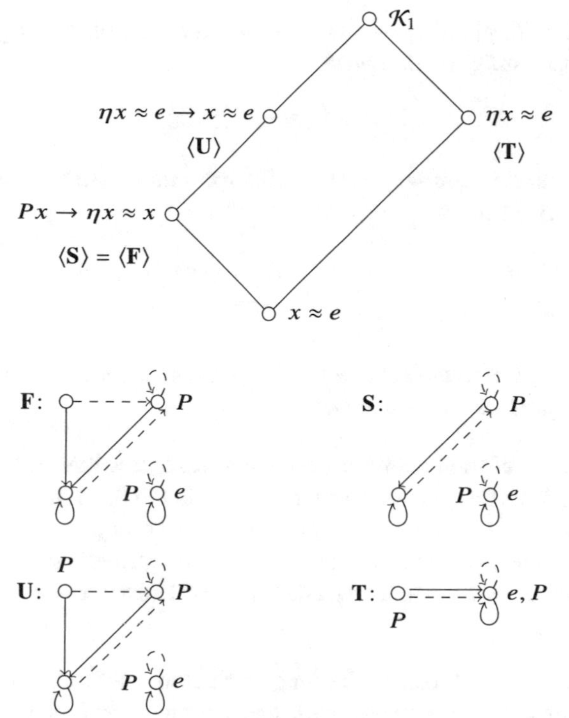

Fig. 7.16 $L_q(\mathcal{K}_1)$ for Running Example 5. Dotted arrows represent η, and solid arrows κ

The quasicritical algebras from $\mathcal{K}_1$ are **S**, **U**, **T** drawn in Fig. 7.16. The free algebra **F** is not quasicritical, since $\mathbf{S} \le \mathbf{F} \le \mathbf{S}^2$. Theorem 7.20(3) applies to ensure that the property (ß) of Lemma 7.19 holds, whence every algebra in $\mathcal{K}_1$ is a subdirect product of its 1-generated subalgebras, and hence not quasicritical unless 1-generated.

7.2 A Case Where Step Six Works: Longstyle

Recall that we have a six-step program to represent a finite pair $(\mathbf{L}, \gamma)$ as $(L_q(\mathcal{K}), \Gamma)$ when possible. For this construction project, the bottlenecks are at steps one and six; steps 2–5 are routine. In this section and Sect. 7.4 we describe the longstyle and shortstyle methods for step six, respectively. The longstyle method was used in Running Examples 2 and 3 and is the basis for Chap. 8 representing linear sums $\mathbf{1} + \mathbf{L}$ as $L_q(\mathcal{K})$.

Recall that as $\langle x \approx e \rangle$ is to be the least subquasivariety, then the laws

$$\eta e \approx e \qquad\qquad P(e)$$

should hold for all function η and predicates P in the language.

Theorem 7.5 *Let* $(\mathbf{L}, \gamma)$ *be a finite pair that has a subsemilattice representation as* $(\mathrm{Sub}(\hat{\mathbf{S}}, \wedge, \hat{0}, H), \Gamma)$ *satisfying the property*

$$h(\hat{x}) = \hat{0} \textit{ implies } \hat{x} = \hat{0}. \qquad (\varpi)$$

Then our construction project, with the interpretation $O(x) \mapsto x \approx e$ *and the additional longstyle axioms*

$$\eta x \approx \eta y \rightarrow x \approx y \qquad \eta x \approx \kappa y \rightarrow x \approx e$$
$$\eta^k x \approx x \rightarrow x \approx e$$

for all the maps η, $\kappa \neq \eta$ *and all* $k > 0$, *always yields a quasivariety representation* $(\mathrm{L}_\mathrm{q}(\mathcal{K}), \Gamma)$ *in a quasivariety with equality.*

The importance of the condition (ϖ) is that it insures that $\eta(\hat{0}) = \hat{0}$ for all the adjoint maps η. Thus the predicate $O(\eta x)$ is assigned to $\hat{0}$, so that $O(x) \leftrightarrow O(\eta x)$ is a law of $\mathcal{K}_0$, whence $\eta x \approx e \leftrightarrow x \approx e$ holds in $\mathcal{K}_1$ for every η.

We have seen that *sometimes* there are shortstyle representations when (ϖ) fails, but an assignment $\hat{a} \rightsquigarrow O(\eta x)$, so that $A(x)$ is equivalent to $\eta x \approx e$, carries additional baggage; see Sect. 7.4.

Example 7.6 To see the effect of the longstyle axioms, it is instructive to consider a prototype longstyle quasivariety with one operation and one predicate, which satisfies $P(\eta x) \leftrightarrow P(\eta^2 x)$. The quasivariety $\mathcal{U}$ has the language η, e, P, and $\approx$, and the following laws:

$$P(e) \qquad \eta e \approx e$$
$$\eta x \approx \eta y \rightarrow x \approx y \qquad \eta^k x \approx x \rightarrow x \approx e \text{ for all } k \geq 1$$
$$P(\eta x) \leftrightarrow P(\eta^2 x).$$

There are four quasicritical structures $\mathbf{U}_1$, $\mathbf{U}_2$, $\mathbf{U}_6$, $\mathbf{U}_7$. The four structures all have the same universe $\{e, 0, 1, 2, \ldots\}$ with $\eta e = e$ and $\eta k = k + 1$ for all $k \geq 0$. They differ in the relation P.

- $P^{\mathbf{U}_1} = \{e\}$
- $P^{\mathbf{U}_2} = \{e, 0, 1, 2, \ldots\}$
- $P^{\mathbf{U}_6} = \{e, 1, 2, \ldots\}$
- $P^{\mathbf{U}_7} = \{e, 0\}$

as illustrated in Fig. 7.17. One can see the embeddings $\mathbf{U}_1 \leq \mathbf{U}_7$ and $\mathbf{U}_2 \leq \mathbf{U}_6$. There are no other dependency relations amongst quasicritical structures, so $\mathrm{L}_\mathrm{q}(\mathcal{U}) \cong \mathbf{3}{\times}\mathbf{3}$. However, there are homomorphisms $\mathbf{U}_1 \rightarrow \mathbf{U}_6$, $\mathbf{U}_1 \rightarrow \mathbf{U}_7$, $\mathbf{U}_6 \rightarrow \mathbf{U}_2$, $\mathbf{U}_7 \rightarrow \mathbf{U}_2$, $\mathbf{U}_1 \rightarrow \mathbf{U}_2$, plus maps to the 1-element structure. Each non-constant homomorphism is the identity map on the set but enlarges the predicate. The homomorphisms determine Γ, since $\Gamma(Q) = \mathbb{H}(Q) \cap \mathcal{U}$. The resulting $(\mathrm{L}_\mathrm{q}(\mathcal{U}), \Gamma)$ is shown in the figure.

It is an interesting exercise to compare $\mathcal{U}$ with the quasivariety $\mathcal{W}$ with the same language but satisfying $P(e)$, $\eta e \approx e$, and $\eta^2 x \approx \eta x$. Then $\mathrm{L_q}(\mathcal{W})$ is a non-distributive lattice with 14 elements. The comparison exhibits some of the difference between longstyle and shortstyle varieties.

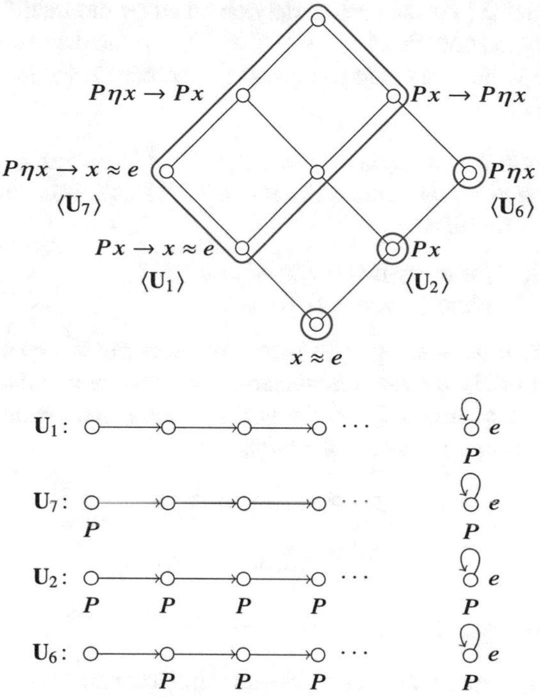

Fig. 7.17 $\mathrm{L_q}(\mathcal{U})$ where $\mathcal{U}$ is the quasivariety of Example 7.6

Running Examples 2 and 3 illustrate the longstyle construction, and indeed they are are subquasivarieties of $\mathcal{U}$ from Example 7.6.

Theorem 7.5 is the finite case of a more general result, Theorem 7.7. The infinite case requires that $1_{\mathbf{S}}$ be compact.

Theorem 7.7 *Suppose* $(\mathbf{L}, \gamma)$ *is represented by* $(\mathrm{S_p}(\mathbf{S}, H), \Gamma)$ *satisfying*

- $1_{\mathbf{S}}$ *is compact.*
- $h(x) = 1_{\mathbf{S}}$ *implies* $x = 1_{\mathbf{S}}$ (ϖ)

for all $x \in S$ *and* $h \in H$. *Then our construction project, with the interpretation* $O(x) \mapsto x \approx e$ *and the additional longstyle axioms from* Theorem 7.5 *yields a quasivariety representation* $(\mathrm{L_q}(\mathcal{K}), \Gamma)$ *in a quasivariety with equality.*

In the infinite case, step 2 is slightly different: we have the dual isomorphism $\mathrm{S_p}(\mathbf{S}, H) \cong^d \mathrm{Con}(\mathbf{K}, +, 0_{\mathbf{S}}, G)$ where $\mathbf{K}$ is the semilattice of compact elements of $\mathbf{S}$, and the maps in G are the adjoints of those in H, given by the same formula $\eta(\hat{s}) = \bigwedge\{\hat{x} \in S : h(\hat{x}) \geq \hat{s}\}$. For details, including the argument that the adjoints map compacts to compacts, see Theorem 14 of [64].

To prove Theorem 7.7, suppose we have $\mathrm{S_p}(\mathbf{S}, H) \cong \mathrm{L_q}(\mathcal{K}_0)$ by our construction (steps 2–5), and let $\mathcal{K}_1$ be the quasivariety obtained by the modifications of Theorem 7.5. We want to show that $\mathrm{L_q}(\mathcal{K}_0) \cong \mathrm{L_q}(\mathcal{K}_1)$, and the next two theorems are in aid of that. For $\mathcal{K}_0$, the next result just modifies Theorem 7.3 to include our stronger conditions on $O(x)$.

Theorem 7.8 *Let $\mathcal{B}$ be an implicational theory in a language $\mathcal{L}$ without equality, and with the restrictions and laws of* Theorem 7.3(1)–(4). *Assume additionally that in $\mathcal{B}$ the predicate O satisfies*

(5) $O(\eta x) \leftrightarrow O(x)$ *for every function symbol* η,
(6) $O(x) \rightarrow P(x)$ *for every predicate symbol* P.

Then every implication holding in a theory extending the theory of $\mathcal{B}$ is equivalent (modulo the laws of $\mathcal{B}$) to a set of implications in only one variable.

Moreover, modulo the quasi-equations of $\mathcal{B}$, every quasi-equation in $\mathcal{L}$ is equivalent to a quasi-equation of one of the forms

$$\underset{1 \leq i \leq k}{\&} P_i(\eta_i(x)) \rightarrow O(x)$$
$$\underset{1 \leq i \leq k}{\&} P_i(\eta_i(x)) \rightarrow Q(\kappa(x))$$

with P_i, $Q \neq O$ but possibly η_i, $\kappa = id$.

The preceding theorem applies to $\mathcal{K}_0$, and thus we can assume that the laws of subquasivarieties of $\mathcal{K}_0$ have the form given there. So let us prove the corresponding statement for $\mathcal{K}_1$.

Theorem 7.9 *Let C be an implicational theory in a language $\mathcal{L}$ with the following restrictions and laws.*

(1) *$\mathcal{L}$ has only unary predicate symbols.*
(2) *$\mathcal{L}$ has only unary function symbols.*
(3) *$\mathcal{L}$ has one constant symbol e.*
(4) *C contains the laws $P(f(e))$ for every predicate P and every formal composition f of functions of $\mathcal{L}$.*

Assume that in C the predicate O satisfies

(5) $O(\eta x) \leftrightarrow O(x)$ *for every function symbol* η,
(6) $O(x) \rightarrow P(x)$ *for every predicate symbol* P.

Let $\mathcal{D}$ be the theory obtained from C by adding equality $\approx$ to the language, along with laws saying that $\approx$ is a congruence relation, converting $O(x) \mapsto x \approx e$ for all instances, and adding the laws:

$$\eta x \approx \eta y \rightarrow x \approx y \qquad\qquad \eta x \approx \kappa y \rightarrow x \approx e$$
$$\eta^k x \approx x \rightarrow x \approx e$$

for all function symbols η, $\kappa \neq \eta$ and all $k > 0$. Then every implication holding in a theory extending the theory of $\mathcal{D}$ is equivalent (modulo the laws of $\mathcal{D}$) to a set of implications in only one variable.

Moreover, modulo the quasi-equations of $\mathcal{D}$, every quasi-equation in $\mathcal{L}$ is equivalent to a quasi-equation of one of the forms

$$\underset{1 \leq i \leq k}{\&} P_i(\eta_i(x)) \rightarrow x \approx e$$
$$\underset{1 \leq i \leq k}{\&} P_i(\eta_i(x)) \rightarrow Q(\kappa(x))$$

with P_i, $Q \neq O$ but possibly η_i, $\kappa = id$.

Note that the laws of $\mathcal{D}$ include the translations of (5) and (6):

$$\eta x \approx e \leftrightarrow x \approx e \qquad\qquad P(e)$$

for all function symbols η and predicates P of the language.

Proof The assumptions guarantee that every atomic formula of $\mathcal{D}$ is equivalent to one of the forms

$$x \approx e \qquad\qquad Q(\kappa(x)) \qquad\qquad x \approx y$$

with $Q \neq O$. So in a quasi-equation $\beta_1 \& \cdots \& \beta_k \rightarrow \alpha$ we may assume that α is one of the above three atomic formulas, and that each β_j is one of the following, where z represents a variable with $z \neq x, y$.

$x \approx e$	$y \approx e$	$z \approx e$
$x \approx y$	$x \approx z$	$y \approx z$
$x \approx \eta y$	$y \approx \eta x$	$z \approx \eta x$
$x \approx \eta z$	$y \approx \eta z$	$z \approx \eta y$
$P(\eta x)$	$P(\eta y)$	$P(\eta z)$

Moreover,

- if any of the first 12 options occurs as a β_j, then we can make a substitution and obtain an equivalent quasi-equation with fewer variables,
- if $P(\eta z)$ occurs as a β_j, where $z \neq x, y$, then via the substitution $z \mapsto e$ we can remove that antecedent to obtain an equivalent quasi-equation in fewer variables.

Thus the original quasi-equation is equivalent, modulo $\mathcal{D}$, to one of the following four forms:

$$\underset{1\le i\le k}{\&} P_i(\eta_i(x)) \to x \approx e$$
$$\underset{1\le i\le k}{\&} P_i(\eta_i(x)) \to Q(\kappa(x))$$
$$\underset{1\le i\le k}{\&} P_i(\eta_i(x)) \to x \approx y$$
$$\underset{1\le i\le \ell}{\&} P_i(\eta_i(x)) \ \& \underset{1\le j\le m}{\&} P_j(\lambda_j(y)) \to x \approx y.$$

However, the third form is equivalent to $\&_{1\le i\le k} P_i(\eta_i(x)) \to x \approx e$ and the fourth is equivalent to the pair

$$\underset{1\le i\le \ell}{\&} P_i(\eta_i(x)) \to x \approx e$$
$$\underset{1\le j\le m}{\&} P_j(\lambda_j(y)) \to e \approx y$$

both of the form given in the theorem. □

In view of Theorems 7.3 and 7.9, it is clear how to translate quasi-equations modulo $\mathcal{K}_0$ into quasi-equations modulo $\mathcal{K}_1$ via $O(x) \mapsto x \approx e$, and vice versa, when the hypotheses of Theorem 7.9 hold. We conclude that $\mathrm{L_q}(\mathcal{K}_0) \cong \mathrm{L_q}(\mathcal{K}_1)$, and this proves Theorem 7.7.

Corollary 7.10 *Suppose* $(\mathbf{L}, \gamma)$ *can be represented as* $(\mathrm{S_p}(\mathbf{S}, H), \Gamma)$. *Then* $(\mathbf{1} + \mathbf{L}, \gamma')$ *can be represented* $(\mathrm{L_q}(\mathcal{K}), \Gamma)$ *for a quasivariety with equality, where* γ' *is* γ *with the new least element as a singleton* γ'*-class.*

Proof For step 1, form $\mathbf{S}' = \mathbf{S} + \mathbf{1}$. Denote the old greatest element as $\hat{0}$, and the new one as $1'$. Extend every $h \in H$ to h' with $h'(1') = 1'$, and add a new map $k(x)$ with $k(x) = \hat{0}$ if $x \in S$, and $k(1') = 1'$. Let $H' = \{h' : h \in H\} \cup \{k\}$. Then $(\mathbf{S}', H')$ satisfies the conditions of Theorem 7.7. □

The quasivariety used in Theorem 7.5 can be regarded as the extreme opposite of that used by Gorbunov and Tumanov for atomistic quasivariety lattices in Theorem 2.64. For that case, the quasivariety consisted of 1-element structures with relations, and there were no operations. In the present case, there are usually operations and relations, and except for the trivial quasivariety the algebra is absolutely free.

We should mention that in [20] there is a different method for converting $\mathcal{K}_0$ to a quasivariety $\mathcal{K}_1$ with equality that requires strong assumptions on the monoid of operators. For example, that method applies when the monoid G of adjoints is a group.

Example 7.11 We illustrate the method of this section with the pair $(\mathbf{L}, \gamma)$ in Fig. 7.18, with γ being the identity map. Reflecting the order in $\mathbf{L}$, we want $\hat{z} \to \hat{a}$, $\hat{z} \to \hat{b}$, $\hat{a} \to \hat{c}$, $\hat{b} \to \hat{c}$.

These maps realize the arrow relations and satisfy the condition (ϖ).

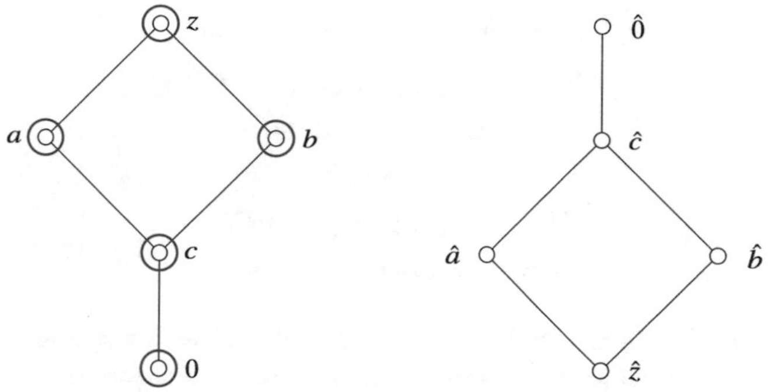

Fig. 7.18 Lattice for Example 7.11, with its companion $\gamma^d(\mathbf{L})$

	h	k	$hk = kh$
$\hat{0}$	$\hat{0}$	$\hat{0}$	$\hat{0}$
$\hat{a}$	$\hat{a}$	$\hat{c}$	$\hat{c}$
$\hat{b}$	$\hat{c}$	$\hat{b}$	$\hat{c}$
$\hat{c}$	$\hat{c}$	$\hat{c}$	$\hat{c}$
$\hat{z}$	$\hat{a}$	$\hat{b}$	$\hat{c}$

The corresponding adjoints are

	η	κ	$\eta\kappa = \kappa\eta$
$\hat{0}$	$\hat{0}$	$\hat{0}$	$\hat{0}$
$\hat{a}$	$\hat{z}$	$\hat{a}$	$\hat{z}$
$\hat{b}$	$\hat{b}$	$\hat{z}$	$\hat{z}$
$\hat{c}$	$\hat{b}$	$\hat{a}$	$\hat{z}$
$\hat{z}$	$\hat{z}$	$\hat{z}$	$\hat{z}$

Note that $\eta^2 = \eta$ and $\kappa^2 = \kappa$.

Predicates are assigned as follows:

$$\begin{aligned}
&\hat{0} \rightsquigarrow O(x), O(\eta x), O(\kappa x)\\
&\hat{a} \rightsquigarrow A(x), A(\kappa x), C(\kappa x)\\
&\hat{b} \rightsquigarrow B(x), B(\eta x), C(\eta x)\\
&\hat{c} \rightsquigarrow C(x)\\
&\hat{z} \rightsquigarrow A(\eta x), B(\kappa x).
\end{aligned}$$

From this it is clear that in $\mathcal{K}_0$ we can replace $A(x)$ by $C(\kappa x)$, and $B(x)$ by $C(\eta x)$.

The quasivariety $\mathcal{K}_0$ has a constant e, predicates O and C, functions η and κ, and the laws

$$O(e) \quad C(e) \quad O(\eta e) \quad O(\kappa e)$$
$$C(\eta^2 x) \leftrightarrow C(\eta x) \quad C(\kappa^2 x) \leftrightarrow C(\kappa x)$$
$$C(\eta\kappa x) \quad C(\kappa\eta x)$$
$$O(x) \rightarrow C(x)$$
$$C(x) \rightarrow C(\eta x) \quad C(x) \rightarrow C(\kappa x)$$
$$C(\eta x) \ \& \ C(\kappa x) \rightarrow C(x).$$

We convert to the quasivariety $\mathcal{K}_1$ by replacing $O(x)$ with $x \approx e$, and laws saying that if any two distinct terms collapse, then everything collapses to e:

$$C(e) \quad \eta e \approx \kappa e \approx e$$
$$C(\eta^2 x) \leftrightarrow C(\eta x) \quad C(\kappa^2 x) \leftrightarrow C(\kappa x)$$
$$C(\eta\kappa x) \quad C(\kappa\eta x)$$
$$C(x) \rightarrow C(\eta x) \quad C(x) \rightarrow C(\kappa x)$$
$$C(\eta x) \ \& \ C(\kappa x) \rightarrow C(x)$$
$$\eta x \approx \eta y \rightarrow x \approx y \quad \kappa x \approx \kappa y \rightarrow x \approx y$$
$$\eta^k x \approx \eta^\ell x \rightarrow x \approx e \text{ for all } k > \ell \geq 0$$
$$\kappa^k x \approx \kappa^\ell x \rightarrow x \approx e \text{ for all } k > \ell \geq 0$$
$$\eta x \approx \kappa y \rightarrow x \approx e$$
$$\eta x \approx \kappa y \rightarrow y \approx e.$$

Note that once we had step 1 with the maps of H satisfying (ϖ), we knew that steps 2–6 would successfully produce a subquasivariety representation $\mathrm{L_q}(\mathcal{K}_1)$, even though the calculations are somewhat tedious.

A slight variation on Corollary 7.10 can occasionally be useful. Suppose we want to represent $(\mathbf{L}, \gamma)$ where $0_\mathbf{L}$ is completely meet irreducible, with upper cover $0_\mathbf{L}^* = c$, the unique atom of $\mathbf{L}$. If $\gamma(c) > c$, then the corollary does not apply directly, and it often happens that the operators on the most natural semilattice representation $\widehat{\mathbf{S}}$ do not satisfy (ϖ). But $\hat{c}$ is a coatom of $\gamma^d(\mathbf{L})$; by adding a new coatom c_1 to $\widehat{\mathbf{S}}$ so that $\hat{c} \leq p_1 \leq c_1$ for all p in the interval $[c, \gamma(c)]$, we may be able to modify the representation to satisfy (ϖ). Thus we are able to complete step 6 and obtain a longstyle representation of the pair $(\mathbf{L}, \gamma) \cong (\mathrm{L_q}(\mathcal{K}_1), \Gamma)$ for a quasivariety with equality.

Figures 7.19 and 7.20 give two examples where this method applies; the details are straightforward and left to the reader. *We do not know whether the pairs* $(\mathbf{L}, \gamma)$ *in either figure can be represented with a locally finite quasivariety with equality!* This is Problem 15 in Chap. 10.

$$\hat{z} \to p_1 \to \hat{c}$$
$$\hat{c} \leftrightarrow c_1$$

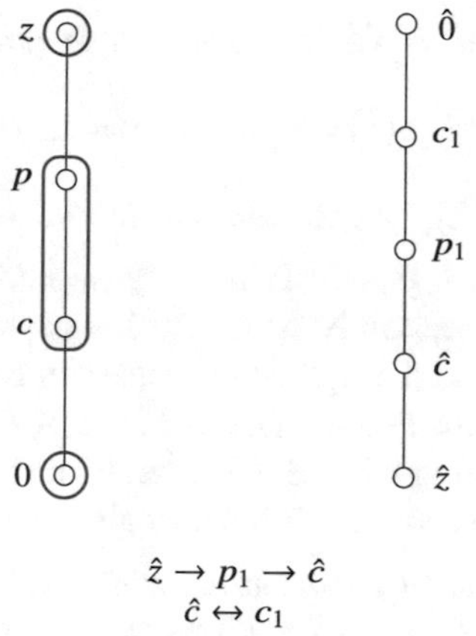

Fig. 7.19 A pair $(\mathbf{L}, \gamma)$ where adding a coatom c_1 to $\widehat{\mathbf{S}}$ allows us to complete step 6

$$\hat{z} \to \hat{b} \to \hat{c}$$
$$\hat{z} \to p_1 \to c_1$$
$$\hat{c} \leftrightarrow c_1$$

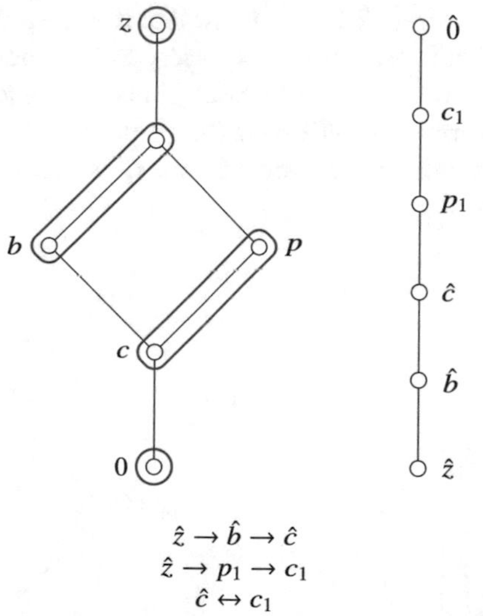

Fig. 7.20 Another pair $(\mathbf{L}, \gamma)$ where adding a coatom c_1 to $\widehat{\mathbf{S}}$ allows us to complete step 6

7.3 Reverse Engineering

So far, we have been focusing on the problem: *Given a pair $(\mathbf{L}, \gamma)$, when can we represent $(\mathbf{L}, \gamma)$ as $(L_q(\mathcal{K}), \Gamma)$?* Another question would be simply to ask: *Can we produce lots of examples of subquasivariety lattices?* The longstyle method of Theorems 7.5 and 7.7 provides one way to do so.

Theorem 7.5 can be restated compactly as follows.

Theorem 7.12 *If* $(\mathbf{S}, \wedge, 1, H)$ *is a finite semilattice with operators satisfying*

$$h(x) = 1 \textit{ implies } x = 1, \qquad (\varpi)$$

then $\mathrm{Sub}(\mathbf{S}, \wedge, 1, H) \cong \mathrm{L_q}(\mathcal{K})$ *for some quasivariety* $\mathcal{K}$ *with equality.*

For once we have $(\mathbf{S}, \wedge, 1, H)$ satisfying (ϖ), we know that the remaining steps 2–6 can be completed successfully using the longstyle method. Thus we obtain $\mathrm{Sub}(\mathbf{S}, \wedge, 1, H) \cong \mathrm{L_q}(\mathcal{K})$, and the equaclosure operator can be read off by inspecting the least elements of the subalgebras. The pair $(\mathrm{Sub}(\mathbf{S}, \wedge, 1, H), \Gamma)$ is naturally some $(\mathbf{L}, \gamma)$, but now we are starting with $(\mathbf{S}, H)$ rather than $(\mathbf{L}, \gamma)$.

The infinite case from Theorem 7.7 requires also that $1_\mathbf{S}$ be compact.

Theorem 7.13 *If* $(\mathbf{S}, H)$ *is an algebraic lattice with operators satisfying* (ϖ) *and such that* $1_\mathbf{S}$ *is compact, then* $\mathrm{S_p}(\mathbf{S}, H) \cong \mathrm{L_q}(\mathcal{K})$ *for some quasivariety* $\mathcal{K}$ *with equality.*

Let us see some examples.

Example 7.14 From Theorem 2.64, we know that if $\mathbf{S}$ is a finite semilattice (with no operators), then $\mathrm{Sub}\ \mathbf{S} \cong \mathrm{L_q}(\mathcal{Q})$ for a quasivariety of 1-element structures. Since (ϖ) is satisfied vacuously, we now know that Sub $\mathbf{S}$ also has a longstyle representation as $\mathrm{L_q}(\mathcal{K})$. The quasivariety $\mathcal{K}$ will have unary predicates and no operations. Its quasicritical structures will have 2 elements, one of which is e. Running Example 1 illustrates this; see Fig. 7.12.

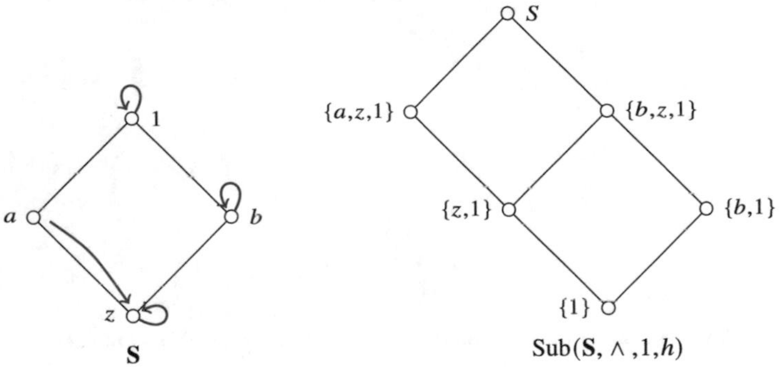

Fig. 7.21 $\mathbf{S}$ is a meet semilattice with the operator h such that $h(a) = z$, $h(x) = x$ otherwise. The lattice $\mathrm{Sub}(\mathbf{S}, \wedge, 1, h)$ has a longstyle representation as a subquasivariety lattice

Example 7.15 The second example is in Fig. 7.21. The semilattice $\mathbf{S}$ is on the left, with the operator h such that $h(a) = z$ and $h(x) = x$ otherwise. Note this is an operator, preserving meets and 1. You can think of h as saying that $a \to z$, i.e., we only want those subsemilattices $\mathbf{T} \leq \mathbf{S}$ such that $a \in T$ implies $z \in T$. The lattice

Sub(**S**, ∧, 1, h) is on the right of the figure. Of course, we already knew that **2**×**3** is a subquasivariety lattice, as it is finite and distributive, but pursuing steps 2–6 would give us a concrete longstyle representation.

We can discern not only the lattice but also the equaclosure operator from Sub **S**. Remember that for a subsemilattice $\mathbf{T} \leq \mathbf{S}$, we have $\Gamma(\mathbf{T}) = \uparrow t_0$ where t_0 is the least element of **T**. Thus if **T** and **T′** have the same least element, then they are in the same Γ-class. Looking at the figure, the subsemilattices in the upper square all have z as their least element, so they are a Γ-class. The remaining subsemilattices are $\uparrow b$ and $\{1\}$, which are Γ-closed. The longstyle subquasivariety lattice will have the same equaclosure operator.

Example 7.16 The third example is in Fig. 7.22. The semilattice **U** has an operator k such that $k(a) = b$ and $k(x) = x$ otherwise. The lattice Sub(**U**, ∧, 1, k) is shown on the right of the figure. The Γ-classes of Sub **U** are determined by whether b, c, z, or 1 is the least element of the subalgebra, and the longstyle subquasivariety lattice will have the corresponding equaclosure operator.

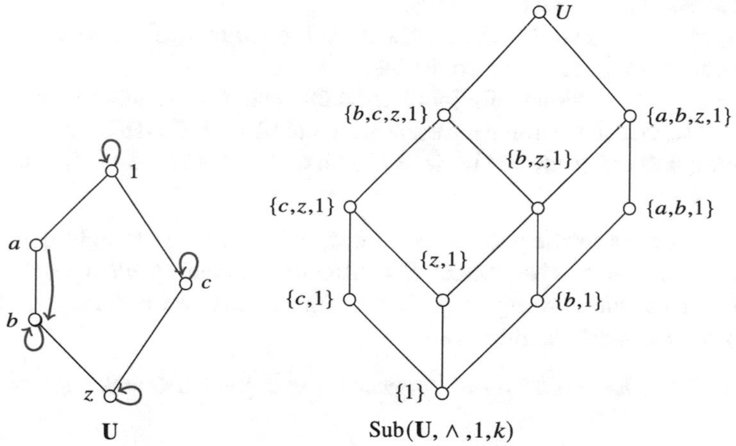

Fig. 7.22 **U** is a meet semilattice with the operator k such that $k(a) = b$, $k(x) = x$ otherwise. The lattice Sub(**U**, ∧, 1, k) has a longstyle representation as a subquasivariety lattice

In general, a semilattice may have many operators satisfying (ϖ), any combination of which may be used. The preceding examples use instances of the following operators. For elements $b, c \in S$, define

$$h(x) = \begin{cases} x & \text{if } x \geq c, \\ x \wedge b & \text{otherwise.} \end{cases}$$

Likewise, one can use Theorem 7.13 for **S** an algebraic lattice with 1_S compact.

This raises an obvious question: *What lattices have a longstyle subquasivariety representation?* That source of examples has yet to be mined.

7.4 A Case Where Step Six Works: Shortstyle

In this section we generalize Running Examples 4 and 5. For this discussion, we need the concept of a strongly quasicritical structure, developed in Appendix A.3. A structure **T** is *quasicritical* if it is finitely generated and $\mathbb{Q}(\mathbf{T})$-subdirectly irreducible. A structure **T** is *strongly quasicritical* if it is quasicritical and satisfies a technical condition (¥) from Appendix A.3.

Since a lattice $\mathrm{L_q}(\mathcal{K})$ is dually algebraic, every subquasivariety $\mathcal{S} \le \mathcal{K}$ is a join of completely join irreducible quasivarieties. The basic facts are these.

- Every strongly quasicritical structure generates a completely join irreducible quasivariety.
- If $\mathcal{Q}$ is completely join irreducible, then $\mathcal{Q} = \mathbb{Q}(\mathbf{T})$ for some strongly quasicritical structure **T**, not necessarily unique.
- The quasivariety generated by a quasicritical structure is join irreducible, but not necessarily completely join irreducible.
- However, if $\mathcal{Q}$ is completely join irreducible and **T** is a quasicritical structure in $\mathcal{Q} \setminus \mathcal{Q}_*$, then **T** is strongly quasicritical and $\mathcal{Q} = \mathbb{Q}(\mathbf{T})$. (Here $\mathcal{Q}_*$ denotes the unique quasivariety covered by $\mathcal{Q}$. By the first part of the proof of Theorem A.10, **T** satisfies (¥).)

Thus every subquasivariety $\mathcal{S} \le \mathcal{K}$ is determined by the strongly quasicritical structures it contains. That means $\mathcal{S}$ is also determined by *all* the quasicritical structures it contains (strong or not), and this is what we normally work with, avoiding the technical condition (¥).

Theorem 7.17 *The following are equivalent for a quasivariety $\mathcal{K}$ and an integer $k > 0$.*

(1) *For every completely join irreducible $\mathcal{Q} \le \mathcal{K}$, there is a k-generated strongly quasicritical structure* **T** *such that $\mathcal{Q} = \mathbb{Q}(\mathbf{T})$.*
(2) *Every subquasivariety $\mathcal{S} \le \mathcal{K}$ is determined, relative to $\mathcal{K}$, by the k-variable quasi-equations it satisfies.*

Proof Assume (1), and let $\mathcal{S} \le \mathcal{K}$. Let $\mathcal{M}$ be the subquasivariety of $\mathcal{K}$ determined by all the k-variable quasi-equations holding in $\mathcal{S}$. Surely $\mathcal{S} \le \mathcal{M}$. If $\mathcal{S} < \mathcal{M}$, then there is a completely join irreducible $\mathcal{Q}$ such that $\mathcal{Q} \le \mathcal{M}$ and $\mathcal{Q} \nleq \mathcal{S}$. Let **T** be a k-generated strongly quasicritical structure generating $\mathcal{Q}$. Since $\mathcal{Q} \nleq \mathcal{S}$, there is a quasi-equation β holding in $\mathcal{S}$ but failing in **T**. Because **T** is k-generated, β can be replaced by a quasi-equation in at most k variables: substitute terms in k variables witnessing the failure of β in **T** to obtain β', a consequence of β failing in **T**. Then β' is a k-variable quasi-equation holding in $\mathcal{S}$ but failing in $\mathcal{M}$, contrary to the definition of $\mathcal{M}$. Thus $\mathcal{S} = \mathcal{M}$, so that (2) holds.

Conversely, assume (2), and let $\mathcal{Q} \leq \mathcal{K}$ be completely join irreducible. Then there is a k-variable quasi-equation γ that holds in $\mathcal{Q}_*$ but not in $\mathcal{Q}$. Then γ fails in some k-generated structure **S** in $\mathcal{Q}$, and hence in some quasicritical subdirect factor **T** of **S**. By the last bullet above, **T** is strongly quasicritical. □

Corollary 7.18 *If a quasivariety $\mathcal{K}$ has the property that every finitely generated subalgebra is a subdirect product of its 1-generated substructures, then every subquasivariety $\mathcal{S} \leq \mathcal{K}$ is determined, relative to $\mathcal{K}$, by the 1-variable quasi-equations it satisfies.*

With these preliminaries out of the way, we return to the shortstyle method.

Caveat *The next argument assumes that we know all the 1-generated structures in $\mathcal{K}$ and need to find out more about other finitely generated structures.* At the end of step 5 we have a quasivariety $\mathcal{K}_0$ with unary operations, unary predicates, and a 1-variable basis for each of its subquasivarieties. Step 6 tries to mimic these properties in a quasivariety $\mathcal{K}_1$ with equality. The shortstyle method is one way to do this, but it requires finding a translation of $\mathcal{K}_0$ into a different language. Sometimes there is no way to do this conversion. But even when it is possible to do the translation, it remains to show that the proposed quasivariety $\mathcal{K}_1$ has the same subquasivariety lattice as $\mathcal{K}_0$. In particular, the subquasivarieties of $\mathcal{K}_1$ should be determined by 1-variable quasi-identities. Put another way, the quasicritical structures in $\mathcal{K}_1$ should be 1-generated. The remainder of this section concerns sufficient conditions to ensure that this happens. Then, in Sect. 7.5 we analyze in more depth what can go wrong with the shortstyle method, which is essentially that $\mathcal{K}_1$ could have quasicritical structures that are not 1-generated.

Recall that if **S** is a substructure of a structure **W**, then a *retraction* of **W** onto **S** is an endomorphism $\sigma : \mathbf{W} \twoheadrightarrow \mathbf{S}$ such that $\sigma^2 = \sigma$. Thus σ is the identity on **S**.

Lemma 7.19 *Let $\mathcal{K}$ be a quasivariety with only unary operations, unary predicates, and a constant e that is fixed by all the operations and such that $P(e)$ holds for every predicate P. Assume $\mathcal{K}$ has the property:*

If **W** is a 1-generated structure in $\mathcal{K}$ and $\mathbf{S} \leq \mathbf{W}$ is a substructure, then **S** is 1-generated and there is a retraction $\sigma : \mathbf{W} \twoheadrightarrow \mathbf{S}$. (ß)

Then every finitely generated structure in $\mathcal{K}$ is a subdirect product of its 1-generated substructures. Hence every quasicritical structure in $\mathcal{K}$ is 1-generated, and 1-variable quasi-equations determine subquasivarieties of $\mathcal{K}$.

Proof Let us prove by induction that, for $m > 1$, if $\mathbf{U} \in \mathcal{K}$ is m-generated and $\mathbf{T} \leq \mathbf{U}$ is generated by $m - 1$ of those generators, then there is a retraction of **U** onto **T**. Moreover, using (ß), we show that the intersection of the kernels of these retractions is Δ. Then every m-generated structure in $\mathcal{K}$ is a subdirect product of its $(m-1)$-generated substructures, and hence by induction of its 1-generated substructures.

Let **U** be generated irredundantly by a set $\{x_1, \ldots, x_m\}$ and let $\mathbf{T} = \mathrm{Sg}(x_2, \ldots, x_m)$. Apply (ß) with $\mathbf{W} = \mathrm{Sg}(x_1)$ and $\mathbf{S} = \mathbf{W} \cap \mathbf{T}$ to obtain a retraction $\sigma_0 : \mathbf{W} \twoheadrightarrow \mathbf{S}$.

(The constant e ensures that $\mathbf{S}$ is nonempty.) This can be extended to a retraction $\sigma : \mathbf{U} \twoheadrightarrow \mathbf{T}$ by defining σ to be the identity on $U \setminus W$. Note that the identity homomorphism is not only the identity elementwise but does not alter the predicates. This makes σ the identity on all of $\mathbf{T}$, whence the only nontrivial pairs in $\ker \sigma$ are from $W^2 \setminus S^2$: that is, if $(u, v) \in \ker \sigma$ and $u \neq v$, then both are in $\mathrm{Sg}(x_1)$ and at least one of them is not in $\mathrm{Sg}(x_2, \ldots, x_m)$.

Repeating the process for each index gives a set of retractions σ_j for $1 \leq j \leq m$ of $\mathbf{U}$ onto $(m-1)$-generated subalgebras with $\bigcap \ker \sigma_j = \Delta$. □

The hypothesis in (ß) that every substructure of a 1-generated structure $\mathbf{W}$ is 1-generated, along with having only unary functions, makes Sub $\mathbf{W}$ a chain, which is slightly more than we need. The following variation, proved by the same argument, removes that requirement, and we will use it for the pair $(\mathbf{W}, \gamma_6)$ in Sect. 7.5.

Suppose $\mathbf{U} \in \mathcal{K}$ *is generated irredundantly by the set* $\{x_1, \ldots, x_m\}$, *and that the following holds for every* i: *with* $\mathbf{W} = \mathrm{Sg}(x_i)$ *and* $\mathbf{T} = \mathrm{Sg}(\{x_j : j \neq i\})$ *and* $\mathbf{S} = \mathbf{W} \cap \mathbf{T}$, *there is a retraction* $\sigma : \mathbf{W} \twoheadrightarrow \mathbf{S}$. $(\beta)'$

In our applications of (ß)$'$, $\mathbf{S}$ will be restricted to only certain possible substructures of $\mathbf{W}$.

One can easily check that the 1-generated structures in $\mathcal{K}_1$ for Running Examples 4 and 5 satisfy (ß); see Figs. 7.15 and 7.16, respectively. More generally, we have this.

Theorem 7.20 *Let* $\mathcal{K}$ *be a quasivariety with only unary operations, unary predicates, and a constant* e *that is fixed by all the operations and such that* $P(e)$ *holds for every predicate* P. *If* $\mathcal{K}$ *satisfies one of the following conditions, then every quasicritical structure in* $\mathcal{K}$ *is 1-generated, and hence 1-variable quasi-equations determine subquasivarieties of* $\mathcal{K}$.

(1) $\mathcal{K}$ *has 1 unary operation and no predicates.*
(2) $\mathcal{K}$ *has 1 unary operation* η, *which satisfies either* $\eta^k x \approx x$ *or* $\eta^k x \approx \eta x$ *for some* $k > 1$, *and* $Px \to P(\eta x)$ *for all predicates* P.
(3) *The operations of* $\mathcal{K}$ *form a left zero semigroup, i.e.,* $\mathcal{K}$ *satisfies* $\kappa\lambda x \approx \kappa x$ *for all* κ, λ, *and there exists an operation* η *such that* $Px \to P(\eta x)$ *for all predicates* P.
(4) *The operations of* $\mathcal{K}$ *form a finite chain under composition, i.e.,* $\eta_1 > \eta_2 > \cdots > \eta_k$ *so that* $\eta_i\eta_j = \eta_{\max(i,j)}$, *and* $Px \to P(\eta_j x)$ *for all* j.

Proof Part (1) is Corollary 6.13 of [63]. The remaining parts are applications of Lemma 7.19, for which we must show that under the given conditions $\mathcal{K}$ satisfies (ß). In each case, since these are substructures, the predicates are inherited from some overstructure $\mathbf{W}$.

For (2), the substructures of a 1-generated structure $\mathbf{W}$ are $\mathbf{W}$, $\{e\}$ and, in case $\mathcal{K}$ satisfies $\eta^k x \approx \eta x$, also $\{\eta x, \eta^2 x, \ldots, \eta^{k-1} x\}$. For this last there is a retraction with $x \mapsto \eta^{k-1} x$.

For (3), let $\mathbf{S} = \mathrm{Sg}(x)$ in $\mathcal{K}$. Let $B = \{\eta x : \eta \in H\}$. The substructures of $\mathbf{S}$, not necessarily distinct, are

- $\{x\} \cup B \cup \{e\}$,
- $B \cup \{e\}$,
- $\{e\}$.

The assumptions are strong enough to ensure that any map $x \mapsto \eta x$ extends to a retraction, while $x \mapsto e$ always yields a retraction.

A couple of comments are in order for this case. There could be infinitely many operations. Moreover, $\eta x = \kappa y$ is allowed for particular elements, but then $\kappa x = \eta x = \kappa y = \eta y$. Also, $\eta x = e$ implies $\kappa x = e$ for all κ; thus the subvarieties $\langle \eta x \approx \kappa y \rangle$ and $\langle \eta x \approx \eta y \rangle$ are both the subvariety with $\tau x \approx e$ for all τ.

For (4), the substructures have the form $\{\eta_j x, \ldots, \eta_k x\}$ for which the map η_j provides the retraction. □

One might hope that (1) could be extended to include predicates, or that the restrictions in (2) could be loosened. The 2-generated quasicritical structure in Fig. 7.23 squelches that hope.

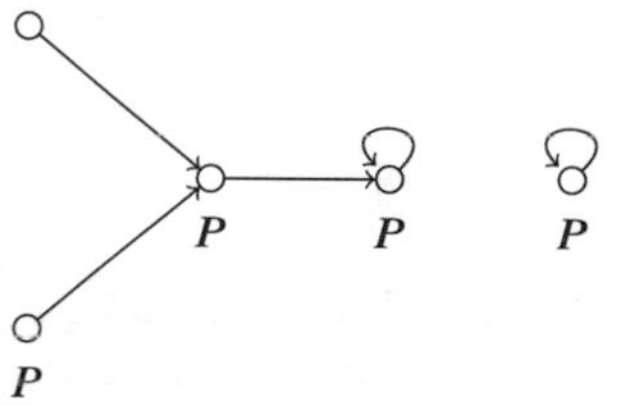

Fig. 7.23 A 2-generated quasicritical structure with one operation and one predicate

One might also seek to strengthen case (4) to semilattices of maps, as used in Sects. 9.3 and 9.4. Perhaps there is a way to do so, but it is not clear how.

Running Example 4 illustrates part (1) of Theorem 7.20, and Running Example 5 uses part (3). The next two extended examples will illustrate (2) and (4), respectively.

Example 7.21 Let us go through the entire six-step program with a pair that uses Theorem 7.20(2).

Consider the lattice $\mathbf{B}_3[a]$ obtained by doubling an atom in the Boolean algebra $\mathbf{2}^3$. In Chap. 3, we saw that $\mathbf{B}_3[a]$ with its least preclop μ fails (K10); this is the example in Fig. 3.6. With a different equaclosure operator γ, there is a representation as Sub($\mathbf{S}$, $\wedge$, 1, H), as diagrammed in Fig. 7.24. Note $\mathrm{J}(\mathbf{B}_3[a]) \subseteq \tau(\mathbf{B}_3[a])$ for this γ.

The adjoint of h is given by

	η
$\hat{0}$	$\hat{b}$
$\hat{a}$	$\hat{z}$
$\hat{b}$	$\hat{b}$
$\hat{c}$	$\hat{b}$
$\hat{z}$	$\hat{z}$

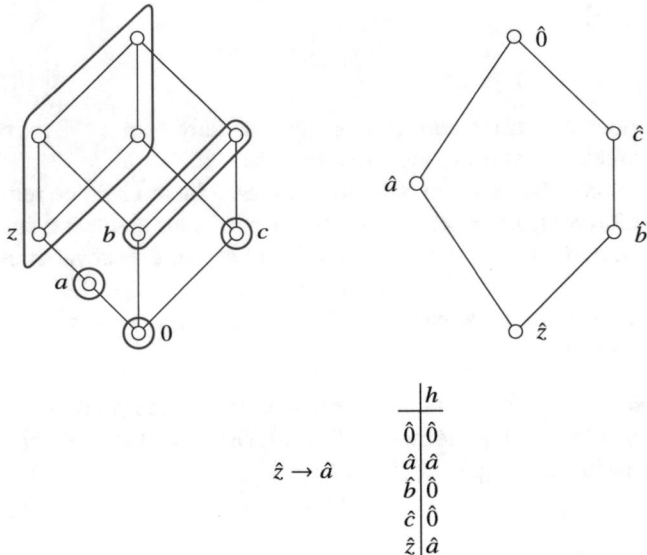

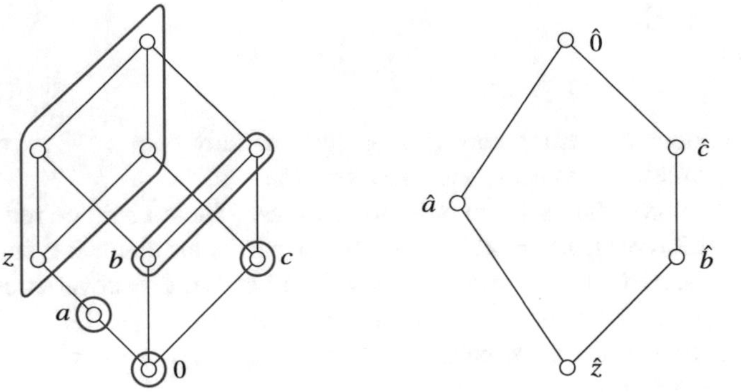

$\hat{z} \to \hat{a}$

	h
$\hat{0}$	$\hat{0}$
$\hat{a}$	$\hat{a}$
$\hat{b}$	$\hat{0}$
$\hat{c}$	$\hat{0}$
$\hat{z}$	$\hat{a}$

Fig. 7.24 Example 7.21

The calculations for step 3 go as follows. The congruences of $\widehat{\mathbf{S}}$ are Δ, $\mathrm{con}(\hat{c}, \hat{0}) = [\hat{c}, \hat{0}]$, $\mathrm{con}(\hat{b}, \hat{c}) = [\hat{b}, \hat{c}]$, $\mathrm{con}(\hat{z}, \hat{b}) = [\hat{z}, \hat{b}][\hat{a}, \hat{0}]$, $\mathrm{con}(\hat{b}, \hat{0}) = [\hat{b}, \hat{c}, \hat{0}]$, $\mathrm{con}(\hat{a}, \hat{c}) = [\hat{z}, \hat{b}][\hat{a}, \hat{c}, \hat{0}]$, $\mathrm{con}(\hat{z}, \hat{c}) = [\hat{z}, \hat{b}, \hat{c}][\hat{a}, \hat{0}]$, $\mathrm{con}(\hat{z}, \hat{a}) = [\hat{z}, \hat{a}][\hat{b}, \hat{c}, \hat{0}]$, ∇ ordered as the dual of $\mathbf{B}_3[a]$. The predicates corresponding to $\widehat{\mathbf{S}} \setminus \{\hat{z}\}$ are O, A, B, and C, assigned as follows:

$$
\begin{aligned}
\hat{0} &\rightsquigarrow O(x)\\
\hat{a} &\rightsquigarrow A(x)\\
\hat{b} &\rightsquigarrow B(x), O(\eta x), B(\eta x), C(\eta x)\\
\hat{c} &\rightsquigarrow C(x)\\
\hat{z} &\rightsquigarrow A(\eta x).
\end{aligned}
$$

Thus $\mathcal{K}_0$ has the operations e, η and predicates O, A, B, C with laws

$$
\begin{array}{rl}
X(e) \qquad X(\eta e) & \text{for } X = O, A, B, C\\
X(\eta^2 x) \leftrightarrow X(\eta x) & \text{for } X = O, A, B, C\\
O(x) \to A(x) & O(x) \to C(x) \to B(x) \leftrightarrow B(\eta x) \leftrightarrow O(\eta x)\\
A(x)\ \&\ B(x) \to O(x) & A(\eta x).
\end{array}
$$

Now for step 4, the join irreducible theories of $\mathrm{ITh}(\mathcal{K}_0)$ are $\langle Ax \rangle$, $\langle Cx \to Ox \rangle$, $\langle Bx \to Cx \rangle$, and $\langle Bx \rangle$, corresponding to the join irreducible congruences of $\widehat{\mathbf{S}}$ from step (3). The lattice of quasivarieties is drawn in Fig. 7.25. Again note that the equaclosure operator Γ agrees with γ from Fig. 7.24.

The shortstyle conversion to a quasivariety $\mathcal{K}_1$ with equality is straightforward. Interpret $O(x) \mapsto x \approx e$, $A(x) \mapsto \eta x \approx x$ and $B(x) \mapsto \eta x \approx e$. The result is a quasivariety with operations η, e, and a predicate C satisfying the laws

$$\eta e \approx e \qquad C(e) \qquad \eta^2 x \approx \eta x$$
$$C(x) \to \eta x \approx e$$

as diagrammed in Fig. 7.26.

It remains to show that **F**, **S**, **I**, and **J** of Fig. 7.26 are the only quasicritical structures in $\mathcal{K}_1$. But we can apply Theorem 7.20(2) to see that every quasicritical structure is 1-generated, and thus the list is complete.

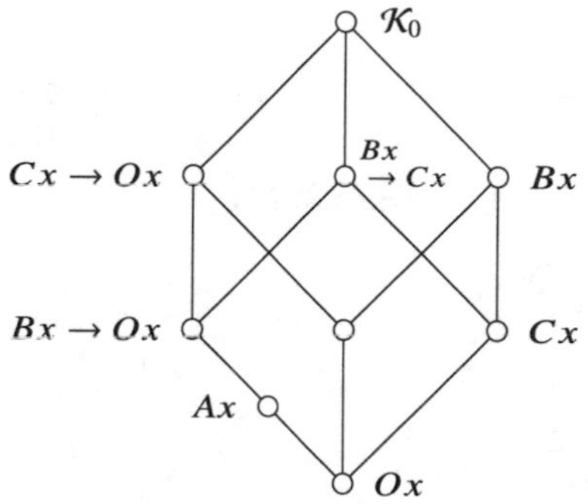

Fig. 7.25 $\mathrm{L_q}(\mathcal{K}_0)$ for Example 7.21. The middle atom in the lower lattice is $Bx \,\&\, (Cx \to Ox)$

Example 7.22 The construction for Example 7.21 can be adapted to show that for every $n \geq 2$, the lattice $\mathbf{B}_n[a]$ can be represented as a subquasivariety lattice. Here $\mathbf{B}_n[a]$ is the lattice obtained by doubling an atom in the lattice $\mathbf{2}^n$. This is a straightforward exercise left to the reader (and a special case of the next example). As in the reverse engineering of Sect. 7.3, the equaclosure operator γ is not specified ahead of time. Let us sketch the method for $\mathbf{B}_4[a]$ as a guide.

The quasivariety $\mathcal{K}_4$ representing $\mathbf{B}_4[a]$ has a unary function η, the constant e, and predicates C, D, and $\approx$. The laws of $\mathcal{K}_4$ are

$$\eta e \approx e \qquad D(e) \qquad C(e) \qquad \eta^2 x \approx \eta x$$
$$D(x) \to C(x) \to \eta x \approx e.$$

The subquasivariety lattice $\mathrm{L_q}(\mathcal{K}_4)$ and quasicritical structures of $\mathcal{K}_4$ are shown in Fig. 7.27.

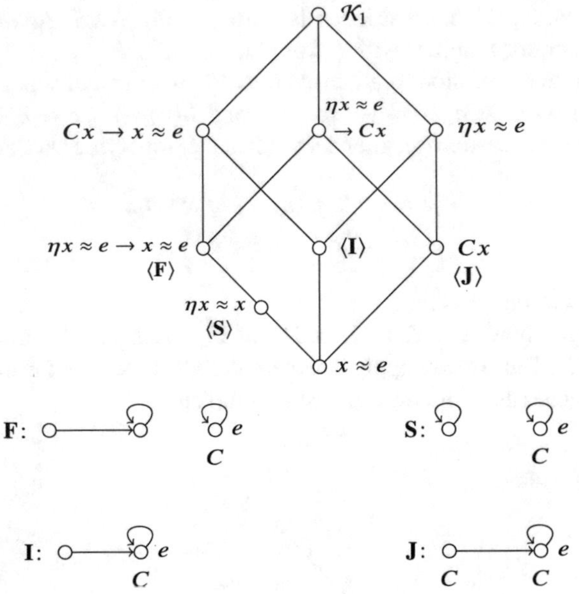

Fig. 7.26 $\mathrm{L_q}(\mathcal{K}_1)$ isomorphic to $\mathbf{B}_3[a]$, Example 7.21

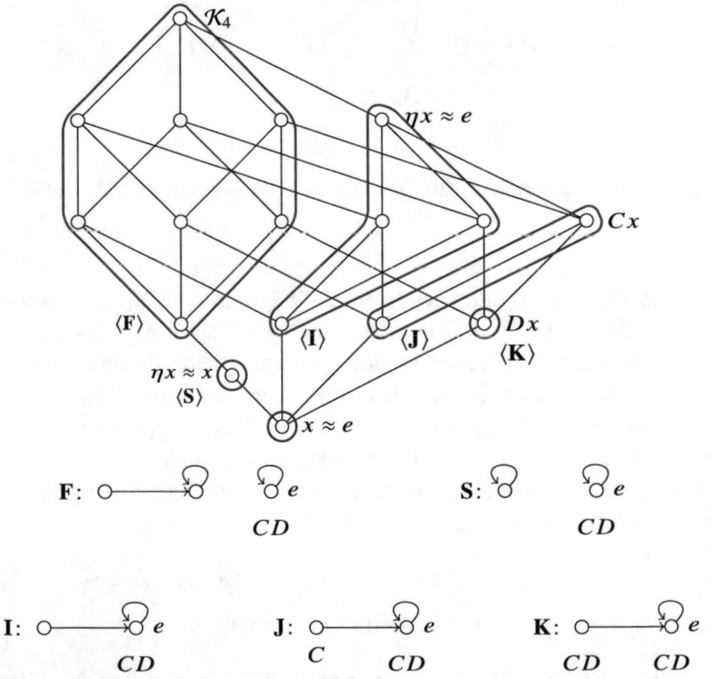

Fig. 7.27 $\mathrm{L_q}(\mathcal{K}_4)$ isomorphic to $\mathbf{B}_4[a]$ (Example 7.22)

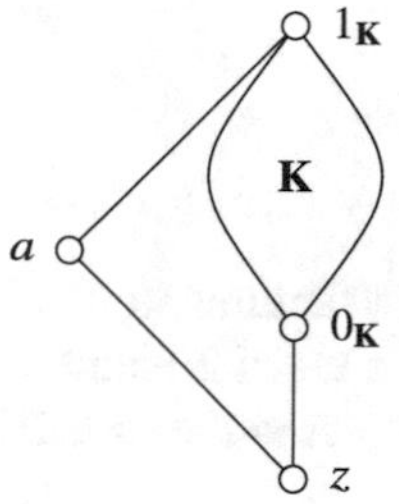

S

	h	η
k	$1_{\mathbf{K}}$	$0_{\mathbf{K}}$
a	a	z
z	a	z

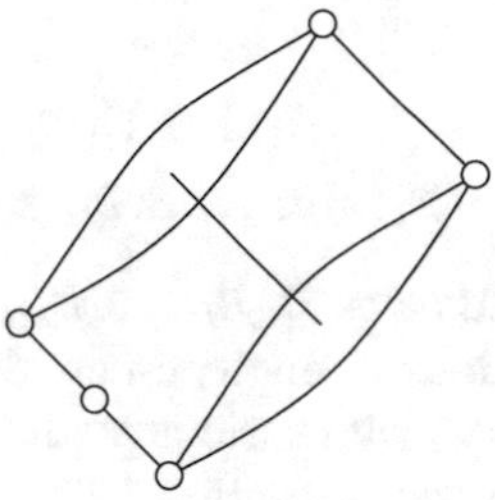

$\mathrm{S_p}(\mathbf{S}, h)$

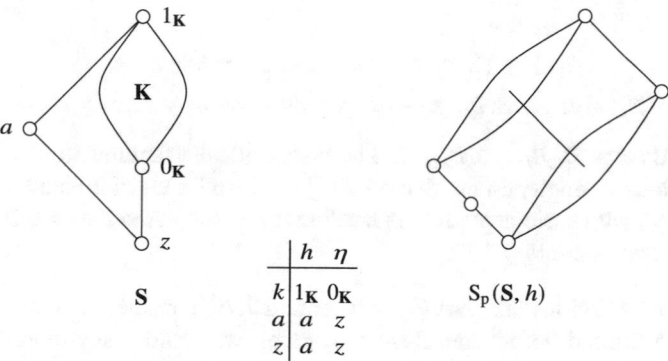

Fig. 7.28 On the left, the lattice **S** and operator h for Example 7.23, where k represents any element of K. On the right is $\mathrm{S_p}(\mathbf{S}, h)$, which is isomorphic to $\mathrm{S_p}(\mathbf{K}) \times \mathbf{2}$ with an atom doubled

Example 7.23 On the other hand, if we let $\mathbf{B}_X = \mathbf{2}^X$ for an infinite set X, and then double an atom, the lattice $\mathbf{B}_X[a]$ is not dually algebraic, and so not a quasivariety lattice. Instead, let us consider a construction that does work, and properly generalizes the previous example.

Recall that for any algebraic lattice $\mathbf{A}$, $\mathrm{S_p}(\mathbf{A})$ is an atomistic, dually algebraic lattice. Let $\mathbf{K}$ be an algebraic lattice whose top element $1_{\mathbf{K}}$ is compact. Form a new lattice $\mathbf{S}$ on $K \cup \{a, z\}$, ordered as on the left side of Fig. 7.28. Then $\mathbf{S}$ is algebraic, in view of the requirement that $1_{\mathbf{K}}$ be compact.

Let h be the operator on $\mathbf{S}$ with

$$h(x) = \begin{cases} a & \text{if } x = z \text{ or } x = a, \\ 1_{\mathbf{K}} & \text{if } x \in K. \end{cases}$$

The h-closed algebraic subsets of $\mathbf{S}$ are

- X for any algebraic subset of $\mathbf{K}$,
- $\{1_{\mathbf{K}}, a\}$,
- $\{a, z\} \cup X$ for any algebraic subset of $\mathbf{K}$.

Thus $\mathrm{S_p}(\mathbf{S}, h)$ is isomorphic to $\mathrm{S_p}(\mathbf{K}) \times \mathbf{2}$ with an atom doubled, as illustrated in Fig. 7.28. Moreover, Theorem 7.20(2) applies to yield a shortstyle quasivariety representation of $\mathrm{S_p}(\mathbf{S}, h)$ as $\mathrm{L_q}(\mathcal{K})$.

The language of $\mathcal{K}$ has a unary function η, the constant e, a predicate A and predicates B for each $b \in K$, and $\approx$. As in the previous example, we can interpret $Ax \mapsto \eta x \approx x$, and the predicate corresponding to $0_{\mathbf{K}}$ as $\eta x \approx e$. The laws of $\mathcal{K}$ are then

$$\eta e \approx e \qquad B(e) \qquad \eta^2 x \approx \eta x$$

$$B(x) \to C(x) \to \eta x \approx e \text{ whenever } b \geq c$$

$$B_1(x) \,\&\, \ldots \,\&\, B_m(x) \to C(x) \text{ whenever } b_1 \vee \ldots \vee b_m \geq c$$

for all predicates $B, B_1, \ldots, B_m, C$. The quasicritical structures in $\mathcal{K}$ are **F** and **S** from previous example, except that all $B(e)$ hold, and a set of 2-element structures with $\eta x = e$, with various predicates holding for x, subject to the restrictions in the laws above; compare Fig. 7.27.

Example 7.24 To illustrate part (4) of Theorem 7.20, consider the lattice $\mathbf{1} + \mathbf{N}_5$ of Fig. 7.29. With a different equaclosure operator, we could easily represent $\mathbf{1} + \mathbf{N}_5$ using the methods of Chap. 8. Instead, let us sketch the six-step program for the equaclosure operator γ in the figure. The arrows, operators, and adjoints are also given in Fig. 7.29.

At step 3, we note that the adjoint maps have the composition table

	η	κ
η	η	κ
κ	κ	κ

so that $\eta > \kappa$, setting up part (4). We are supposed to use the opposite composition $\star$, but the maps commute.

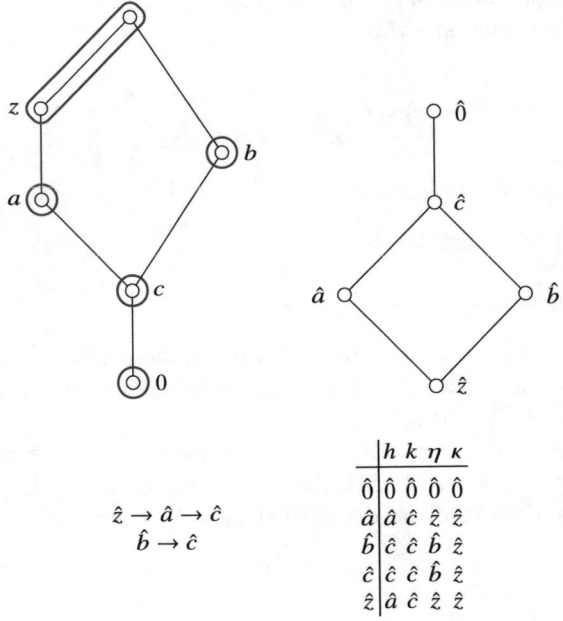

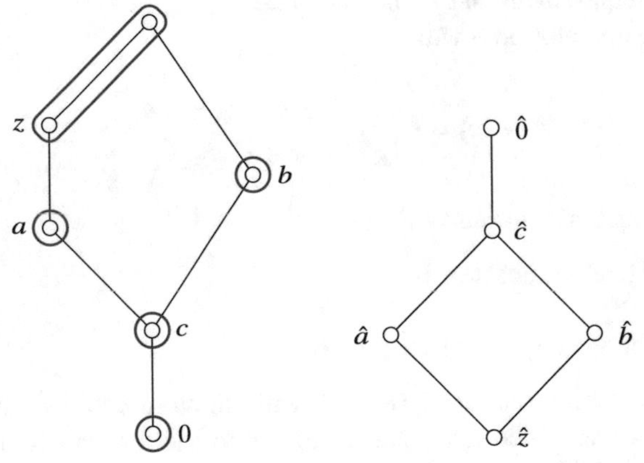

$\hat{z} \to \hat{a} \to \hat{c}$
$\hat{b} \to \hat{c}$

	h	k	η	κ
$\hat{0}$	$\hat{0}$	$\hat{0}$	$\hat{0}$	$\hat{0}$
$\hat{a}$	$\hat{a}$	$\hat{c}$	$\hat{z}$	$\hat{z}$
$\hat{b}$	$\hat{c}$	$\hat{c}$	$\hat{b}$	$\hat{z}$
$\hat{c}$	$\hat{c}$	$\hat{c}$	$\hat{b}$	$\hat{z}$
$\hat{z}$	$\hat{a}$	$\hat{c}$	$\hat{z}$	$\hat{z}$

Fig. 7.29 Example 7.24 with a chain of operations

Then for the quasivariety $\mathcal{K}_0$ without equality, there are operations η, κ, the constant e, and predicates A, B, C satisfying, in addition to the standard laws involving e and the operations,

$$O(x) \leftrightarrow O(\eta x) \leftrightarrow O(\kappa x) \qquad Bx \leftrightarrow B(\eta x) \leftrightarrow C(\eta x)$$
$$A(\eta x) \qquad A(\kappa x) \qquad B(\kappa x) \qquad C(\kappa x)$$
$$Cx \to Ax \qquad Cx \to Bx \qquad Ax \,\&\, Bx \to Cx.$$

To convert to a quasivariety $\mathcal{K}_1$ with equality, we convert

$$Ox \mapsto x \approx e \qquad Ax \mapsto \eta x \approx x \qquad Bx \mapsto \eta x \approx \kappa x$$

which entails $Cx \mapsto \kappa x \approx x$. The laws of $\mathcal{K}_0$ then translate to the following laws for $\mathcal{K}_1$:

$$\eta e \approx e \qquad \kappa e \approx e$$
$$\eta x \approx e \to x \approx e \qquad \kappa x \approx e \to x \approx e$$
$$\eta^2 x \approx \eta x \qquad \kappa^2 x \approx \kappa x$$
$$\eta\kappa x \approx \kappa x \approx \kappa\eta x \qquad .$$

The quasicritical algebras **F**, **S**, **T**, **D** of Fig. 7.30 provide models.

The conversion from $\mathcal{K}_0$ to $\mathcal{K}_1$ was successful, meaning that the representation indicated by Fig. 7.30 is correct so far as 1-variable quasi-identities, or 1-generated quasicritical algebras, are concerned. Now we invoke Theorem 7.20(4) to say that *all* the quasicritical algebras of $\mathcal{K}_1$ are 1-generated, and conclude that the representation of the figure is complete!

7.5 Shortstyle Representations Revisited

Now let us analyze what can go wrong with shortstyle representations. The six-step program starts with a pair $(\mathbf{L}, \gamma)$ and constructs a quasivariety $\mathcal{K}_0$ in a language without equality such that $(\mathbf{L}, \gamma) \cong (\mathrm{L_q}(\mathcal{K}_0), \Gamma)$. The language of $\mathcal{K}_0$ has unary operations, unary predicates, and a constant. All subquasivarieties of $\mathcal{K}_0$ are determined by quasi-equations in 1 variable. Step 6 attempts to translate $\mathcal{K}_0$ into a quasivariety $\mathcal{K}_1$ with equality such that $\mathrm{L_q}(\mathcal{K}_0)$ embeds into $\mathrm{L_q}(\mathcal{K}_1)$. The translation requires interpreting some predicates of $\mathcal{K}_0$ in terms of equality, e.g., $Ox \mapsto x \approx e$ and say $Ax \mapsto \eta x \approx x$. Admittedly, we lack a systematic method for these translations. If

(1) the lattices of 1-variable implicational theories of $\mathcal{K}_0$ and $\mathcal{K}_1$ are isomorphic under the translation, and
(2) each subquasivariety of $\mathcal{K}_1$ is determined by quasi-equations in 1 variable,

then $(\mathrm{L_q}(\mathcal{K}_0), \Gamma) \cong (\mathrm{L_q}(\mathcal{K}_1), \Gamma)$, and the program has succeeded.

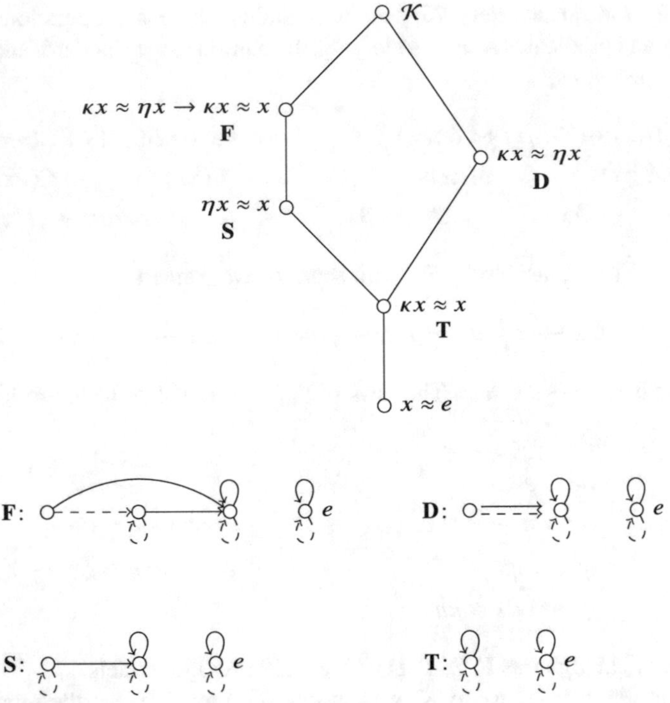

Fig. 7.30 $\mathrm{L_q}(\mathcal{K}_1)$ for Example 7.24. Dotted arrows represent η, and solid arrows κ

So two things can go wrong. First, the translation for 1-variable quasi-identities may not work. This happens fairly often and is rather obvious when it occurs.

Example 7.25 We saw that Running Example 3 has a longstyle representation. Under the most natural translation, the atom on the left of the proposed $\mathrm{L_q}(\mathcal{K}_1)$ is generated by the algebra in Fig. 7.31. This algebra is quasicritical and has a nontrivial subalgebra, so it cannot generate an atom of $\mathrm{L_q}(\mathcal{K}_1)$. Attempts to fix this, without going to the longstyle representation, just lead to worse problems. Perhaps the pair in Running Example 3 has a locally finite representation, but if so, our method does not find it.

Fig. 7.31 Structure for Example 7.25 that cannot generate an atom in the subquasivariety lattice

The second problem that the proposed $\mathcal{K}_1$ may contain subquasivarieties with laws in more than 1 variable is more subtle and devious. Sometimes we can use

Theorem 7.20 to show that every quasicritical structure in $\mathcal{K}$ is 1-generated, but that has limited applicability. If $\mathcal{K}_1$ contains a quasicritical structure $\mathbf{T}$ that is not 1-generated, then $\mathbf{T}$ generates a subquasivariety with no analogue in $L_q(\mathcal{K}_0)$. That can happen too easily, and it is a nearly irremediable problem: the program has failed for this pair, unless we try a different representation in step 1. This is best illustrated by an example.

Example 7.26 Consider the pair $(\mathbf{W}, \mu)$ from Sect. 6.2.1. Since $J(\mathbf{W}) \subseteq \tau(\mathbf{W})$ for μ, we can use the representation from Sect. 5.2; see Fig. 7.32. (The general methods of Chap. 6 need not give the simplest representation of a finite lower bounded lattice; for present purposes, the simpler the better.) Note that since $h(\hat{z}) = \hat{a}$ and $k(\hat{z}) = \hat{c}$, we can use the meet $h(\hat{z}) \wedge k(\hat{z}) = \hat{a} \wedge \hat{c} = \hat{d}$ to enforce $\hat{z} \to \hat{d}$, and do not need another map for that purpose. On the other hand, H should include the composition $\ell = hk = kh$.

The adjoint maps, with $\ell' = \zeta$, and the operation table for H'^{opp} are

	η	κ	ζ
$\hat{0}$	$\hat{c}$	$\hat{b}$	$\hat{z}$
$\hat{a}$	$\hat{z}$	$\hat{b}$	$\hat{z}$
$\hat{b}$	$\hat{z}$	$\hat{b}$	$\hat{z}$
$\hat{c}$	$\hat{c}$	$\hat{z}$	$\hat{z}$
$\hat{d}$	$\hat{z}$	$\hat{z}$	$\hat{z}$
$\hat{z}$	$\hat{z}$	$\hat{z}$	$\hat{z}$

$\star$	η	κ	ζ
η	η	ζ	ζ
κ	ζ	κ	ζ
ζ	ζ	ζ	ζ

As always, steps 2–5 are routine (though long), yielding $\mathcal{K}_0$ that has the operations e, η, κ, ζ and predicates O, A, B, C, D with laws

$$X(e) \quad X(fe) \quad \text{for } f = \eta, \kappa, \zeta$$
$$B(x) \leftrightarrow O(\kappa x) \leftrightarrow A(\kappa x) \leftrightarrow B(\kappa x)$$
$$A(\eta x) \quad B(\eta x) \quad C(\kappa x) \quad D(\eta x) \quad D(\kappa x) \quad X(\zeta x)$$
$$O(x) \to A(x) \to B(x) \qquad O(x) \to C(x) \to D(x) \qquad A(x) \to D(x)$$
$$B(x) \,\&\, D(x) \to A(x) \qquad B(x) \,\&\, C(x) \to O(x)$$
$$X(\eta(\eta x)) \leftrightarrow X(\eta x) \qquad X(\eta(\kappa(x)) \leftrightarrow X(\zeta(x)) \qquad X(\eta(\zeta x)) \leftrightarrow X(\zeta x)$$
$$X(\kappa(\eta x)) \leftrightarrow X(\zeta x) \qquad X(\kappa(\kappa(x)) \leftrightarrow X(\kappa(x)) \qquad X(\kappa(\zeta x)) \leftrightarrow X(\zeta x)$$
$$X(\zeta(\eta x)) \leftrightarrow X(\zeta x) \qquad X(\zeta(\kappa(x)) \leftrightarrow X(\zeta(x)) \qquad X(\zeta(\zeta x)) \leftrightarrow X(\zeta x)$$

for $X = O, A, B, C, D$.

The lattice of quasivarieties $L_q(\mathcal{K}_0)$ is drawn in Fig. 7.33. Note that $\Gamma\langle B(x) \to A(x)\rangle = \mathcal{K}_0$.

Now convert to $\mathcal{K}_1$ using a shortstyle representation, choosing the translations

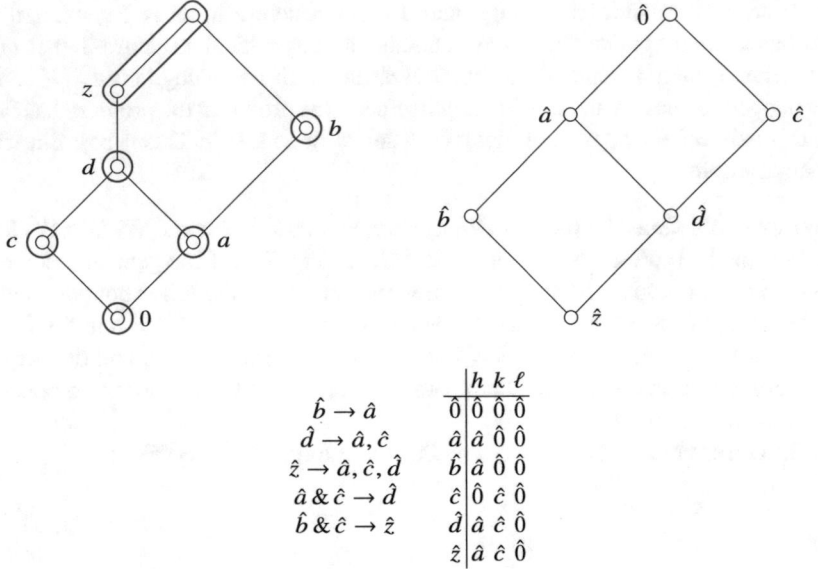

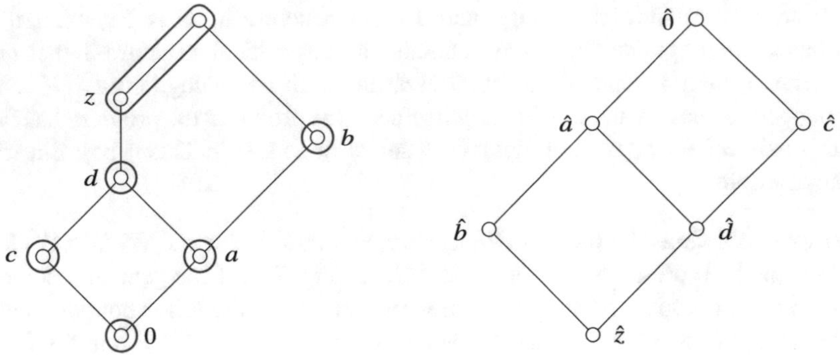

$$\begin{aligned} \hat{b} &\to \hat{a} \\ \hat{d} &\to \hat{a}, \hat{c} \\ \hat{z} &\to \hat{a}, \hat{c}, \hat{d} \\ \hat{a} \,\&\, \hat{c} &\to \hat{d} \\ \hat{b} \,\&\, \hat{c} &\to \hat{z} \end{aligned}$$

	h	k	ℓ
$\hat{0}$	$\hat{0}$	$\hat{0}$	$\hat{0}$
$\hat{a}$	$\hat{a}$	$\hat{0}$	$\hat{0}$
$\hat{b}$	$\hat{a}$	$\hat{0}$	$\hat{0}$
$\hat{c}$	$\hat{0}$	$\hat{c}$	$\hat{0}$
$\hat{d}$	$\hat{a}$	$\hat{c}$	$\hat{0}$
$\hat{z}$	$\hat{a}$	$\hat{c}$	$\hat{0}$

Fig. 7.32 Step 1 for Example 7.26: $(\mathbf{W}, \mu)$

$$\begin{aligned} O(x) &\mapsto x \approx e \\ B(x) &\mapsto \kappa x \approx e \\ C(x) &\mapsto \eta x \approx e \\ A(x) &\mapsto \kappa x \approx e \,\&\, D(x). \end{aligned}$$

Moreover, since $O(\zeta x)$ is a law of $\mathcal{K}_0$, we can replace all instances of ζx by e. That leaves us with a language having operations η, κ, e and predicates D, $\approx$.

The laws from $\mathcal{K}_0$ convert to:

$$\begin{array}{rll} \eta e \approx e & \kappa e \approx e & D(e) \\ \eta^2 x \approx \eta x & \kappa^2 x \approx \kappa x & \\ \eta \kappa x \approx e & \kappa \eta x \approx e & \\ D(\eta x) & D(\kappa x) & \\ \eta x \approx e \to D(x) & \kappa x \approx e \,\&\, \eta x \approx e \to x \approx e. & \end{array}$$

As an exercise, one can derive that $\mathcal{K}_1$ satisfies these quasi-equations:

$$\begin{aligned} \eta x \approx x &\to \kappa x \approx e \\ \kappa x \approx x &\to \eta x \approx e \\ \eta x \approx \kappa y &\to \eta x \approx e. \end{aligned}$$

But a quick check (using the homomorphic images of **F**) shows that we still have too many 1-generated quasicritical structures: 7 when we need 4, the number of join irreducibles in **W**. To that end, we add two new laws:

$$\eta x \approx e \longrightarrow \kappa x \approx x$$
$$D(x)\,\&\,\kappa x \approx e \longrightarrow \eta x \approx x$$

leaving the 4 quasicritical structures **F**, **H**, **I**, and **J** of Fig. 7.34.

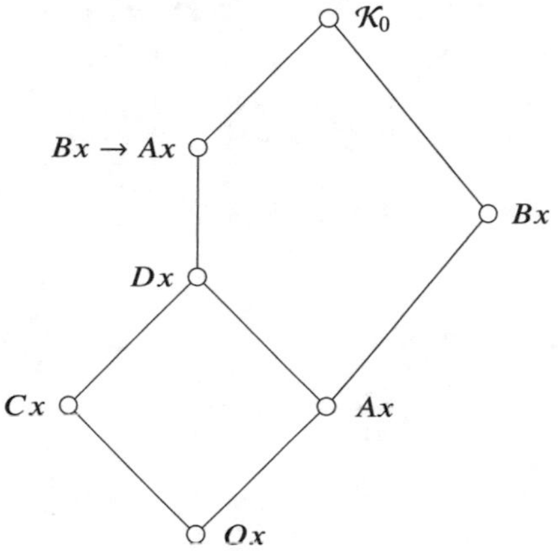

Fig. 7.33 $\mathrm{L_q}(\mathcal{K}_0)$ for Example 7.26: $(\mathbf{W}, \mu)$

It appears that we have almost reached the goal. Then, in checking for 2-generated quasicritical structures, we find the structure **X** in Fig. 7.35. Alas, **X** is a subdirect product of **F** and **J**; as long as both of those are in the quasivariety $\mathcal{K}_1$, then **X** comes with them.

One might hope that a different choice of the added laws might eliminate this problem, but none of those options is plausible. In short, the shortstyle method has failed for $(\mathbf{W}, \mu)$. The mediumstyle method of Sect. 7.6 will fix this, representing $(\mathbf{W}, \mu)$ as $(\mathrm{L_q}(Q), \Gamma)$ for a non-locally finite quasivariety.

Example 7.27 Our second example is similar, so we will only sketch it. The pair $(\mathbf{M}, \gamma)$ of Fig. 7.36 has $\mathrm{J}(\mathbf{M}) \subseteq \tau(\mathbf{M})$. Except for the labeling, it has the same semigroup of operators as $(\mathbf{W}, \mu)$, and hence the same adjoints. Following the prescription, we obtain a quasivariety $\mathcal{M}_0$, then interpret

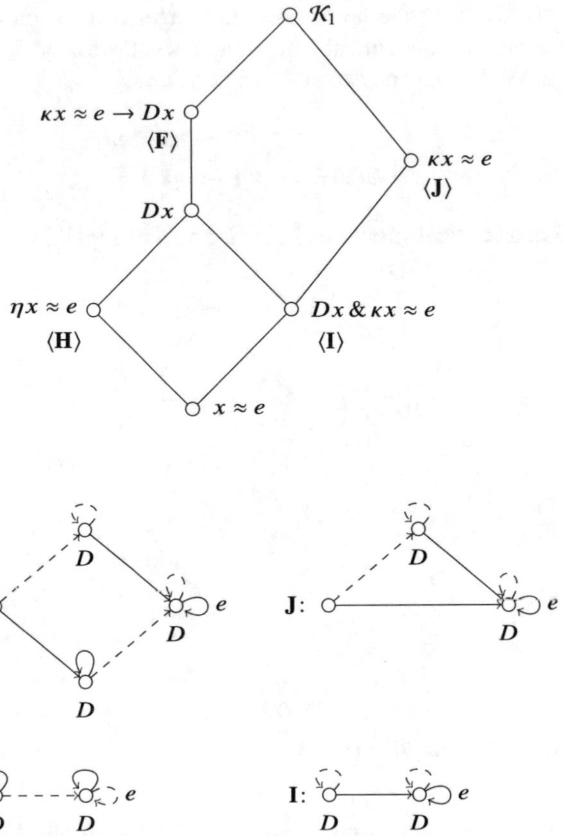

Fig. 7.34 Proposed $\mathrm{L_q}(\mathcal{K}_1)$ for the pair $(\mathbf{W}, \mu)$. It is not correct: the quasivariety generated by the structure **X** in Fig. 7.35 is among the missing. Dotted arrows represent η, and solid arrows κ

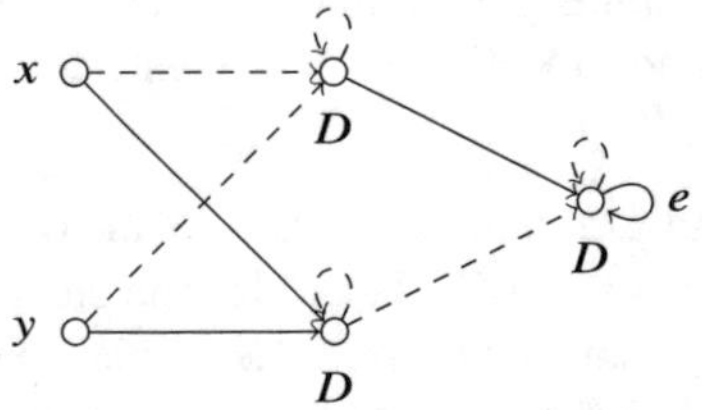

Fig. 7.35 A 2-generated quasicritical structure **X** in $\mathcal{K}_1$. Dotted arrows represent η, and solid arrows κ

$$
\begin{aligned}
O(x) &\mapsto x \approx e \\
A(x) &\mapsto \eta x \approx x \\
B(x) &\mapsto \kappa x \approx x \\
C(x) &\mapsto \kappa x \approx e \\
D(x) &\mapsto \eta x \approx e
\end{aligned}
$$

to obtain the putative representation $(\mathrm{L_q}(\mathcal{M}), \Gamma)$ of Fig. 7.37. The laws of $\mathcal{M}$ are

$$
\begin{gathered}
\eta e \approx e \qquad \kappa e \approx e \\
\eta^2 x \approx \eta x \qquad \kappa^2 x \approx \kappa x \qquad \eta\kappa x \approx e \qquad \kappa\eta x \approx e \\
\eta x \approx x \to \kappa x \approx e \qquad \kappa x \approx x \to \eta x \approx e \\
\kappa x \approx e \,\&\, \eta x \approx e \to x \approx e.
\end{gathered}
$$

The free structure **F** on one generator is the same, except this time there is no unary predicate, so $\mathbf{F} = \mathbf{S}\times\mathbf{T}$. Again the quasivariety generated by the quasicritical algebra **X** of Fig. 7.35 (without the predicates) is missing. Thus the shortstyle program fails for $(\mathbf{M}, \gamma)$. Again, the mediumstyle method of Sect. 7.6 will fix this.

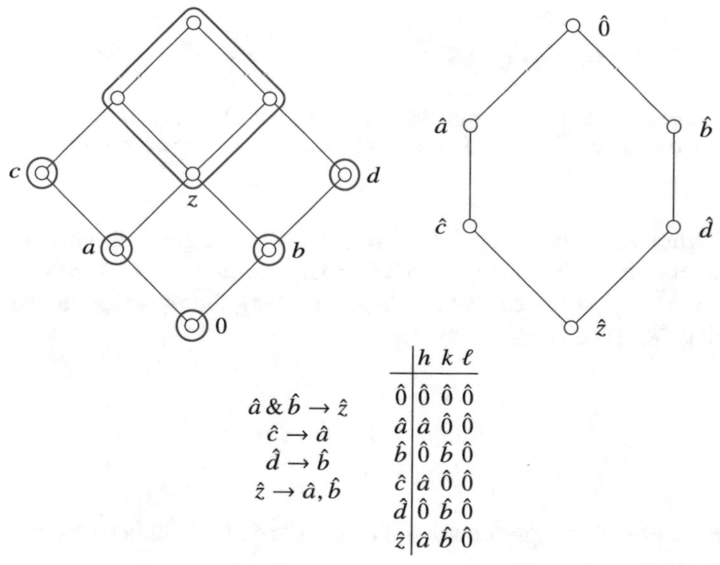

$\hat{a} \,\&\, \hat{b} \to \hat{z}$
$\hat{c} \to \hat{a}$
$\hat{d} \to \hat{b}$
$\hat{z} \to \hat{a}, \hat{b}$

	h	k	ℓ
$\hat{0}$	$\hat{0}$	$\hat{0}$	$\hat{0}$
$\hat{a}$	$\hat{a}$	$\hat{0}$	$\hat{0}$
$\hat{b}$	$\hat{0}$	$\hat{b}$	$\hat{0}$
$\hat{c}$	$\hat{a}$	$\hat{0}$	$\hat{0}$
$\hat{d}$	$\hat{0}$	$\hat{b}$	$\hat{0}$
$\hat{z}$	$\hat{a}$	$\hat{b}$	$\hat{0}$

Fig. 7.36 Step 1 for Example 7.27: $(\mathbf{M}, \gamma)$

Example 7.28 After all the attention given to the lattice **W** in Chap. 6, it behooves us to show that, with at least one of its equaclosure operators, it is a subquasivariety

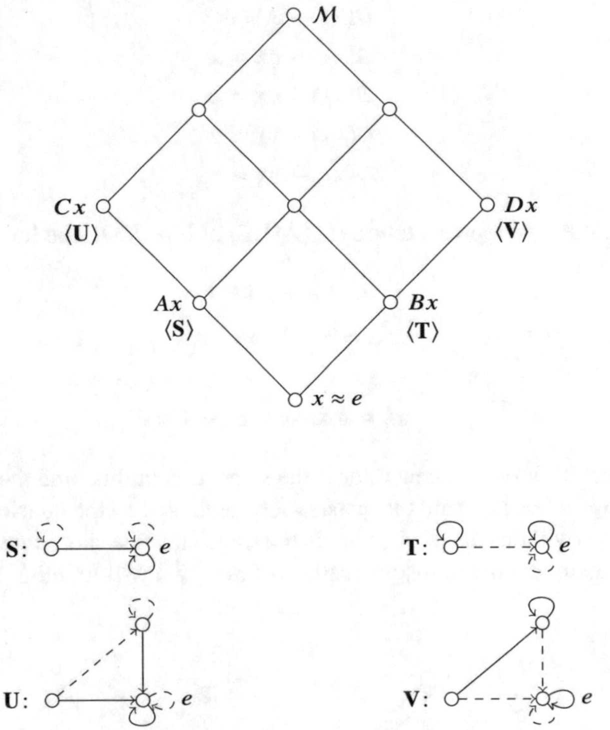

Fig. 7.37 Proposed $L_q(\mathcal{M})$ for the pair $(\mathbf{M}, \gamma)$ of Example 7.27. It is not correct: the quasivariety generated by the algebra **X** in Fig. 7.35 is missing. Dotted arrows represent η, and solid arrows κ

lattice. Figure 7.38 gives step 1 for $(\mathbf{W}, \gamma_6)$, with a slightly different semilattice representation than given in Sect. 6.2.6. Again we have $\mathrm{J}(\mathbf{W}) \subsetneq \tau(\mathbf{W})$, and this time $kh = k^2 = \ell$ is the constant $\hat{0}$ map. Following the prescription, we obtain a quasivariety $\mathcal{K}_0$, then choose to interpret

$$O(x) \mapsto x \approx e$$
$$A(x) \mapsto B(x) \,\&\, \kappa x \approx e$$
$$C(x) \mapsto \kappa x \approx e$$

to obtain the proposed representation $(L_q(\mathcal{K}_1), \Gamma)$ of Fig. 7.39 in terms of η, κ, B, $\approx$, and e. The laws of $\mathcal{K}_1$ are

$$\eta e \approx e \qquad \kappa e \approx e \qquad B(e)$$

$$\eta^2 x \approx \eta x \qquad \kappa^2 x \approx e \qquad \eta\kappa x \approx \kappa x \qquad \kappa\eta x \approx e$$

$$\eta x \approx x \to \kappa x \approx e \qquad \kappa x \approx x \to \eta x \approx e$$

$$\eta x \approx e \leftrightarrow x \approx e$$

$$B(\kappa x) \qquad B(\eta x) \leftrightarrow B(x).$$

More will be added momentarily.

If the representation of Fig. 7.39 is to be correct, then $\mathcal{K}_1$ must be the quasivariety generated by the structures **F**, **G**, **H**, **I** in that figure. The plan is to identify additional laws that hold in those four structures, add them to the list, and then employ Lemma 7.19 using the alternate form with (ß)′.

The extra laws for $\mathcal{K}_1$ are

$$\kappa x \approx e \to \eta x \approx x$$

$$\eta x \approx \kappa x \to x \approx e$$

$$\kappa x \approx \eta y \to y \approx \kappa x$$

$$\kappa x \approx \kappa y \to \kappa x \approx e$$

(Section 2.2 of [63] gives a general method for finding quasi-equations to eliminate unwanted quasicritical structures, and we can use them so long as they hold in **F**, **G**, **H**, and **I**.)

Recall the alternate condition of Lemma 7.19:

(ß)′ *Suppose* $\mathbf{U} \in \mathcal{K}$ *is generated irredundantly by the set* $\{x_1, \ldots, x_m\}$*, and that the following holds for every i: with* $\mathbf{W} = \mathrm{Sg}(x_i)$ *and* $\mathbf{T} = \mathrm{Sg}(\{x_j : j \neq i\})$ *and* $\mathbf{S} = \mathbf{W} \cap \mathbf{T}$*, there is a retraction* $\sigma : \mathbf{W} \twoheadrightarrow \mathbf{S}$.

To see that this holds in the current situation, we note that $\mathbf{S} = \mathbf{W} \cap \mathbf{T}$ can only be $\{e\}$ or $\{\eta x, e\}$, and there are retractions $\sigma : \mathbf{W} \twoheadrightarrow \mathbf{S}$ in both these cases, since **W** must be isomorphic to one of our four structures. We conclude that $\mathcal{K}_1$ has only 1-generated quasicritical structures, meaning only those four, and the representation of Fig. 7.39 is correct.

7.6 Mediumstyle Representations

Our general plan has been to construct quasivarieties $\mathcal{K}_1$ consisting of structures whose algebra reduct depends on the adjoint monoid H' from step 2. In the longstyle method, the structures in $\mathcal{K}_1$ have operators that act freely, whereas in the shortstyle method the operators act compactly. For example, if $\eta \in H'$ satisfies $\eta^2 = \eta$, then for the longstyle method we have $x \neq \eta x \neq \eta^2 x \neq \cdots$, but we posit, for all predicates P, that $P(\eta x)$ holds iff $P(\eta^2 x)$ holds. However, for the shortstyle method, we require that $\eta x \approx \eta^2 x$. For a particular representation problem $(\mathbf{L}, \gamma)$, when neither of these

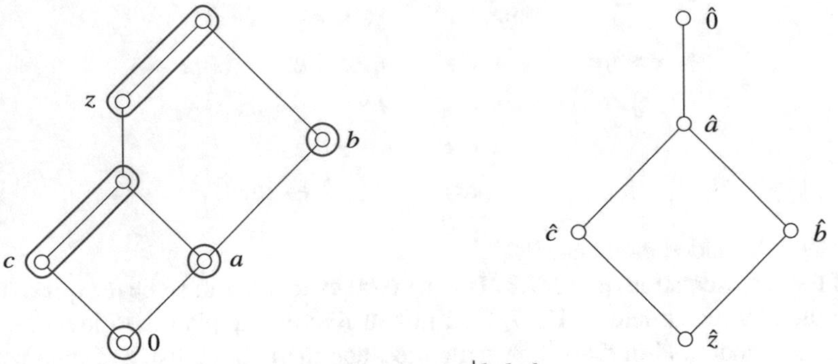

$\hat{b} \to \hat{a}$
$\hat{z} \to \hat{a}, \hat{c}$
$\hat{b} \,\&\, \hat{c} \to \hat{z}$

	h	k	ℓ
$\hat{0}$	$\hat{0}$	$\hat{0}$	$\hat{0}$
$\hat{a}$	$\hat{a}$	$\hat{0}$	$\hat{0}$
$\hat{b}$	$\hat{a}$	$\hat{a}$	$\hat{0}$
$\hat{c}$	$\hat{c}$	$\hat{0}$	$\hat{0}$
$\hat{z}$	$\hat{c}$	$\hat{a}$	$\hat{0}$

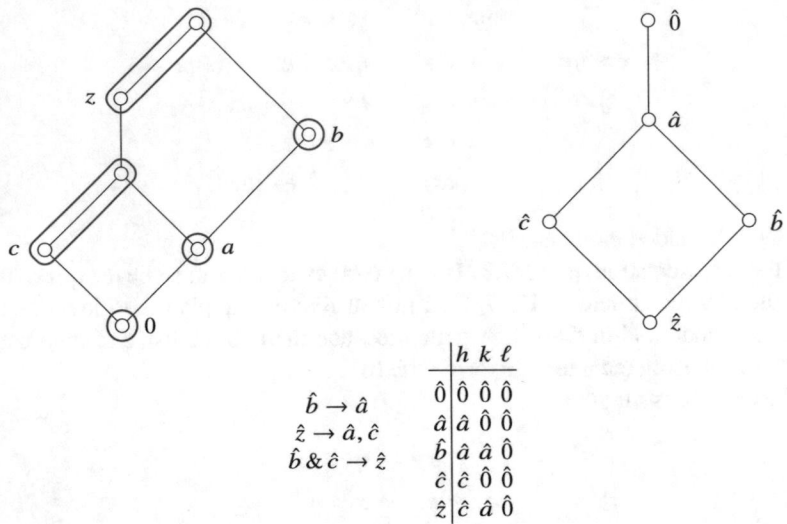

Fig. 7.38 Step 1 for Example 7.28: $(\mathbf{W}, \gamma_6)$

extremes works, we might try something with features of both, especially when more than one operator is involved. This we term a *mediumstyle* representation. It is best illustrated by examples.

Example 7.29 The lattice $\mathbf{B}_3 = \mathbf{2}^3$, with the equaclosure operator shown in Fig. 7.40, seems difficult to represent as a quasivariety lattice. Neither the longstyle nor short-style representation works. (The details are omitted here.) But there is a combination that does yield a representation.

Let Q be the quasivariety generated by the 3 algebras in Fig. 7.41. The language has 2 unary operations, η and κ, the constant e, and only the predicate $\approx$. The algebras $\mathbf{F}, \mathbf{A}, \mathbf{B}$ satisfy the following laws, for all $\ell, m \geq 1$.

$$\eta e \approx e \approx \kappa e \qquad \eta\kappa x \approx \kappa\eta x$$

$$\eta^m x \approx x \to x \approx e \qquad \kappa^m x \approx x \to x \approx e$$

$$\eta^\ell \kappa^m x \approx x \to x \approx e \qquad \eta^\ell x \approx \kappa^m x \to x \approx e$$

$$\eta^2 x \approx e \to \eta x \approx e \qquad \kappa^2 x \approx e \to \kappa x \approx e$$

$$\eta x \approx \eta y \;\&\; \kappa x \approx \kappa y \to x \approx y \quad .$$

Now $\mathbf{F}$ is the free algebra $\mathbf{F}_Q(1)$, and its Q-congruence lattice is $\mathbf{1} + \mathbf{2} \times \mathbf{2}$ exactly as in the second lattice of Fig. 7.40. Its Q-homomorphic images are $\mathbf{E}$ (the 1-element algebra), $\mathbf{A}$, $\mathbf{B}$, $\mathbf{C}$ (the subalgebra of $\mathbf{A} \times \mathbf{B}$ generated by (x, x), see Fig. 7.42), and $\mathbf{F}$. Moreover, the k-generated free algebra $\mathbf{F}_Q(k)$ consists of k copies of $\mathbf{F}$ glued over $\{e\}$.

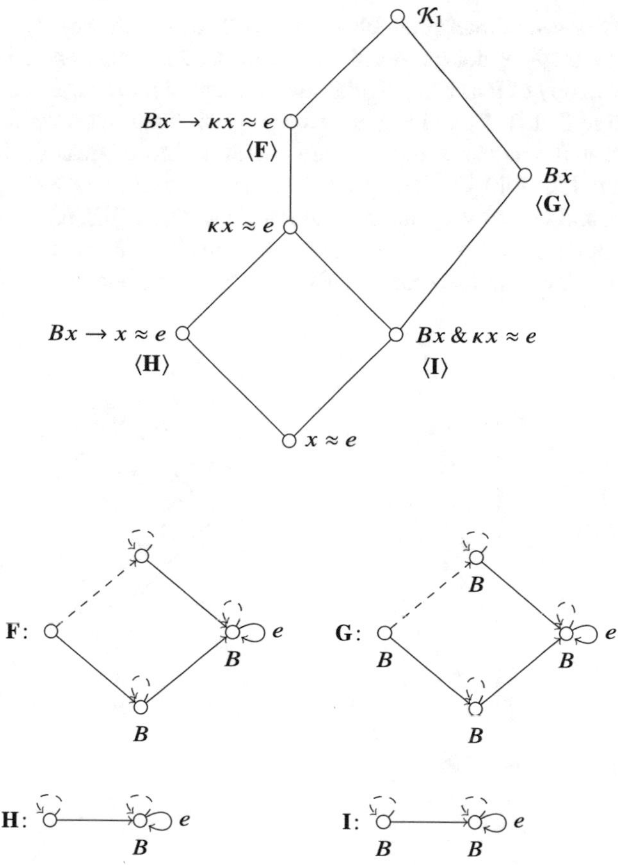

Fig. 7.39 $\mathrm{L_q}(\mathcal{K}_1)$ for Example 7.28: $(\mathbf{W}, \gamma_6)$. This one is correct! Dotted arrows represent η, and solid arrows κ

The k-generated algebras in Q are Q-homomorphic images of $\mathbf{F}_Q(k)$. They consist of say $m \leq k$ components, each isomorphic to one of the following forms, glued over $\{e\}$:

- **A**,
- **B**,
- **C**,
- **F**,
- $\mathbf{C}' = \mathbf{C} \setminus \{(x, x)\}$,
- a finitely generated subalgebra of **F**.

Now $\mathbf{C}$ and $\mathbf{C}'$ are subdirect products of $\mathbf{A}$ and $\mathbf{B}$, of which they contain copies as subalgebras, so they are not quasicritical. Likewise, a finitely generated subalgebra of $\mathbf{F}$ contains a copy of $\mathbf{F}$ as a subalgebra, so neither are those algebras quasicritical.

We conclude that $\mathbf{A}$, $\mathbf{B}$, and $\mathbf{F}$ are the only quasicritical algebras in $\mathcal{Q}$, and thus $\mathrm{L_q}(\mathcal{Q})$ is $\mathbf{2}^3$ with the indicated equaclosure operator. The subquasivariety $\mathbb{Q}(\mathbf{F})$ corresponds to z in Fig. 7.40, $\mathbb{Q}(\mathbf{A}) = \langle \kappa x \approx e \rangle$ corresponds to a, $\mathbb{Q}(\mathbf{B}) = \langle \eta x \approx e \rangle$ corresponds to b, and $\langle \eta\kappa(x) \approx e \rangle$ corresponds to c. Note that $\mathbb{Q}(\mathbf{F}, \mathbf{A})$ is determined by the quasi-equation $\eta x \approx e \rightarrow x \approx e$, and symmetrically $\mathbb{Q}(\mathbf{F}, \mathbf{B})$ is $\kappa x \approx e \rightarrow x \approx e$. In these calculations, we use that $\mathcal{Q}$ satisfies $\eta\kappa x \approx \kappa\eta x$ and $\eta x \approx e \,\&\, \kappa x \approx e \rightarrow x \approx e$.

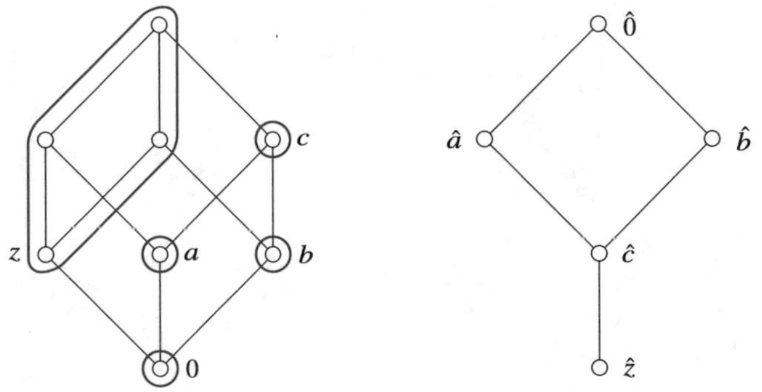

Fig. 7.40 $(\mathbf{2}^3, \gamma)$ for Example 7.29

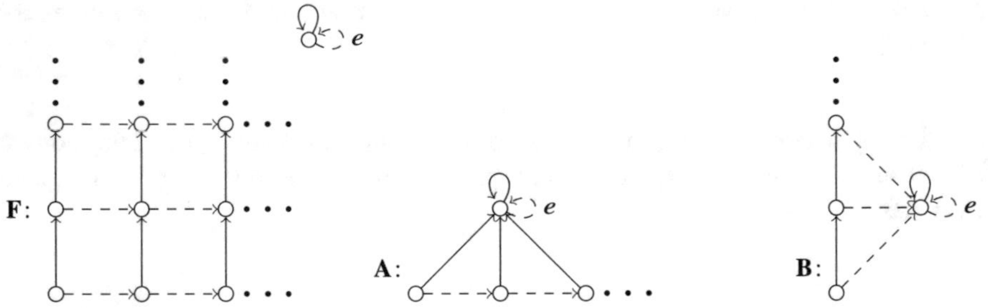

Fig. 7.41 Algebras **F**, **A**, **B** generating the quasivariety $\mathcal{Q}$ of Example 7.29. Dotted arrows represent η, and solid arrows κ

Example 7.30 Next we look at the lattice $\mathbf{M} = \mathbf{3}\times\mathbf{3}$ with the equaclosure operator of Fig. 7.43. This is the same pair for which the shortstyle method failed in Example 7.27 (Fig. 7.36).

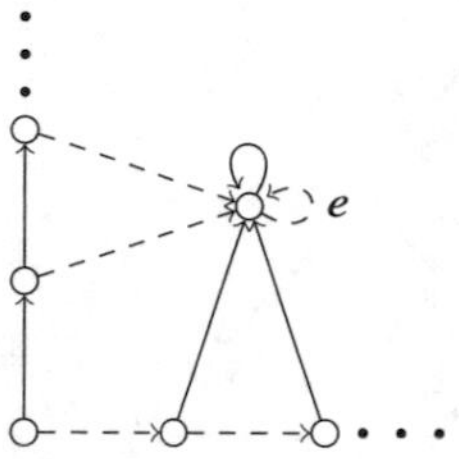

Fig. 7.42 The algebra $\mathbf{C} \leq \mathbf{A} \times \mathbf{B}$ generated by (x, x)

Let $\mathcal{M}$ be the quasivariety generated by the structures **U** and **V** in Fig. 7.44. The language has 2 unary operations, η and κ, the constant e, and predicates A, B, and $\approx$. The structures **U** and **V** satisfy the following laws, for all $m \geq 1$.

$$\begin{aligned} &\eta e \approx e \approx \kappa e && A(e),\ B(e) \\ &A(\eta x) && B(\kappa x) \\ &Ax \longrightarrow \kappa x \approx e && Bx \longrightarrow \eta x \approx e \\ &\eta^m x \approx x \longrightarrow x \approx e && \kappa^m x \approx x \longrightarrow x \approx e \\ &\eta x \approx \kappa y \longrightarrow \eta x \approx e && \eta x \approx \eta y \ \&\ \kappa x \approx \kappa y \longrightarrow x \approx y. \end{aligned}$$

It follows that $\eta\kappa x \approx e \approx \kappa\eta x$ and that $\eta x \approx e \ \&\ \kappa x \approx e \longrightarrow x \approx e$. Note that **U** has a substructure $\mathbf{S} = \{\eta x, \eta^2 x, \ldots\} \cup \{e\}$ for which As holds for all $s \in S$. Similarly, **V** has a substructure **T** in which Bt holds for all $t \in T$. These **S**, **T** are $\mathcal{M}$-homomorphic images of **U**, **V**, respectively.

The 1-generated $\mathcal{M}$-free structure **F**, the substructure of $\mathbf{U} \times \mathbf{V}$ generated by (x, x), is also shown in Fig. 7.44. As in the previous example, the k-generated free structure $\mathbf{F}_{\mathcal{M}}(k)$ consists of k copies of **F** glued over $\{e\}$.

Again, the k-generated structures in $\mathcal{M}$ are $\mathcal{M}$-homomorphic images of $\mathbf{F}_{\mathcal{M}}(k)$. They consist of say $m \leq k$ components, each isomorphic to one of the following forms, glued over $\{e\}$:

- **U**,
- **V**,
- **S**,
- **T**,
- **F**.

Note that $\mathbf{F}' = \mathbf{F} \setminus \{(x, x)\}$ is already of this form, being $\mathbf{S} \cup \mathbf{T}$ glued over $\{e\}$.

Thus **U**, **V**, **S**, and **T** are the only quasicritical algebras in $\mathcal{M}$, and these give $\mathrm{L_q}(\mathcal{M})$ as $\mathbf{3} \times \mathbf{3}$ with the indicated equaclosure operator.

Example 7.31 Now let us double the middle element in $\mathbf{M} = \mathbf{3} \times \mathbf{3}$, as in Fig. 7.45. Denote the new lattice as $\mathbf{M}[m]$. We omit the details of the six-step program, and only sketch the results, which are similar to the previous two examples.

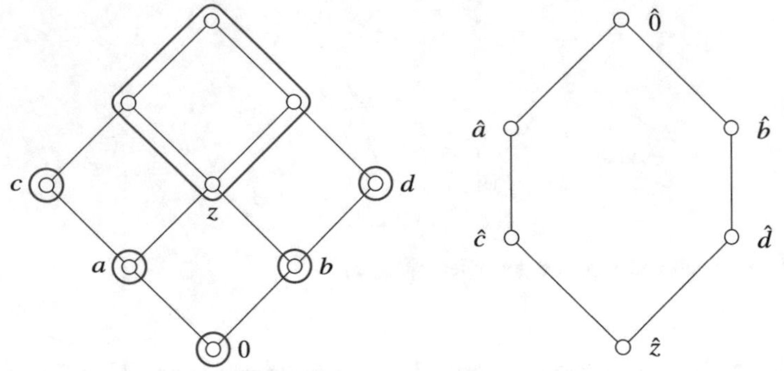

Fig. 7.43 ($\mathbf{M}, \gamma$) for Example 7.30

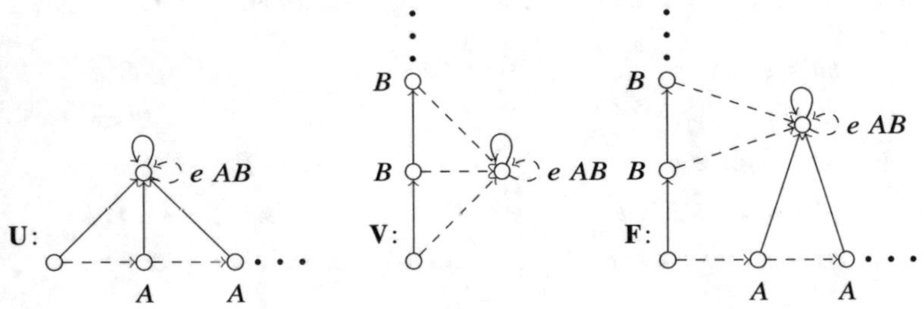

Fig. 7.44 Structures **U**, **V** generating the quasivariety $\mathcal{M}$ of Example 7.30 and the 1-generated $\mathcal{M}$-free structure $\mathbf{F} \le \mathbf{U} \times \mathbf{V}$ generated by (x, x). Dotted arrows represent η, and solid arrows κ

Let $\mathcal{N}$ be the quasivariety generated by the structures **U** and **V** in Fig. 7.46. The language has 2 unary operations, η and κ, the constant e, and predicates E and $\approx$. The structures **U** and **V** satisfy the following laws, for all $m \ge 1$.

$$\eta e \approx e \approx \kappa e \qquad E(e)$$
$$E(\eta x) \qquad E(\kappa x)$$
$$\eta^m x \approx x \to x \approx e \qquad \kappa^m x \approx x \to x \approx e$$
$$\eta x \approx \kappa y \to \eta x \approx e \qquad \eta\kappa x \approx e \approx \kappa\eta x$$
$$\eta x \approx \eta y \;\&\; \kappa x \approx \kappa y \to x \approx y .$$

It follows that $\eta x \approx e \;\&\; \kappa x \approx e \to x \approx e$. Note that **U** has a substructure $\mathbf{S} = \{\eta x, \eta^2 x, \dots\} \cup \{e\}$ for which Es holds for all $s \in S$, and likewise **V** has a substructure **T** in which Et holds for all $t \in T$. These **S**, **T** are $\mathcal{N}$-homomorphic images of **U**, **V**, respectively.

The 1-generated $\mathcal{N}$-free structure **F**, the substructure of $\mathbf{U} \times \mathbf{V}$ generated by (x, x), is also shown in Fig. 7.46. As in the previous example, the k-generated free structure $\mathbf{F}_{\mathcal{N}}(k)$ consists of k copies of **F** glued over $\{e\}$.

Again, the k-generated structures in $\mathcal{N}$ are $\mathcal{N}$-homomorphic images of $\mathbf{F}_{\mathcal{N}}(k)$. They consist of say $m \leq k$ components, each isomorphic to one of the following forms, glued over $\{e\}$:

- **U**,
- **V**,
- **S**,
- **T**,
- **F**,
- $\mathbf{F}'$ obtained from **F** by adding $E(x, x)$.

However, $\mathbf{F}'$ is not quasicritical since $\mathbf{S}, \mathbf{T} \leq \mathbf{F}' \leq \mathbf{S} \times \mathbf{T}$, while $\mathbf{F}'' = \mathbf{F} \setminus \{(x, x)\}$ is $\mathbf{S} \cup \mathbf{T}$ glued over $\{e\}$.

Thus **U**, **V**, **S**, **T**, and **F** are the only quasicritical algebras in $\mathcal{N}$, and these give $\mathrm{L_q}(\mathcal{N})$ as $\mathbf{M}[m]$ with the indicated equaclosure operator.

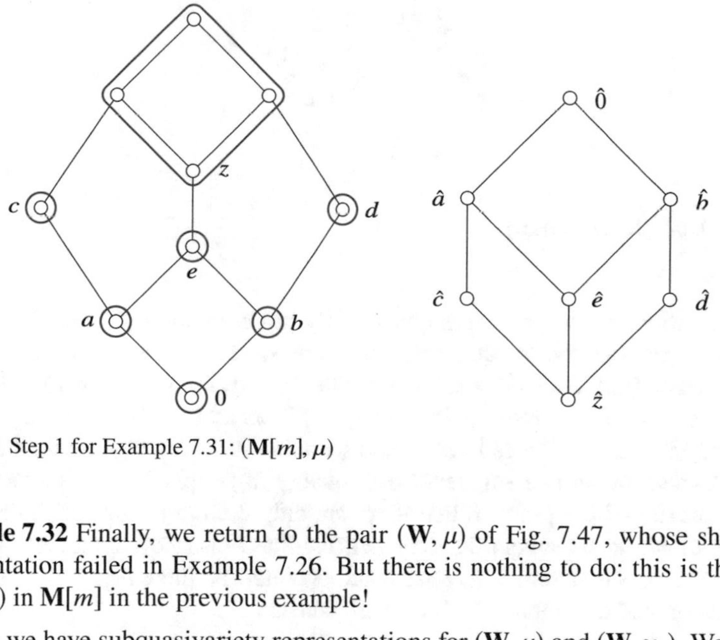

Fig. 7.45 Step 1 for Example 7.31: $(\mathbf{M}[m], \mu)$

Example 7.32 Finally, we return to the pair $(\mathbf{W}, \mu)$ of Fig. 7.47, whose shortstyle representation failed in Example 7.26. But there is nothing to do: this is the ideal $\downarrow(a \vee d)$ in $\mathbf{M}[m]$ in the previous example!

Thus we have subquasivariety representations for $(\mathbf{W}, \mu)$ and $(\mathbf{W}, \gamma_6)$. We do not know whether any of the remaining pairs $(\mathbf{W}, \gamma_j)$ for $1 \leq j \leq 5$ can be represented as subquasivariety lattices with equality!

This chapter has been concerned with the problem of how to represent a pair $(\mathbf{L}, \gamma)$ as $(\mathrm{L_q}(\mathcal{K}), \Gamma)$. In our approach, the first step is to represent $(\mathbf{L}, \gamma)$ as $(\mathrm{S_p}(\mathbf{S}, H), \Gamma)$ for an algebraic (and often finite) lattice **S** with operators. If we can do so, then

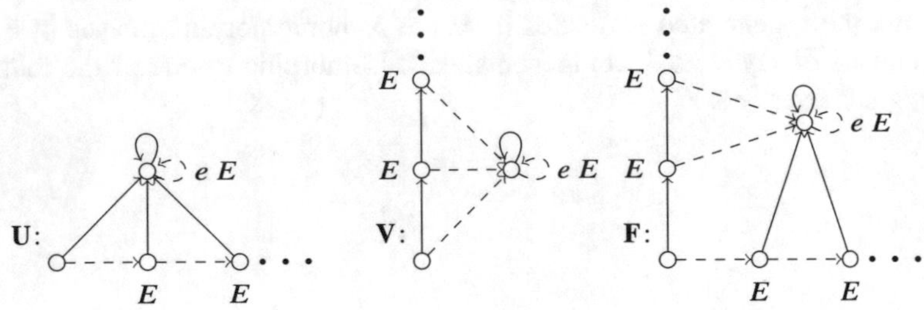

Fig. 7.46 Structures **U**, **V** generating the quasivariety $\mathcal{N}$ of Example 7.31 and the 1-generated $\mathcal{N}$-free structure $\mathbf{F} \leq \mathbf{U} \times \mathbf{V}$ generated by (x, x). Dotted arrows represent η, and solid arrows κ

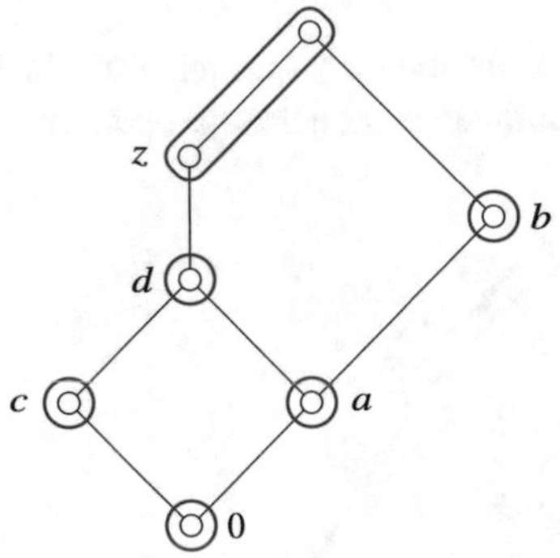

Fig. 7.47 Example 7.32: $(\mathbf{W}, \mu)$

it is straightforward to produce a quasivariety representation $(\mathrm{L_q}(\mathcal{K}_0), \Gamma)$ in a language without equality. In some circumstances, we know how to convert this to a representation $(\mathrm{L_q}(\mathcal{K}_1), \Gamma)$ in a language with equality. On the other hand, there are infinite lattices that are isomorphic to an $\mathrm{S_p}(\mathbf{S}, H)$ but admit no equaclosure operator satisfying (I8). Such lattices have a representation in a language without equality, but no representation in a language with equality (Example 3.8). It is a measure of the complexity of the problem that there are only 2 large classes of finite lattices that are known to be subquasivariety lattices: finite distributive lattices and finite atomistic lattices that admit an equaclosure operator. So there is plenty of room for investigation and development of new methods here.

Chapter 8
Lattices $\mathbf{1} + \mathbf{L}$ as $\mathrm{L_q}(\mathcal{K})$

Фролов. Хороший вечер.
Маша. Да ... Тишина и прохлада. Хочется сказать какую-нибудь глупость.
Фролов. В чём же дело?
Маша. Не умею. Чувствую, а сказать не умею. – Александр Вампилов, Прощание в июне

On n'est point toujours une bête pour l'avoir été quelquefois. – Denis Diderot

D. Pigozzi and G. Tardos proved that for every dually algebraic lattice $\mathbf{L}$, the linear sum $\mathbf{1}+\mathbf{L}$ is isomorphic to a lattice of subvarieties $\mathrm{L_v}(\mathcal{V})$ for a variety $\mathcal{V}$ of algebras (with equality) [99]. In this chapter we consider the possibility of similar results for quasivarieties, in which case $\mathbf{L}$ needs to also be join semidistributive. That is, we seek conditions under which we can show that $\mathbf{1}+\mathbf{L} \cong \mathrm{L_q}(\mathcal{K})$ for some quasivariety $\mathcal{K}$ of structures in a language with equality. By Corollary 7.10, if $\mathbf{L} \cong \mathrm{S_p}(\mathbf{S}, H)$, then $\mathbf{1}+\mathbf{L} \cong \mathrm{L_q}(\mathcal{K})$ for some $\mathcal{K}$ with equality. As a general plan for the chapter, we use the methods of Sect. 7.2 to obtain a longstyle representation of $\mathbf{1}+\mathbf{L}$ as $\mathrm{L_q}(\mathcal{K})$, with particular attention to the case where $\mathbf{L}$ itself does not admit an equaclosure operator (so that Corollary 7.10 does not apply).

We weakly conjecture:

(A) For any finite join semidistributive lattice, the linear sum $\mathbf{1}+\mathbf{L}$ has a longstyle representation as a subquasivariety lattice.
(B) More generally, if $\mathbf{L}$ is dually algebraic and satisfies the Jónsson-Kiefer Property, the linear sum $\mathbf{1}+\mathbf{L}$ has a longstyle representation as $\mathrm{L_q}(\mathcal{K})$.

A counterexample to either conjecture would involve necessary conditions for representation not localized at 0. Note that when $\mathbf{L}$ is dually algebraic, then $\mathbf{1}+\mathbf{L}$ has an equaclosure operator γ with just two γ-classes, $\{0\}$ and L.

K. Adaricheva et al., *A Primer of Subquasivariety Lattices*, CMS/CAIMS Books in Mathematics 3, https://doi.org/10.1007/978-3-030-98088-7_8

8.1 The Leaf Lattice and Generalizations

The lattice $\mathrm{Co}(\mathbf{4})$ of convex subsets of a 4-element chain does not support an equaclosure operator, so it cannot be represented as a subquasivariety lattice. As $\mathrm{Co}(\mathbf{4})$ is not lower bounded, neither is any lattice containing it as a sublattice lower bounded. In this section, we show that the *leaf lattice* $\mathbf{1}+\mathrm{Co}(\mathbf{4})$ of Fig. 8.1 is isomorphic to $\mathrm{L_q}(Q)$ for a quasivariety Q with equality, even though it contains $\mathrm{Co}(\mathbf{4})$. By Theorem 1.25, there is no locally finite quasivariety $\mathcal{K}$ of finite type such that $\mathbf{1} + \mathrm{Co}(\mathbf{4}) \cong \mathrm{L_q}(\mathcal{K})$.

The construction in this section is a modification of one from Adaricheva and Gorbunov [16]. An earlier representation of the leaf lattice as a lattice of H-closed algebraic subsets $\mathrm{S_p}(\mathbf{L}, H)$ for an algebraic lattice with operators can be found in Adaricheva and Nation [20]. That paper also showed that the lattice $(\mathbf{2}\times\mathbf{2}) + \mathrm{Co}(\mathbf{4})$, which is not lower bounded, is a subquasivariety lattice, while leaving the question of $\mathbf{1} + \mathrm{Co}(\mathbf{4})$ open.

We use Theorem 7.7 to represent the leaf lattice as $\mathrm{L_q}(\mathcal{K})$. Thus we will represent $\mathbf{1}+\mathrm{Co}(\mathbf{4})$ as $\mathrm{S_p}(\mathbf{S}, H)$ in such a way that $1_\mathbf{S}$ is compact and the condition ϖ is satisfied, that is, $h(x) = 1_\mathbf{S}$ implies $x = 1_\mathbf{S}$, which then leads to a longstyle quasivariety representation.

Note that $\mathbf{1} + \mathrm{Co}(\mathbf{4})$ supports only one equaclosure operator, with 0 as one γ-class and the rest of the lattice as the other. The minimal nontrivial join covers are $b \le a \vee c$, $c \le b \vee d$ and both $b, c \le a \vee d$. The arrow relations to satisfy are $a, b, c, d \to z$ and $z \leftrightarrow z_1$. The lattice $\mathbf{S}$ is drawn in Fig. 8.2; it is algebraic. The monoid of operators is generated by:

	p	m	f	g
$\hat{0}$	$\hat{0}$	$\hat{0}$	$\hat{0}$	$\hat{0}$
z_1	z_1	z_1	z_1	$\hat{z}$
a_i	a_{i+1}	a_{i-1}	z_1	$\hat{z}$
b_i	b_{i+1}	b_{i-1}	z_1	$\hat{z}$
c_i	c_{i+1}	c_{i-1}	z_1	$\hat{z}$
d_i	d_{i+1}	d_{i-1}	z_1	$\hat{z}$
$\hat{z}$	$\hat{z}$	$\hat{z}$	z_1	$\hat{z}$

Now it is child's play to produce an infinite sequence of subquasivariety lattices that are not lower bounded. For $k \ge 2$, let $\mathbf{L}_k$ be the join semilattice with 0 generated by $\{a_0, b_0, \ldots, a_{k-1}, b_{k-1}\}$ subject to the relations that $a_i \le b_i \vee a_{i+1}$ and $a_i \le b_i \vee b_{i+1}$, where in $\mathbf{L}_k$ the subscripts are taken modulo k. Note that $\mathbf{L}_2 \cong \mathrm{Co}(\mathbf{4})$ with $a_0 = b, b_0 = a, a_1 = c$, and $b_1 = d$. These lattices are atomistic, not lower bounded because of the D-cycle $a_0 \mathrel{D} a_1 \mathrel{D} \ldots \mathrel{D} a_{k-1} \mathrel{D} a_0$, and join semidistributive because we can represent $\mathbf{1}+\mathbf{L}_k$ as $\mathrm{S_p}(\mathbf{S}, f, g, m_k, p_k)$. Here, $\mathbf{S}$ is the same lattice as Fig. 8.2, labeled as in Fig. 8.3, and the operators are given by the following table:

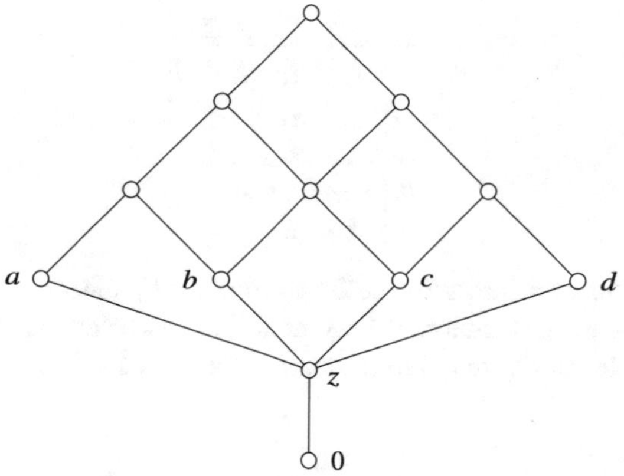

Fig. 8.1 The leaf lattice $\mathbf{1} + \mathrm{Co}(\mathbf{4})$ is isomorphic to $\mathrm{L_q}(\mathcal{Q})$ for a quasivariety $\mathcal{Q}$, but not to $\mathrm{L_q}(\mathcal{K})$ for any locally finite quasivariety $\mathcal{K}$ of finite type (Sect. 8.1)

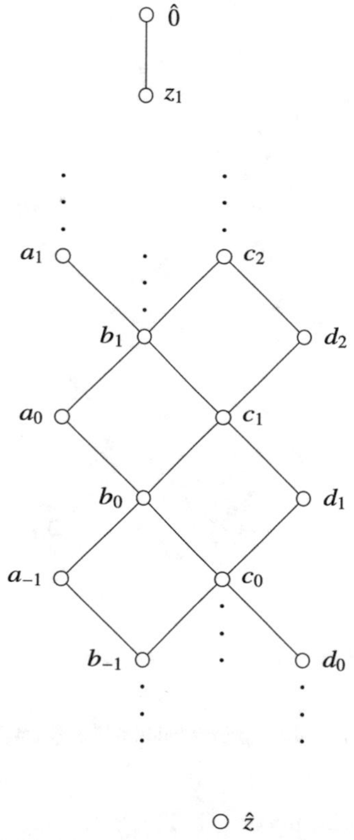

Fig. 8.2 The lattice **S** used to represent $\mathbf{1} + \mathrm{Co}(\mathbf{4})$ (Sect. 8.1)

	p_k	m_k	f	g
$\hat{0}$	$\hat{0}$	$\hat{0}$	$\hat{0}$	$\hat{0}$
z_1	z_1	z_1	z_1	$\hat{z}$
a_i	a_{i+k}	a_{i-k}	z_1	$\hat{z}$
b_i	b_{i+k}	b_{i-k}	z_1	$\hat{z}$
$\hat{z}$	$\hat{z}$	$\hat{z}$	z_1	$\hat{z}$

where in $\mathbf{S}$ the subscripts are taken in $\mathbb{Z}$ (not modulo k). Checking the details is left to the reader. Since the operators satisfy ϖ and the top element $\hat{0}$ is compact in $\mathbf{S}$, there is a longstyle quasivariety representation for each $\mathbf{1}+\mathbf{L}_k$.

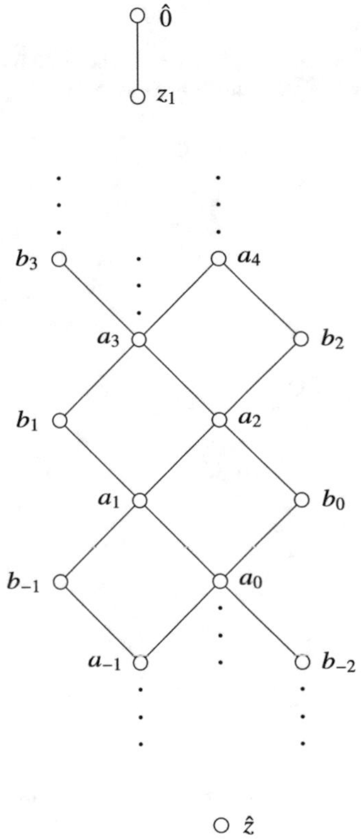

Fig. 8.3 The lattice $\mathbf{S}$ relabeled for the representation of $\mathbf{L}_k$ (Sect. 8.1)

Another type of leaf lattice appears in Fig. 8.4.

Fig. 8.4 Leaf lattice with natural equaclosure operator

8.2 Examples of Representations for 1 + L

This section contains more representations of lattices $\mathbf{1}+\mathbf{L}$ as $L_q(\mathcal{K})$, with particular attention to cases where $\mathbf{L}$ itself may not be a subquasivariety lattice. For example, $\mathbf{L}$ need not support an equaclosure operator, or it may not be atomic.

8.2.1 The Lattice 1 + Co(2 × 2)

The lattice $\mathrm{Co}(\mathbf{2}\times\mathbf{2})$ of convex subsets of the ordered set $\mathbf{2}\times\mathbf{2}$ is lower bounded and subdirectly reducible, but it does not admit an equaclosure operator; see Theorem 4.9. So we consider the lattice $\mathbf{K} = \mathbf{1}+\mathrm{Co}(\mathbf{2}\times\mathbf{2})$ drawn in Fig. 8.5. The minimal nontrivial join covers in $\mathbf{K}$ are $b \leq a \vee d$ and $c \leq a \vee d$. We need to use different right-hand sides to realize these join covers in the lattice Sub $\mathbf{T}$, say $b = a \wedge d$ and $c = a' \wedge d'$. The remaining arrow relations needed are $a, b, c, d \to z$ and $a \leftrightarrow a'$ and $d \leftrightarrow d'$ and $\hat{z} \leftrightarrow z_1$. Thus we can use the following operators on the lattice $\mathbf{T}$ of Fig. 8.6.

	h	k	f	g
$\hat{0}$	$\hat{0}$	$\hat{0}$	$\hat{0}$	$\hat{0}$
z_1	z_1	z_1	z_1	$\hat{z}$
a	a'	$\hat{z}$	z_1	$\hat{z}$
d	$\hat{z}$	d'	z_1	$\hat{z}$
a'	a	$\hat{z}$	z_1	$\hat{z}$
d'	$\hat{z}$	d	z_1	$\hat{z}$
b	$\hat{z}$	$\hat{z}$	z_1	$\hat{z}$
c	$\hat{z}$	$\hat{z}$	z_1	$\hat{z}$
$\hat{z}$	$\hat{z}$	$\hat{z}$	z_1	$\hat{z}$

8.2.2 The Lattice 1 + J

The lattice known as $\mathbf{J}$ is bounded but does not admit an equaclosure operator (Sect. 4.1.1). So we consider the lattice $\mathbf{N} = \mathbf{1} + \mathbf{J}$ drawn in Fig. 8.7. The minimal nontrivial join covers are $p \leq q \vee r$ and $r \leq s \vee t$. There will be a number of arrow relations: for all $i \in \mathbb{Z}$,

$$t_i \to p_i \to q_i \to \hat{z}$$
$$r_i \to s_i \to \hat{z}$$
$$\hat{z} \leftrightarrow z_1$$
$$x_i \leftrightarrow x_{i+1} \text{ for } x = p, q, r, s, t.$$

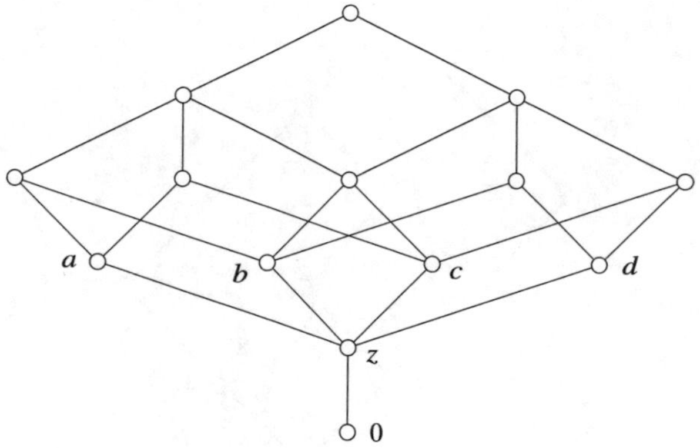

Fig. 8.5 The lattice **K** = **1** + Co(**2** × **2**) of Sect. 8.2.1

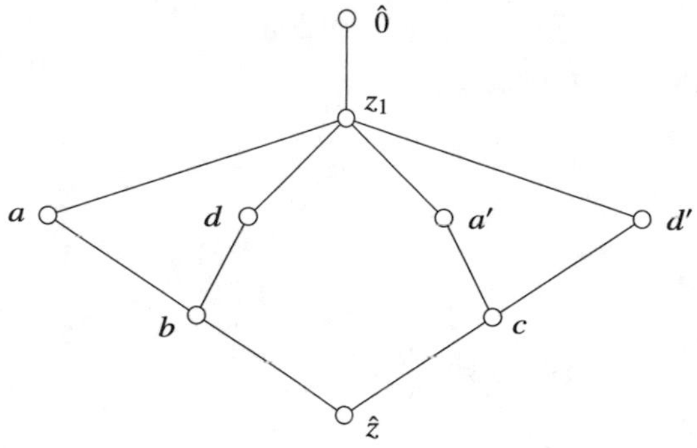

Fig. 8.6 The lattice **T** for **K** = **1** + Co(**2** × **2**) of Sect. 8.2.1 and **P** = **1** + **H** of Sect. 8.2.5

We can use the following operators on the lattice **U** of Fig. 8.8.

	p	m	h	f	g
$\hat{0}$	$\hat{0}$	$\hat{0}$	$\hat{0}$	$\hat{0}$	$\hat{0}$
z_1	z_1	z_1	z_1	z_1	$\hat{z}$
p_i	p_{i+1}	p_{i-1}	q_i	z_1	$\hat{z}$
q_i	q_{i+1}	q_{i-1}	q_i	z_1	$\hat{z}$
r_i	r_{i+1}	r_{i-1}	s_i	z_1	$\hat{z}$
s_i	s_{i+1}	s_{i-1}	s_i	z_1	$\hat{z}$
t_i	t_{i+1}	t_{i-1}	p_{i+1}	z_1	$\hat{z}$
$\hat{z}$	$\hat{z}$	$\hat{z}$	$\hat{z}$	z_1	$\hat{z}$

There may be another way to do this one, but this version is fairly natural.

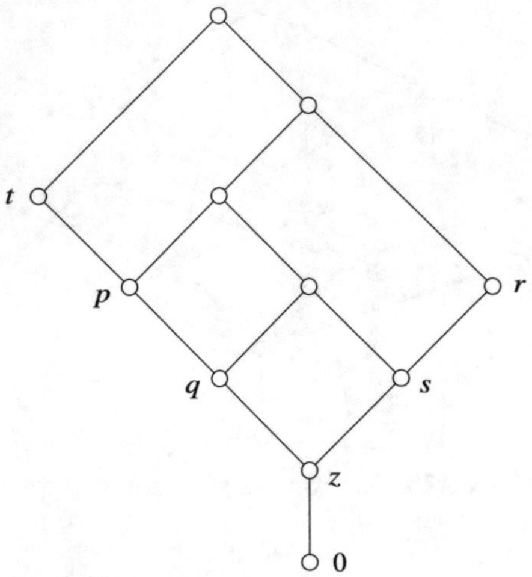

Fig. 8.7 The lattice $\mathbf{N} = \mathbf{1} + \mathbf{J}$ of Sect. 8.2.2

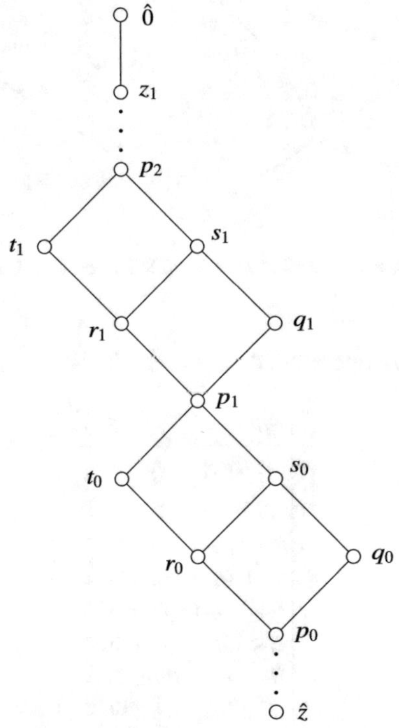

Fig. 8.8 The lattice $\mathbf{U}$ for $\mathbf{N} = \mathbf{1} + \mathbf{J}$ of Sect. 8.2.2

8.2.3 The Lattice $\mathbf{1} + (\mathbf{2} \times \mathbf{3})$

Now let us do an example of $\mathbf{1} + \mathbf{D}$ for a small distributive lattice, to set the pattern for the next section. Consider the lattice $\mathbf{O} = \mathbf{1} + (\mathbf{2} \times \mathbf{3})$ drawn in Fig. 8.9. There are no minimal nontrivial join covers, and the arrow relations are just the inclusions from the picture plus $\hat{z} \leftrightarrow z_1$.

The operators h_j on the representing lattice $\mathbf{V}$ of Fig. 8.10 do the job:

$$h_j(x) = x \vee j \qquad h_j(\hat{0}) = \hat{0} \qquad h_j(\hat{z}) = \hat{z}$$

for $j = b, c, d$ and $x = a, b, c, d, e, z_1$, along with the usual maps f and g that map to z_1 and $\hat{z}$, respectively.

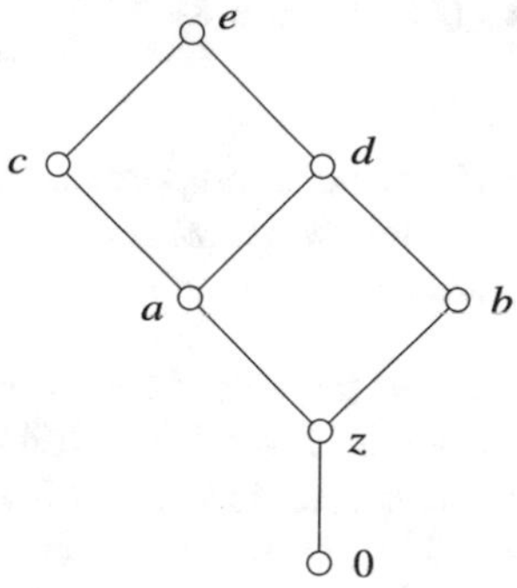

Fig. 8.9 The lattice $\mathbf{O} = \mathbf{1} + (\mathbf{2} \times \mathbf{3})$ of Sect. 8.2.3

8.2.4 $\mathbf{1} + \mathcal{O}(\mathbf{P})$

In Chap. 9, Theorem 9.12, we will prove that for any dually algebraic, distributive lattice $\mathbf{D}$, the lattice $\mathbf{1}+\mathbf{D}$ is isomorphic to $\mathrm{L_q}(\mathcal{K})$ for a quasivariety of structures with equality. Likewise, Theorem 9.18 says that any lattice of the form $\mathcal{O}(\mathbf{P})$ is isomorphic to some $\mathrm{L_q}(\mathcal{Q})$, whenever $\mathbf{P}$ has only finitely many minimal elements (so that the least element of $\mathcal{O}(\mathbf{P})$ is dually compact). The result in this subsection is weaker than either of those, but the proof is too nice to omit.

The preceding example shows us exactly how to do a restricted version, namely to represent $\mathbf{1}+\mathbf{D}$ when $\mathbf{D}$ is distributive, algebraic, and dually algebraic. First, recall the classic characterization of lattices $\mathcal{O}(\mathbf{P})$ of order ideals of an ordered set [31, 59].

Lemma 8.1 *The following are equivalent for a complete lattice* $\mathbf{D}$.

(1) $\mathbf{D}$ *is distributive, algebraic, and dually algebraic.*

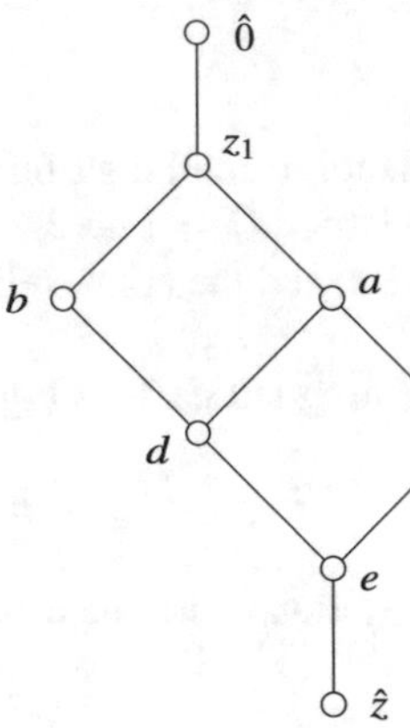

Fig. 8.10 The lattice **V** for $\mathbf{N} = \mathbf{1} + (\mathbf{2} \times \mathbf{3})$ of Sect. 8.2.3

(2) **D** *is distributive, dually algebraic, and upper continuous.*
(3) **D** *is distributive, algebraic, and lower continuous.*
(4) $\mathbf{D} \cong \mathcal{O}(\mathbf{P})$ *for some ordered set* **P**.

Proof The lattice $\mathcal{O}(\mathbf{P})$ consists of the closed sets of a finitary closure operator $\downarrow$, and the union of order ideals is an order ideal. Thus $\mathcal{O}(\mathbf{P})$ is a distributive, algebraic lattice. The complement of an order ideal is an order filter of **P**, whence the lattice of order filters satisfies $\mathcal{F}(\mathbf{P}) \cong^d \mathcal{O}(\mathbf{P})$, with both lattices ordered by set containment. Meanwhile, $\mathcal{F}(\mathbf{P}) \cong \mathcal{O}(\mathbf{P}^d)$; that makes $\mathcal{F}(\mathbf{P})$ algebraic and hence $\mathcal{O}(\mathbf{P})$ dually algebraic. Thus (4) implies (1).

Algebraic lattices are upper continuous, so (1) implies (2). We will show that (2) implies (4). The argument that (1) implies (3) implies (4) is of course dual.

Assume that **D** is distributive, dually algebraic, and upper continuous. Because it is dually algebraic, every element of D is a join of completely join irreducible elements. Let **P** be the set of completely join irreducible elements of **D**, with the order inherited from **D**. We want to show that these elements are in fact completely join prime.

So let $p \in P$ and suppose $p \leq \bigvee B$ for some $B \subseteq D$. Now, using upper continuity and then distributivity,

$$\begin{aligned} p &= p \wedge \bigvee B \\ &= p \wedge \Big(\bigvee_{F \subseteq B \text{ finite}} \bigvee F \Big) \\ &= \bigvee_{F \subseteq B \text{ finite}} \Big(p \wedge \bigvee F \Big) \\ &= \bigvee_{F \subseteq B \text{ finite}} \bigvee_{f \in F} (p \wedge f) \end{aligned}$$

As p is completely join irreducible, we conclude that $p = p \wedge f$, i.e., $p \leq f$ for some $f \in B$. Thus the elements of **P** are completely join prime. Combining this with the fact that every element of **D** is a join of elements in **P**, we see that the map $\varphi : \mathbf{D} \to \mathcal{O}(\mathbf{P})$ *via* $\varphi(x) = \downarrow x \cap P$ is an isomorphism. □

Note that the lattices of Lemma 8.1 satisfy complete distributive laws, including that $x \vee (\bigwedge Y) = \bigwedge_{y \in Y}(x \vee y)$.

Theorem 8.2 *Let* **D** *be a distributive, algebraic, and dually algebraic lattice. Then* **1** + **D** *is isomorphic to* $\mathrm{L_q}(\mathcal{K})$ *for a quasivariety of structures with equality.*

Proof Let $\mathbf{S} = \mathbf{1} + (\mathbf{1} + \mathbf{D})^d$ with the labeling conventions as in Fig. 8.10 for $\hat{0}$, z_1, and $\hat{z}$. Clearly **S** is an algebraic lattice. We use the standard operators f, g, and the h_j for $j \in D$, where $h_j(x) = x \vee j$ (with the join taken in **S**) for $x \in S \setminus \{\hat{0}, \hat{z}\}$, while $h_j(\hat{0}) = \hat{0}$ and $h_j(\hat{z}) = \hat{z}$. (It would suffice to use the operators h_j with j completely join irreducible in **S**.) Check that these are indeed operators, i.e., preserve arbitrary meets and directed joins, using the complete distributivity noted above. Moreover, the H-closed algebraic subsets of **S** are

- $\{\hat{0}\}$,
- $\{\hat{0}, \hat{z}, z_1\}$,
- $\{\hat{0}, \hat{z}, z_1\} \cup \uparrow_{\mathbf{S}} x$ for every $x \in S$ with $1_{\mathbf{D}} \leq x < z_1$.

Thus $\mathbf{1} + \mathbf{D} \cong \mathrm{S_p}(\mathbf{S}, H)$, and since ϖ is satisfied and $\hat{0}$ is compact in **S**, there is a longstyle representation as subquasivarieties. □

8.2.5 The Lattice 1 + H, the Hexagon

Returning to simple examples, recall that the hexagon **H** does not support an equaclosure operator. Let $\mathbf{P} = \mathbf{1} + \mathbf{H}$, labeled as in Fig. 8.11. The minimal nontrivial join covers are the same as for $\mathbf{1} + \mathrm{Co}(\mathbf{2} \times \mathbf{2})$: $b, c \leq a \vee d$. For the arrow relations, besides $a, b, c, d \to z$ and $a \leftrightarrow a'$ and $d \leftrightarrow d'$ and $\hat{z} \leftrightarrow z_1$, include $b \to a$ and $c \to d$. So we can reuse the lattice **T** of Fig. 8.6, and the operators for $\mathbf{1} + \mathrm{Co}(\mathbf{2} \times \mathbf{2})$, plus two additional operators for the last two arrows. (In other words, since $\mathbf{1} + \mathbf{H} \leq \mathbf{1} + \mathrm{Co}(\mathbf{2} \times \mathbf{2})$, we use additional maps to get the sublattice: if $H \leq K$, then $\mathrm{S_p}(\mathbf{S}, K) \leq \mathrm{S_p}(\mathbf{S}, K)$. Those extra maps are the point of this example.) Again the condition ϖ is satisfied, and there is a longstyle subquasivariety representation of **1** + **H**.

	ℓ	m
$\hat{0}$	$\hat{0}$	$\hat{0}$
z_1	z_1	z_1
a	a	$\hat{z}$
d	z_1	$\hat{z}$
a'	$\hat{z}$	z_1
d'	$\hat{z}$	d'
b	a	$\hat{z}$
c	$\hat{z}$	d'
$\hat{z}$	$\hat{z}$	$\hat{z}$

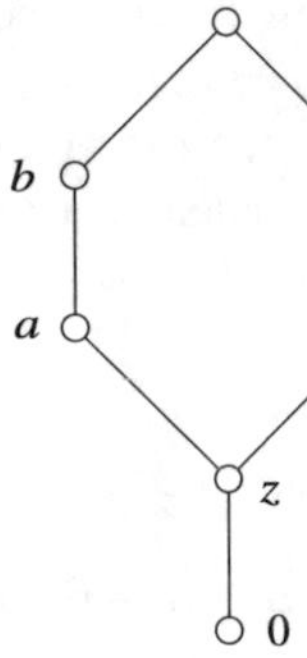

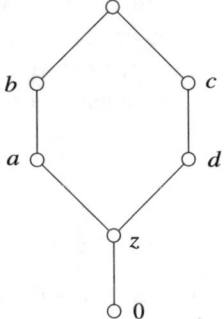

Fig. 8.11 The lattice $\mathbf{P} = \mathbf{1} + \mathbf{H}$ of Sect. 8.2.5

Next, having done the distributive case, one might want to prove that $\mathbf{1}+\mathbf{L}$ *is a subquasivariety lattice when* $\mathbf{L}$ *is a finite, lower bounded lattice. The constructions in the next two Sects. 8.2.6 and 8.2.7, do that for some special cases. They are included to aid those who might work on this problem, but other readers can skip to Sect. 8.3 without loss of continuity.*

8.2.6 $\mathbf{1}+\mathbf{L}$ with L Subdirectly Irreducible, Rank 1

Recall that a finite lattice $\mathbf{L}$ is *lower bounded of rank* k if $\mathrm{J}(\mathbf{L}) \subseteq \mathrm{D_k}(\mathbf{L})$. The finite lower bounded lattices of rank k form a pseudovariety, denoted $\mathcal{LB}(k)$. Distributive lattices are of rank 0, and, for example, $\mathrm{Co}(\mathbf{2}\times\mathbf{2})$ and the hexagon $\mathbf{H}$ are of rank 1. See [21, 46, 89, 91] for discussions of the theory of $\mathcal{LB}(k)$.

In this section we will prove the next theorem.

Theorem 8.3 *Let* $\mathbf{L}$ *be a finite lattice that is subdirectly irreducible and lower bounded lattice of rank* 1. *Then* $\mathbf{1}+\mathbf{L}$ *is isomorphic to* $\mathrm{L_q}(\mathcal{K})$ *for a quasivariety of structures with equality.*

Doubtlessly, Theorem 8.3 has limited value, the intent being only to offer some support for Conjecture (A) on page 217. As an added benefit, though, we get to describe all finite, subdirectly irreducible, lower bounded lattices of rank 1.

A finite lattice $\mathbf{L}$ is subdirectly irreducible and in $\mathcal{LB}(1)$ if and only if $\mathbf{L} \cong \mathbf{2}$ or $\mathrm{J}(\mathbf{L})$ contains precisely one non-join-prime element, say x, and $x \mathrel{\mathrm{D}} y$ for every other $y \in \mathrm{J}(\mathbf{L})$. (Here we use the convention that $0 \notin \mathrm{J}(\mathbf{L})$.) For example, the pentagon is subdirectly irreducible, the hexagon is not.

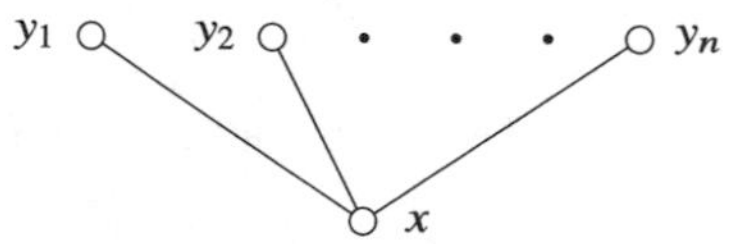

Fig. 8.12 $(\mathrm{J}(\mathbf{L}), \overline{\mathrm{D}})$ for $\mathbf{L}$ subdirectly irreducible and in $\mathcal{LB}(1)$

Thus the ordered set $(\mathrm{J}(\mathbf{L}), \overline{\mathrm{D}})$ looks like Fig. 8.12, and the elements of Y are join prime, i.e., in $\mathrm{D}_0(\mathbf{L})$. Unfortunately that suppresses some information, *viz.*, which subsets of Y form minimal nontrivial join covers of x. If we let $Y_1, \ldots, Y_m$ denote the minimal nontrivial join covers of x, then we have that

(1) $\mathrm{J}(\mathbf{L}) = \{x\} \cup Y$,
(2) $Y = Y_1 \cup \cdots \cup Y_m$,
(3) each Y_i is an antichain,
(4) $x \nleq y$ for all $y \in Y$,
(5) if $i \neq j$, then $Y_i \ll Y_j$ does not hold.

(Recall that $A \ll B$ means that for each $a \in A$ there exists $b \in B$ such that $a \leq b$. If A and B are distinct join covers of x and $A \ll B$, then B is not minimal.)

Whenever x and Y satisfy these conditions, we can construct a lattice that is subdirectly irreducible and in $\mathcal{LB}(1)$, with these conditions reflecting the structure of $\mathbf{L}$. Explicitly, let $Y_1, \ldots, Y_m$ be sets with $2 \leq |Y_i| < \infty$ for all i, and let $Y = \bigcup Y_i$, and let $x \notin Y$. Let $\mathbf{P}$ be a partial order on $Y \cup \{x\}$ such that conditions (2)–(5) above are satisfied. Thus the sets Y_i need not be pairwise disjoint, but no Y_i refines another Y_j, which means, in particular, $Y_i \nsubseteq Y_j$. Define $\mathbf{G}(x, Y_1, \ldots, Y_m, \mathbf{P})$ to be the join semilattice with 0 generated by $\{x\} \cup Y$ subject to the relations of $\mathbf{P}$ and $x \leq \bigvee Y_i$ for all i. These lattices are hard to draw, but easy to imagine and exemplify. For an easy one, the Boolean algebra with an atom doubled, $\mathbf{B}_3[a]$, is isomorphic to $\mathbf{G}(x, \{y_1, y_2\}, \{y_1, y_3\}, y_1 \leq x)$, as shown in Fig. 8.13.

Let Δ denote the antichain order on $\{x\} \cup Y$.

Lemma 8.4 *A finite lattice* $\mathbf{L}$ *is subdirectly irreducible and lower bounded of rank* 1 *if and only if* $\mathbf{L}$ *is isomorphic to* $\mathbf{G}(x, Y_1, \ldots, Y_m, \mathbf{P})$ *for a set of parameters satisfying* (1)–(5). *Moreover, for any set of parameters satisfying* (1)–(5), *we have* $\mathbf{G}(x, Y_1, \ldots, Y_m, \mathbf{P}) \leq \mathbf{G}(x, Y_1, \ldots, Y_m, \Delta)$.

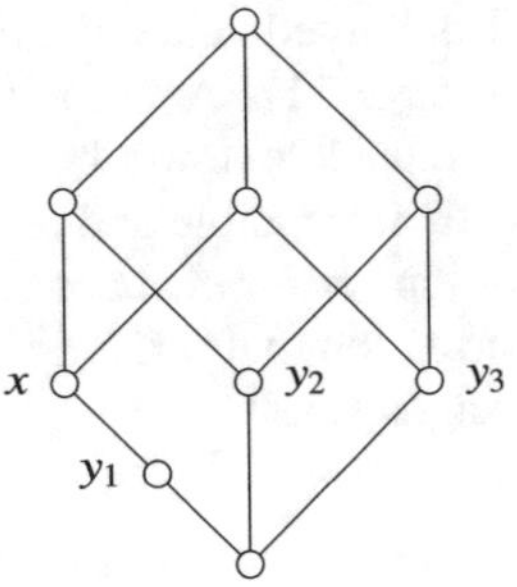

Fig. 8.13 $\mathbf{B}_3[a]$ as $\mathbf{G}(x, \{y_1, y_2\}, \{y_1, y_3\}, y_1 \leq x)$

In these terms, $\{x\} \cup Y$ is J(**L**), **P** is the order on J(**L**), and $x \leq \bigvee Y_i$ are the minimal nontrivial join covers of **L**. The embedding of the second part just maps $x \mapsto x \vee \bigvee\{y_j : y_j \leq x\}$ and $y_i \mapsto y_i \vee \bigvee\{y_k : y_k \leq y_i\}$ for each i. For an easy example, embed the pentagon into $\mathbf{G}(x, \{y_1, y_2\}, \Delta)$.

Corollary 8.5 *If* **L** *is a finite lattice in* $\mathcal{LB}(1)$*, then it is a subalgebra of a direct product of lattices of the form* $\mathbf{G}(x, Y_1, \ldots, Y_m, \Delta)$.

There will be one subdirect factor for each join irreducible element that is minimal in $(\mathrm{J}(\mathbf{L}), \overline{\mathrm{D}})$. That is because congruences on a finite lattice **L** correspond to D-closed subsets of J(**L**). For Day's description of lattice congruences in terms of the D-relation, see [46, Section II.3] or [90, Chapter 10]. Section III.4 of [46] is also relevant with regard to subdirectly irreducible, finite, lower bounded lattices.

Thus our first task is to represent the lattices $\mathbf{1} + \mathbf{G}(x, Y_1, \ldots, Y_m, \Delta)$. Let **F** denote the free meet semilattice with 1 generated by $\{x\} \cup Y$ subject to the relations $x = \bigwedge Y_i$ for $1 \leq i \leq m$, denoting the largest element of **F** by z_1. Then form the lattice $\mathbf{S} = \{\hat{z}\} + \mathbf{F} + \{\hat{0}\}$. This is illustrated for two examples in Fig. 8.14. Note that x covers $\hat{z}$, since $y \geq x$ for every $y \in Y$, because each y is in some minimal nontrivial join cover Y_j of x, reflecting the subdirect irreducibility of the lattice.

We are temporarily assuming that $\{x\} \cup Y$ is an antichain, so we need only to enforce the arrow relations $x \to \hat{z} \leftrightarrow z_1$. For that purpose, we can employ the operators f and g used throughout this section.

Next, we start adding relations to **P**. Consider additional relations $y \leq y'$ with y, $y' \in Y$. Necessarily y and y' come from different Y_i and Y_j. To enforce $y' \to y$ we use operators

$$k(u) = \begin{cases} \hat{0} & \text{if } u = \hat{0} \\ z_1 & \text{if } u = z_1 \\ y & \text{if } u = y' \\ \hat{z} & \text{otherwise.} \end{cases}$$

It is not hard to check that this works, because for any $T \subseteq S$, we have $k(\bigwedge T) = \hat{z} = \bigwedge k(T)$ unless $T \subseteq \{y', z_1, \hat{0}\}$.

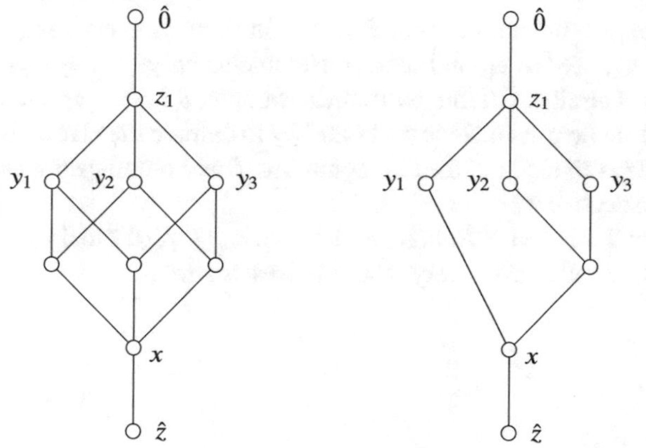

Fig. 8.14 The lattices **S** for $\mathbf{1}+\mathbf{G}(x, \{y_1, y_2, y_3\}, \Delta)$ and $\mathbf{1}+\mathbf{G}(x, \{y_1, y_2\}, \{y_1, y_3\}, \Delta)$, respectively

A little more care is required for inclusions $y \leq x$ with $y \in Y_i$, say. Note that for each j, the condition $y \leq x \leq \bigvee Y_j$ implies that $y \leq y'$ for some $y' \in Y_j$, since the elements of Y are join prime. That observation allows us to use the following map:

$$\ell(u) = \begin{cases} \hat{0} & \text{if } u = \hat{0} \\ \hat{z} & \text{if } u = \hat{z} \\ y & \text{if } u = \bigwedge_{\mathbf{F}} Z \text{ with } Z \subseteq Y \text{ and } y \leq z \text{ in } \mathbf{L} \text{ for some } z \in Z \\ z_1 & \text{otherwise.} \end{cases}$$

Thus we can add operators for each inclusion in **P**, and so represent $\mathbf{1} + \mathbf{G}(x, Y_1, \ldots, Y_m, \mathbf{P})$ as $\mathrm{Sub}(\mathbf{S}, \wedge, \hat{0}, H)$ with **S** finite and H satisfying ϖ. That proves Theorem 8.3.

Example 8.6 Let $\mathbf{M}[m]$ be $\mathbf{3}\times\mathbf{3}$ with the middle point doubled, and form $\mathbf{1}+\mathbf{M}[m]$ as in Fig. 8.15. As in the figure, we take $\mathbf{S}' = \{\hat{z}\} + \mathbf{F} + \{\hat{0}\}$ where **F** is the meet semilattice with top z_1, bottom x, coatoms y_1, y_2, y_3, y_4, subject to the relations $y_1 \wedge y_4 = x$ and $y_2 \wedge y_3 = x$.

We need to enforce the arrow relations

$$\begin{array}{cc} y_3 \longrightarrow y_1 & y_4 \longrightarrow y_2 \\ x \longrightarrow y_1 & x \longrightarrow y_2 \\ y_1, y_2 \longrightarrow z_1 & z_1 \leftrightarrow \hat{z} \end{array}$$

For $y_3 \longrightarrow y_1$ we use the map k_1 that fixes $\hat{0}$ and z_1, maps y_3 to y_1, and sends everything else (including y_1) to $\hat{z}$. The map k_2 to enforce $y_4 \longrightarrow y_2$ is symmetric.

For $x \longrightarrow y_1$ use the map ℓ_1 that fixes $\hat{0}$ and $\hat{z}$, maps everything in the intervals $[x, y_1]$ and $[x, y_3]$ of $\mathbf{S}'$ to y_1, and sends the elements $z_1, y_2, y_4, y_2 \wedge y_4$ to z_1. The map ℓ_2 for $x \longrightarrow y_2$ is similar. (In this particular example, $\ell_1(y_3) = y_1$ and $\ell_2(y_4) = y_2$, so those maps could be used instead of k_1 and k_2 to enforce the arrows in the first line.)

For the arrows in the last line, we again use f and g collapsing everything but $\hat{0}$ to z_1 and $\hat{z}$, respectively.

Thus we get $\mathbf{1}+\mathbf{M}[m] \cong \mathrm{Sub}(\mathbf{S}', \wedge, \hat{0}, k_1, k_2, \ell_1, \ell_2, f, g)$ satisfying ϖ, which then yields a longstyle subquasivariety lattice representation.

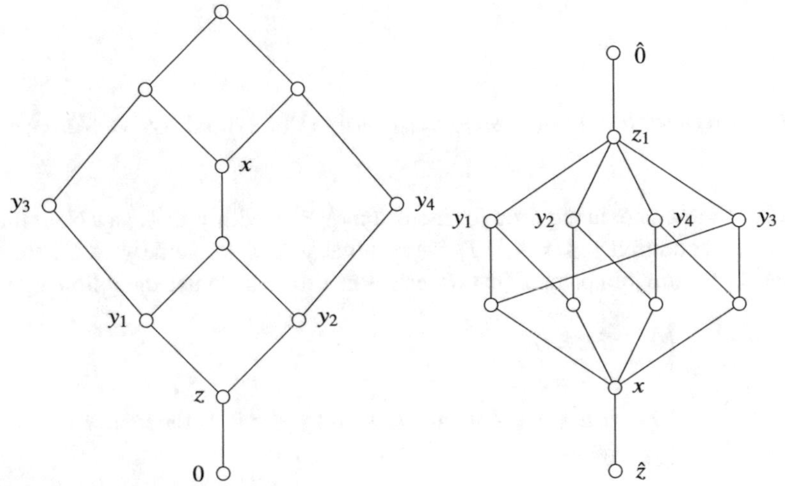

Fig. 8.15 $\mathbf{1}+\mathbf{M}[m]$ and $\mathbf{S}'$ to represent $\mathbf{1}+\mathbf{M}[m]$ of Example 8.6

Aside A straightforward adaptation of the construction of Theorem 8.3 yields a nice bonus.

Theorem 8.7 *Every finite, subdirectly irreducible, lower bounded lattice of rank* 1 *is isomorphic to* $\mathrm{Sub}(\mathbf{S}, \wedge, 1, H)$ *for a finite semilattice with operators.*

Proof Consider the least preclop μ on such a lattice $\mathbf{L}$ given by Theorem 4.10,

$$\mu(t) = \begin{cases} 1 & \text{if } t \geq x, \\ t & \text{otherwise,} \end{cases}$$

where x is the unique non-join-prime join irreducible in $\mathrm{J}(\mathbf{L})$. Then $\mathrm{J}(\mathbf{L}) \subseteq \tau(\mathbf{L})$, so we take $\widehat{\mathbf{S}} = \mu^d(\mathbf{L})$. In the semilattice $\widehat{\mathbf{S}}$ there is no z_1. Remove the operators f, g from the preceding construction, and wherever z_1 appears in the definition of an operator k or ℓ, replace it by $\hat{0}$. Now one can check that the construction yields the desired representation of $\mathbf{L}$ as $\mathrm{Sub}(\mathbf{S}, \wedge, 1, H)$. However, the condition ϖ, which held for **1** + **L**, need not be satisfied, and we may not obtain a representation of **L** as $\mathrm{L_q}(\mathcal{K})$. □

8.2.7 1 + L with L Subdirectly Reducible, Rank 1

Representing **1** + **L** as $\mathrm{L_q}(\mathcal{K})$ when **L** is subdirectly reducible, or in $\mathcal{LB}(k)$ for some $k > 1$, or not lower bounded at all, gets much more involved. Let us content ourselves with two very simple cases.

Case 1 The next theorem covers *some* cases where J(**L**) has a unique non-join-prime element but is not subdirectly irreducible.

Theorem 8.8 *Let* **L** *be a finite lattice with exactly one join irreducible element* x *that is not join prime. Let* $Y = \{y \in \mathrm{J}(\mathbf{L}) : x \mathrel{D} y\}$*, and let* $E = \mathrm{J}(\mathbf{L}) \setminus (Y \cup \{x\})$*. Assume that* x *is incomparable to* e *for all* $e \in E$*. Then* **1** + **L** *can be represented as* $\mathrm{L_q}(\mathcal{K})$ *for some quasivariety* $\mathcal{K}$*.*

Indeed, the prescription is almost the same as before. We take **F** to be the semilattice freely generated by J(**L**), subject to the relations $x = \bigwedge Y_i$ whenever $x \le \bigvee Y_i$ is a minimal nontrivial join cover in **L**. Then let $\mathbf{S} = \mathbf{F} + \{\hat{0}\}$. The arrow relations are of the forms $y \to y'$, $x \to y$, $y \to e$, $e \to y$, $t \to z_1$, and $z_1 \leftrightarrow \hat{z}$. Though x is no longer the unique atom in **S**, these can be handled using the operators f, g, k, ℓ defined exactly as before. This gives **1** + **L** as $\mathrm{Sub}(\mathbf{S}, \wedge, 1, H)$ satisfying ϖ, which again yields a longstyle representation as $\mathrm{L_q}(\mathcal{K})$.

Example 8.9 For the lattice **1** + **K** in Fig. 8.16, we have $Y = \{y_1, y_2\}$ and $E = \{y_3\}$. Besides the arrows involving z, we need to enforce $y_1 \to y_3$ and $y_2 \to y_3$. The prescription gives maps k_1 and k_2 doing that.

However, there is no analogue of Theorem 8.8 for this case. The lattice **K** fails (K9) and (K10) and so cannot be represented as $\mathrm{S_p}(\mathbf{S}, H)$.

(We have also found representations as $\mathrm{L_q}(\mathcal{K})$ for some lattices **1** + **L** where J(**L**) has only one non-join-prime element, but relaxing the hypotheses of Theorem 8.8 to allow $e \le x$ or $x \le e$ for some $e \in E$. Some changes are required, and we have not done enough cases to establish the general pattern.)

Case 2 Now let us consider a class of lattices generalizing the hexagon, where the lattice is in $\mathcal{LB}(1)$ but J(**L**) contains more than one non-join-prime element.

Let $X = \{x_1, \dots, x_n\}$ and Y be disjoint finite sets with $|Y| \ge 2$, and let **Q** be an order on $X \cup Y$ such that Y is an antichain and no relation $x \le y$ holds in **Q**. Define the lattice $\mathbf{H}(x_1, \dots, x_n, Y, \mathbf{Q})$ to be the join semilattice with 0 generated by $X \cup Y$ subject to the inclusions of **Q** and $x_i \le \bigvee Y$ for $1 \le i \le n$. For example, $\mathrm{Co}(\mathbf{2} \times \mathbf{2})$ is $\mathbf{H}(b, c, \{a, d\}, \Delta)$, while the hexagon is $\mathbf{H}(b, c, \{a, d\}, a \le b,\ d \le c)$. These are lower bounded lattices of rank 1, but for $n > 1$ they are not subdirectly irreducible.

Theorem 8.10 *Every lattice of the form* $\mathbf{1} + \mathbf{H}(x_1, \dots, x_n, Y, \mathbf{Q})$ *is isomorphic to* $\mathrm{L_q}(\mathcal{K})$ *for a quasivariety of structures with equality.*

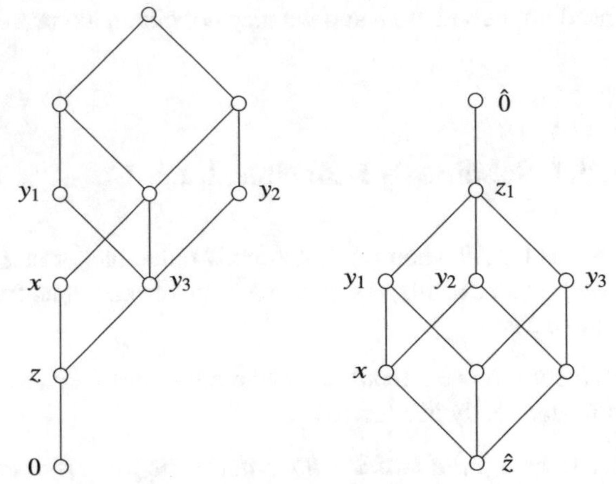

Fig. 8.16 $\mathbf{1}+\mathbf{K}$ and $\mathbf{S}$ to represent $\mathbf{1}+\mathbf{K}$ of Example 8.9

Proof For each i, let Y_i be a copy of Y, and let $\mathbf{P}_i$ be the lattice representing $\mathbf{1}+\mathbf{G}(x_i, Y, \Delta)$, as constructed in the proof of Theorem 8.3. String $\mathbf{P}_1, \ldots, \mathbf{P}_n$ together in parallel, identifying $\hat{0}$, z_1 and $\hat{z}$, as in Fig. 8.6. We need to find operators to enforce the arrow relations

- $y_i \leftrightarrow y_j$ whenever $i \neq j$ (for different copies of the same element $y \in Y$),
- $x_i \longrightarrow x_j$ whenever $x_j \leq x_i$ in $\mathbf{Q}$,
- $x_i \longrightarrow y_i$ whenever $y \leq x_i$ in $\mathbf{Q}$,
- $x_i \longrightarrow \hat{z}$ and $y \longrightarrow \hat{z}$ for all i and $y \in Y$,
- $\hat{z} \leftrightarrow z_1$. □

Note that there are no relations $x_i \leq y$ or $y \leq y'$ in $\mathbf{Q}$. The standard operators f and g take care of the last two items.

To enforce $y_i \leftrightarrow y_j$ for different copies of the same element $y \in Y$, let

$$h(t) = \begin{cases} \hat{0} & \text{if } t = \hat{0} \\ z_1 & \text{if } t = z_1 \\ y_j & \text{if } t = y_i \\ y_i & \text{if } t = y_j \\ \hat{z} & \text{otherwise.} \end{cases}$$

To enforce a relation $x_i \longrightarrow x_j$ from $\mathbf{Q}$, we use

$$k(t) = \begin{cases} \hat{0} & \text{if } t = \hat{0} \\ z_1 & \text{if } t = z_1 \\ t_j & \text{if } t = t_i \in P_i \\ \hat{z} & \text{otherwise.} \end{cases}$$

To enforce a relation $x_i \to y_i$ from $\mathbf{Q}$, we use

$$\ell(t) = \begin{cases} \hat{0} & \text{if } t = \hat{0} \\ z_1 & \text{if } t = z_1 \\ y_i & \text{if } t = t_i \in P_i \text{ and } x_i \leq t_i \leq y_i \text{ in } \mathbf{P}_i \\ z_1 & \text{if } t = t_i \in P_i \text{ and } t_i \nleq y_i \text{ in } \mathbf{P}_i \\ \hat{z} & \text{otherwise.} \end{cases}$$

These results provide some limited support for Conjectures (A) and (B) on page 217, but we are far from the general case, if indeed the conjectures are true.

8.3 An Age-Old Question Answered

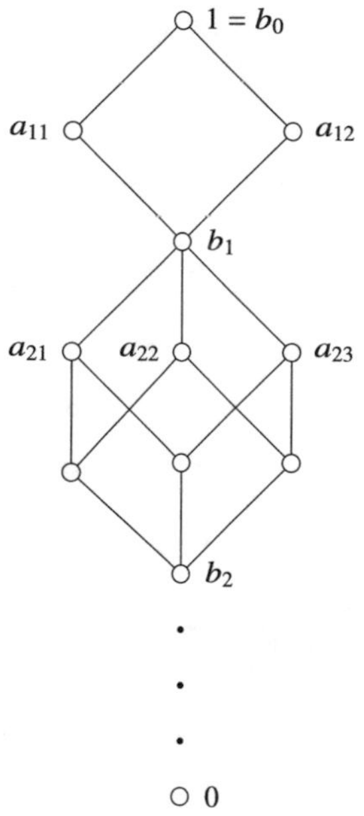

Fig. 8.17 The tower lattice $\mathbf{T}$ of Sect. 8.3. The elements b_j ($j \in \omega$) form a descending chain, and for $j \geq 1$ the interval $[b_j, b_{j-1}]$ is the Boolean lattice $\mathbf{B}_{j+1}$

The tower lattice $\mathbf{T}$ is drawn in Fig. 8.17. We take the set of operators h_{ij} for $1 \leq i, j \leq i + 1$, where

$$h_{ij}(x) = \begin{cases} 1 & \text{if } x \geq b_i \\ a_{ij} & \text{otherwise.} \end{cases}$$

We claim that $\mathbf{P} = \mathrm{S_p}(\mathbf{T}, H)$ is the ascending linear sum of lattices Sub $\mathbf{B}_n$. Indeed, the proper H-closed algebraic subsets of $\mathbf{T}$ are of the form $\mathbf{S}\dot{+}\mathbf{B}_k\dot{+}\dots\dot{+}\mathbf{B}_2$ where $\mathbf{S}$ is any subsemilattice of $\mathbf{B}_k$; here we use $\dot{+}$ to denote the glued linear sum.

To see this, consider any element x with $b_{k+1} \leq x < b_k$. Then $h_{ij}(x) = a_{ij}$ for all $i \leq k$ and $j \leq i+1$, so that the entire interval $[b_k, 1]$ is contained in $\mathrm{Sg}(x)$. On the other hand, $h_{ij}(x) = 1$ whenever $i > k$.

Now $\mathbf{P}$ satisfies no lattice identities, because subsemilattice lattices satisfy no identities [48]. Moreover, we know that $\mathbf{1}+\mathbf{P}$ is a subquasivariety lattice by Corollary 7.10. But clearly it does not contain a copy of the free lattice $\mathrm{FL}(X)$ for $|X| \geq 3$; indeed, $\mathbf{1}+\mathbf{P}$ contains no infinite antichain, while FL(3) does.

Problems 16 and 17 of [1] asked whether every subquasivariety lattice $\mathrm{L_q}(\mathcal{K})$ that satisfies no lattice identities must have a sublattice isomorphic to FL(3). The lattice $\mathbf{1}+\mathbf{P}$ shows that the answer is negative.

Chapter 9
Representing Distributive Dually Algebraic Lattices

In this dark, when we all talk at once, some of us must learn to whistle. – Walt Kelly

И казалось, что ещё немного - и решение будет найдено, и тогда начнется новая, прекрасная жизнь; и обоим было ясно, что до конца ещё далеко-далеко и что самое сложное и трудное только ещё начинается. – Антон Чехов, Дама с собачкой

Years ago, Tumanov proved that every finite distributive lattice is isomorphic to a lattice of subquasivarieties ([112], our Theorem 2.66). This chapter deals with dually algebraic, distributive lattices.

Theorem 9.1 is that every distributive dually algebraic lattice can be represented as $\mathrm{S_p}(\mathbf{S}, H)$ with $\mathbf{S}$ an algebraic lattice and H a monoid of operators. As a consequence, every linear sum $\mathbf{1} + \mathbf{D}$ with $\mathbf{D}$ distributive and dually algebraic is isomorphic to a lattice of subquasivarieties $\mathrm{L_q}(\mathcal{K})$ with equality, which is Corollary 9.12.

Theorem 9.18 is that every distributive lattice that is both algebraic and dually algebraic, and has its least element dually compact, is isomorphic to $\mathrm{L_q}(\mathcal{K})$ for some quasivariety $\mathcal{K}$ with equality.

On the other hand, Example 2.61 was that the chain $(\omega+1)^d$, which is distributive, algebraic, and dually algebraic, can be represented as $\mathrm{S_p}(\mathbf{S}, H)$ but is not isomorphic to any subquasivariety lattice $\mathrm{L_q}(\mathcal{K})$ in a language with equality, because the lattice is not atomic, as required by Theorem 2.60.

9.1 Distributive Dually Algebraic Lattices as $\mathrm{S_p}(\mathbf{S}, H)$

In this section we prove the following result.

Theorem 9.1 *Every distributive dually algebraic lattice can be represented as* $\mathrm{S_p}(\mathbf{S}, H)$ *with* $\mathbf{S}$ *an algebraic lattice and* H *a monoid of operators.*

K. Adaricheva et al., *A Primer of Subquasivariety Lattices*, CMS/CAIMS Books in Mathematics 3, https://doi.org/10.1007/978-3-030-98088-7_9

Let **D** be an algebraic distributive lattice, and **P** the join semilattice of its compact elements, so that **D** is isomorphic to the ideal lattice $\mathcal{I}(\mathbf{P})$. We are going to show that $\mathbf{D}^d$ is representable as $\mathrm{S_p}(\mathcal{I}(\mathbf{K}), H)$ for a semilattice **K**, with **P** and **K** related by a semilattice homomorphism $\mu : \mathbf{K} \to \mathbf{P}$. We note in advance that **K** and **P** are both join semilattices with 0; **P** is distributive, though **K** need not be so.

Recall that a $(\vee, 0)$-semilattice **Q** is *distributive* if whenever $u \le x \vee y$ in **Q**, there exist $a, b \in Q$ such that $a \le x$, $b \le y$, and $u = a \vee b$. This is equivalent to saying that the lattice of ideals $\mathcal{I}(\mathbf{Q})$ is a distributive lattice. We will eventually need a lemma of E. T. Schmidt [106].

Lemma 9.2 *Let* **Q** *be a distributive semilattice with* 0*, and let F be a finite subset of Q. Then there exists a finite distributive subsemilattice* **E** *of* **Q** *with $F \subseteq E \subseteq Q$.*

The next lemma is also useful.

Lemma 9.3 *Let* **K** *be a* $(\vee, 0)$*-semilattice and* $\mathbf{S} = \mathcal{I}(\mathbf{K})$ *its ideal lattice. Consider a map* $h : \mathbf{S} \to \mathbf{S}$*, and let* $h_0 : \mathbf{K} \to \mathbf{S}$ *be the restriction of h to* **K**.

(1) *The map h preserves nonempty directed joins if and only if*

  (a) *h_0 is order preserving, and*
  (b) $h(x) = \bigvee\{h_0(k) : k \in K$ *and* $k \le x\}$ *for all $x \in S$.*

(2) *Assume that h satisfies the property of* (1)*. Then h preserves arbitrary meets if and only if*

  (a) $h(1_S) = 1_S$*, and*
  (b) *if x, y_α $(\alpha \in A)$ are elements of K such that $x \le h_0(y_\alpha)$ for all α, then there exists $z \in K$ such that $z \le y_\alpha$ for all α and $x \le h_0(z)$.*

Note that an operator h need not preserve non-directed joins nor need it map compact elements to compact elements. Thus the proof of the lemma requires a slight bit of care.

Proof Surely the conditions in (1) are necessary for h to preserve directed joins, since for any $x \in L$, $\{k \in K : k \le x\}$ is a directed set.

Conversely, assume that h satisfies the conditions of (1). The fact that h_0 is order preserving, along with (1b), implies that h is order preserving. So it remains to show that $h(\bigvee d_i) \le \bigvee h(d_i)$ whenever $D = \{d_i : i \in I\}$ is a directed subset of S. Since **S** is algebraic, it suffices to show that for a compact element k, $k \le h(\bigvee d_i)$ implies $k \le \bigvee h(d_i)$.

Assume $k \le h(\bigvee d_i) = \bigvee\{h_0(c) : c \in K \text{ and } c \le \bigvee d_i\}$. By the compactness of k, there is a finite subset $c_1, \ldots, c_m$ such that $k \le h_0(c_1) \vee \cdots \vee h_0(c_m)$. By the compactness of c_j and the directedness of D, each $c_j \le d_{i_j}$ for some $i_j \in I$. Let d_ℓ be such that $d_{i_j} \le d_\ell$ for $1 \le j \le m$, so that $e = c_1 \vee \cdots \vee c_m \le d_\ell$. Then $k \le h_0(c_1) \vee \ldots \vee h_0(c_m) \le h_0(e)$ since h_0 preserves order, and $h_0(e) \le h(d_\ell) \le \bigvee_{i \in I} h(d_i)$, as desired.

If h preserves arbitrary meets, then of course $h(1_S) = 1_S$ (the empty meet). Suppose that h also satisfies property (1). In particular, when y_α $(\alpha \in A)$ are compact elements, then $\bigwedge_{\alpha \in A} h_0(y_\alpha) \leq h(\bigwedge_{\alpha \in A} y_\alpha) = \bigvee \{h_0(k) : k \in K \text{ and } k \leq y_\alpha \text{ for all } \alpha\}$. Thus whenever x is compact and $x \leq h_0(y_\alpha)$ for all α, there is a finite subset $k_1, \ldots, k_m$ such that each $k_j \leq \bigwedge y_\alpha$ and $x \leq h_0(k_1) \vee \cdots \vee h_0(k_m)$. Taking $z = k_1 \vee \cdots \vee k_m$ we have $z \leq \bigwedge y_\alpha$ and $x \leq h_0(z)$ since h_0 preserves order. Thus (2) holds.

Conversely, assume that h satisfies (1) and (2). We want to show that for any set $\{u_\alpha : \alpha \in A\} \subseteq S$ we have $\bigwedge h(u_\alpha) \leq h(\bigwedge u_\alpha)$. So consider any compact element k such that $k \leq \bigwedge h(u_\alpha)$, i.e., $k \leq h(u_\alpha)$ for all $\alpha \in A$. Arguing as before, by virtue of (1), for each α there exists $y_\alpha \in K$ such that $y_\alpha \leq u_\alpha$ and $k \leq h_0(y_\alpha)$. By (2), there exists $z \in K$ such that $z \leq y_\alpha$ for all α and $k \leq h_0(z)$. Then $z \leq \bigwedge y_\alpha \leq \bigwedge u_\alpha$, whence $h_0(z) \leq h(\bigwedge u_\alpha)$. Therefore, $k \leq h(\bigwedge u_\alpha)$, as desired. □

We assume that the distributive algebraic lattice $\mathbf{D}$ is given, with $\mathbf{P}$ its semilattice of compact elements, so that $\mathbf{D} = \mathcal{I}(\mathbf{P})$. We want to show that there is a $(\vee, 0)$-semilattice $\mathbf{K}$, and a set H of operators on $\mathcal{I}(\mathbf{K})$, such that $\mathbf{D} \cong^d S_p(\mathcal{I}(\mathbf{K}), H)$. Throughout, we let $\mathbf{S} = \mathcal{I}(\mathbf{K})$, which is of course an algebraic lattice.

First we describe a set of six conditions that a $(\vee, 0)$-semilattice might satisfy and show that when $\mathbf{K}$ satisfies them we get $\mathcal{I}(\mathbf{P}) \cong^d S_p(\mathcal{I}(\mathbf{K}), H)$, which is Theorem 9.5. Afterwards, we prove that for every distributive semilattice $\mathbf{P}$ there exists such a $\mathbf{K}$. The properties we consider are:

(C0) $\mathbf{K}$ has a largest element 1_K iff $\mathbf{P}$ has a largest element 1_P, i.e., iff 1_D is compact in $\mathbf{D}$.
(C1) There is a surjective join homomorphism $\mu : \mathbf{K} \to \mathbf{P}$.
(C2) $\mu(x) = 0_P$ iff $x = 0_K$.
(C3) Let $\mathbf{K}^\dagger = \mathbf{K} \setminus \{0_K, 1_K\}$, the partial semilattice obtained by removing 0_K and 1_K if there is one. (References to 1_K in the sequel should be omitted if $\mathbf{K}$ has no largest element.) There is a map $f : (K^\dagger)^2 \to K^\dagger$ such that for all u, k, $\ell \in K^\dagger$,

 (a) $f(u, k) \nleq u$,
 (b) $k \leq f(u, k) = f(u, f(u, k))$,
 (c) $k \leq \ell$ implies $f(u, k) \leq f(u, \ell)$,
 (d) $f(u, k) = k \leq \ell$ implies $f(u, \ell) = \ell$,
 (e) $\mu(f(u, k)) = \mu(k)$,
 (f) $\mu(\ell) \leq \mu(k) < 1_P$ implies $\ell \leq f(\ell, k)$.

(C4) There is a subset $B \subseteq K^\dagger$ such that B join-generates $K^\dagger$, i.e., every $u \in K^\dagger$ is a join of finitely many members of B. Moreover, there is a map $g : B \times K \to K$ such that for all $b \in B$ and $k, \ell \in K$,

$$k \leq b \vee \ell \text{ iff } \ell \geq g(b, k).$$

Note $g(b, k) \leq k$.
(C5) $\mu(k) = 1_P$ implies $k = 1_K$ or there exist $u_1, \ldots, u_n, v_1, \ldots, v_n$ such that

- $u_i \leq v_i$ and $\mu(u_i) = \mu(v_i) < 1_P$ for each i,
- $\bigvee u_i = k$ and $\bigvee v_i = 1_K$.

Observe that conditions (C3b)–(C3d) say that, for each u, the range of $f(u,*)$ is an order filter of fixed points of $f(u,*)$. Condition (C4) is reminiscent of Wehrung's definition of *fermentability* [114].

If such a semilattice $\mathbf{K}$ exists, then the monoid H will be generated by two types of operators. In accordance with Lemma 9.3(1), we define the maps on $\mathbf{K}$ and extend them naturally to $\mathbf{S}$. For each u in $K^\dagger$, define z_u by extending the map

$$z_u(x) = \begin{cases} x & \text{if } x = f(u,x) \text{ or } x = 1_K, \\ 0_K & \text{otherwise} \end{cases}$$

for $x \in K$. Note $z_u(u) = 0_K$ by (C3a).

For each $u \in K$, define ℓ_u by $\ell_u(x) = x \vee u$.

Lemma 9.4 *If* (C0)–(C5) *hold, then each* z_u $(u \in K^\dagger)$ *and each* ℓ_u $(u \in K)$ *is an operator.*

Proof By Lemma 9.3(1), both types of maps preserve nonempty directed joins, and both types preserve 1_K. It remains to check that the property of Lemma 9.3(2b) holds for z_u and ℓ_u.

Let x, y_α $(\alpha \in A)$ be such that $x \le z_u(y_\alpha)$ for all α. We claim that $f(u,x)$ has the property of (2b). Without loss of generality $x > 0$ and thus $z_u(y_\alpha) = y_\alpha = f(u, y_\alpha)$ for all α. But $x \le y_\alpha$ implies $f(u,x) \le f(u,y_\alpha) = y_\alpha$ for all α, while $x \le f(u,x) = z_u(f(u,x))$ also holds. So in Lemma 9.3(2b), we can take $z = f(u,x)$.

First we show that ℓ_b preserves arbitrary meets for each $b \in B$ from condition (C4). Assume x, y_α $(\alpha \in A)$ are such that $x \le \ell_b(y_\alpha) = y_\alpha \vee b$ holds for all α. By property (C4), then $y_\alpha \ge g(b,x)$ for all α, while $x \le b \vee g(b,x) = \ell_b(g(b,x))$. Thus we can take $z = g(b,x)$ for Lemma 9.3(2b).

Clearly ℓ_{0_K} and ℓ_{1_K} (when 1_K exists) preserve meets, so consider $u \in K^\dagger$. Then there exist $b_1, \ldots, b_m \in B$ such that $u = b_1 \vee \cdots \vee b_m$, whence the map $\ell_u = \ell_{b_1} \circ \cdots \circ \ell_{b_m}$ also preserves arbitrary meets. □

Theorem 9.5 *If the semilattice* $\mathbf{K}$ *and the maps* μ, f, g *satisfy* (C0)–(C5), *then with* $\mathbf{S} = \mathcal{I}(\mathbf{K})$ *and* H *the monoid generated by the operators* z_u, ℓ_u *above, we have* $\mathrm{S_p}(\mathbf{S}, H) \cong^d \mathcal{I}(\mathbf{P})$.

Thus, assuming we can find a semilattice $\mathbf{K}$ with the desired properties, we want to establish a dual isomorphism between H-closed algebraic subsets of $\mathbf{S}$ and ideals of the semilattice $\mathbf{P}$. The proof of Theorem 9.5 requires a series of lemmas.

Let $\mathbf{S}$ be an algebraic lattice with a monoid of operators H. Recall that an element $x \in S$ is *fully invariant* if $h(x) \ge x$ for every $h \in H$. The fully invariant elements form a complete sublattice of $\mathbf{S}$ (Theorem 1.22). Every algebraic subset has a least element, and the fully invariant elements of $\mathbf{S}$ are precisely those elements that can be the least element of an H-closed algebraic subset. Moreover, in view of the operators ℓ_u for every compact u, the H-closed algebraic subsets of our particular $\mathbf{S}$ are exactly the principal filters $\uparrow x$ with x fully invariant. We record this thusly.

Lemma 9.6 *If* $\mathbf{S} = \mathcal{I}(\mathbf{K})$ *and* ℓ_u *is in* H *for every* $u \in K$*, then* $A \subseteq S$ *is an* H*-closed algebraic subset if and only if* $A = \uparrow x_0$ *for some fully invariant element* $x_0 \in S$*. Hence* $\mathrm{S_p}(\mathbf{S}, H) \cong^d \Phi$*, the sublattice of* $\mathbf{S}$ *consisting of all fully invariant elements.*

It is a *dual* isomorphism because filters reverse inclusion.

Proof Assume that $\ell_u \in H$ for every compact u. Let $x \in A$ where A is an H-closed algebraic subset, and consider any $y \geq x$. Then $\ell_u(x) = x \vee u$ is in A for every compact $u \leq y$. Since these form a directed set, $y = \bigvee_{u \leq y} \ell_u(x)$ is in A. Thus $\uparrow x \subseteq A$. □

So, given $\mathbf{K}$ satisfying (C0)–(C5), it remains to identify the fully invariant elements of $\mathbf{S} = \mathcal{I}(\mathbf{K})$. The elements of $\mathbf{S}$ are ideals of $\mathbf{K}$, and in view of Lemma 9.3, for each $X \in S$ and $h \in H$ we have that $h(X)$ is the ideal generated by $\{h(k) : k \in X\}$. An ideal X is fully invariant exactly when $X \subseteq h(X)$ for every $h \in H$. As $\ell_u(k) \geq k$ for every $u, k \in K$, we concentrate on the operators z_u with $u \in K^\dagger$.

Lemma 9.7 *An ideal* X *of* $\mathbf{K}$ *is a fully invariant element of* $\mathbf{S}$ *if and only if either* $1_K \in X$ *(in which case* $X = K$ *itself) or for every* $u \in K^\dagger$ *and* $k \in X$*, there exists* $k' \in X$ *such that* $k \leq k' = f(u, k')$*.*

Proof Because $z_u(k) \in \{k, 0_K\}$ for every $k \in K$, we have $z_u(X) \leq X$ for any ideal. The condition of the lemma says that $k \leq k' = z_u(k')$, and so avoids proper containment. □

For each ideal $J \in \mathcal{I}(\mathbf{P})$, let $I_J = \mu^{-1}(J) = \{k \in K : \mu(k) \in J\}$. By property (C1), every I_J is an ideal of $\mathbf{K}$.

Lemma 9.8 *If* (C0)–(C5) *hold, then each* I_J *is a fully invariant element of* $\mathbf{S}$.

Proof If $J = P$ itself, then $I_J = K$ which is fully invariant, so assume $J \subset P$; in particular, $1_P \notin J$. It suffices to show that for every $u \in K^\dagger$, and every k with $0 < k \in I_J$, there is an element k' such that $k \leq k' \in I_J$ and $z_u(k') = k'$. By assumptions (C3b) and (C3e), $k' = f(u, k)$ has these properties. □

Lemma 9.9 *If* X *is a fully invariant element of* $\mathbf{S}$*, then* $X = I_J$ *for some ideal* J *of* $\mathbf{P}$.

Proof Assume that X is fully invariant, and let J be the ideal of $\mathbf{P}$ generated by $\{\mu(k) : k \in X\}$. Clearly $X \subseteq I_J$; we want to show $I_J \subseteq X$.

If $J = \{0_P\}$, then $X = \{0_K\} = I_J$ by condition (C2), so assume $J > \{0_P\}$.

Next suppose $1_P \notin J$. Let $0_K < k \in K$ and let $\mu(k) = p < 1_P$, and consider an element $\ell \in K$ with $\mu(\ell) \leq p$. We want to show that $\ell \in X$. Now by condition (C3f) we have $\ell \leq f(\ell, k)$, while by Lemma 9.7 there exists $w \in X$ with $k \leq w = f(\ell, w)$. Thus $\ell \leq f(\ell, k) \leq f(\ell, w) = w \in X$ using (C3c), whence $\ell \in X$.

Finally, assume $1_P \in J$, so that $\mu(k) = 1_P$ for some $k \in X$. We want to show $1_K \in X$. By condition (C5), either $k = 1_K$, in which case we are done, or there exist $u_1, \ldots, u_n, v_1, \ldots, v_n$ such that $u_i \leq v_i$ and $\mu(u_i) = \mu(v_i) < 1_P$ for each i, with $\bigvee u_i = k$ and $\bigvee v_i = 1_K$. Now we repeat the previous argument for each i. As

$\mu(u_i) = \mu(v_i) < 1_P$, by (C3f) we have $v_i \leq f(v_i, u_i)$. Meanwhile $u_i \leq k$ implies $u_i \in X$, whence by Lemma 9.7 there exists $w_i \in X$ such that $u_i \leq w_i = f(v_i, w_i)$. Combining yields $v_i \leq f(v_i, u_i) \leq f(v_i, w_i) = w_i \in X$, whence $v_i \in X$. But then $1_K = \bigvee v_i \in X$, as desired. □

Corollary 9.10 *If* (C0)–(C5) *hold, then* X *is a fully invariant element of* $\mathbf{S}$ *if and only if* $X = I_J$ *for some ideal* J *of* $\mathbf{P}$.

Now we are set up to prove Theorem 9.5. We have that $\mathbf{D} = \mathcal{I}(\mathbf{P})$, and the fully invariant elements of $\mathbf{S} = \mathcal{I}(\mathbf{K})$ form a lattice dually isomorphic to $\mathcal{I}(\mathbf{P})$, via the map $J \mapsto \uparrow I_J$. It remains to observe that the map is surjective and $J \leq J'$ if and only if $I_J \leq I_{J'}$, using the assumptions that μ is surjective and order preserving. Lemma 9.6 then says that this map is a dual isomorphism.

To complete the proof of Theorem 9.1, we must show that, given the semilattice $\mathbf{P}$, such a $\mathbf{K}$ satisfying (C0)–(C5) exists. By Lemma 9.2, every finite subset of $\mathbf{P}$ is contained in some finite distributive subsemilattice $\mathbf{E} \leq \mathbf{P}$. As the penultimate step for finding $\mathbf{K}$, we will construct for each finite distributive $\mathbf{E} \leq \mathbf{P}$ a semilattice $\mathbf{K_E}$ and maps μ, $f_\mathbf{E}$, $g_\mathbf{E}$ that satisfy (C0)–(C5) with respect to $\mathbf{E}$.

We may assume that $0_P \in E$, in symbols $0_E = 0_P$, since the type is $(\vee, 0)$-semilattices. Also $\mathbf{E}$ has a largest element 1_E, which may or may not be 1_P. Let $E^\dagger = E \setminus \{0_P, 1_P\}$. Take $\mathbf{K_E} = (\mathbf{E}^\dagger \times \omega) \cup \{0_P, 1_P\}$ but only adding 1_P if $\mathbf{P}$ has a largest element 1_P. This is a join semilattice with

$$(p, i) \vee (q, j) = \begin{cases} (p \vee q, i \vee j) & \text{if } p \vee q < 1_P, \\ 1_P & \text{otherwise,} \end{cases}$$

and 0_P, 1_P behaving as usual.

Now define

$$\begin{aligned} \mu(0_P) &= 0_P, \\ \mu((q, i)) &= q, \\ \mu(1_P) &= 1_P. \end{aligned}$$

Clearly this is a join homomorphism.

For property (C3), for each (q, i) define

$$f_\mathbf{E}((q, i), (r, j)) = (r, \max(i + 1, j)).$$

Note that $f_\mathbf{E}((q, i), (r, j)) = (r, j)$ iff $j \geq i + 1$. With that observation, one can check properties (C3a)–(C3f).

Our next lemma is well-known.

Lemma 9.11 *Let* $\mathbf{G}$ *be a distributive lattice satisfying the descending chain condition. For each pair* x, $z \in G$ *there is a least element* $z \setminus x$ *with the property*

$$x \vee y \geq z \text{ iff } y \geq z \setminus x.$$

Although $\mathbf{K_E}$ need not be distributive, because we have collapsed the top, each principal ideal $\downarrow(p,i)$ is distributive, and that is enough to yield condition (C4). Set $B_{\mathbf{E}} = K_{\mathbf{E}} \setminus \{1_P\}$, which contains 0_P and all (p,i) with $p \in E^{\dagger}$. Remember that we want to find $g_{\mathbf{E}}(b,k)$ on $B_{\mathbf{E}} \times K_{\mathbf{E}}$ such that

$$k \leq b \vee \ell \text{ iff } \ell \geq g_{\mathbf{E}}(b,k).$$

Clearly $g_{\mathbf{E}}(0_P,k) = k$. For $b = (p,i)$, check that the following map works:

$$g_{\mathbf{E}}((p,i),k) = \begin{cases} 0_P & \text{if } k = 0_P, \\ (1_P \setminus p, 0) & \text{if } k = 1_P \text{ and } p \vee r = 1_P \text{ for some } r < 1_P, \\ 1_P & \text{if } k = 1_P \text{ and } p \vee r = 1_P \text{ implies } r = 1_P, \\ (q \setminus p, j \setminus i) & \text{if } k = (q,j). \end{cases}$$

Property (C5) holds trivially for $\mathbf{K_E}$, since $\mu(k) = 1_P$ only for $k = 1_P$.

Now for $\mathbf{K}$ we take $\sum \mathbf{K_E}/\theta$, where

- $\sum \mathbf{K_E}$ is the direct sum of the semilattices $\mathbf{K_E}$ over all finite distributive 0-subsemilattices $\mathbf{E}$ of $\mathbf{P}$, that is, all finitely nonzero members of the direct product. An element of the direct sum will be denoted $\langle x_{\mathbf{E}} \rangle$,
- θ is the join semilattice congruence that collapses all x in the direct sum such that $x_{\mathbf{E}} = 1_P$ for some $\mathbf{E}$. This congruence class is an order filter in the direct sum, which we denote as 1_K.

It remains to define μ, f, and g on $\mathbf{K}$ and check properties (C0)–(C5).

For an element $k \in \mathbf{K}$, let $\mu(k) = \bigvee \mu(x_{\mathbf{E}})$. Since μ is a join homomorphism on each component and θ is a join-congruence, μ is a join homomorphism.

The definition of $f(u,k)$ is componentwise, extending the maps $f_{\mathbf{E}}$ on each $\mathbf{K_E}$ that were defined above. For $u, v \in K$, define

$$u \wedge_{\mathbf{E}} v = \bigvee \{e \in E : e \leq u \text{ and } e \leq v\}$$

which makes sense even when $u \notin E$ and/or $v \notin E$. Consider u, $k \in K^{\dagger}$, so that neither is the zero vector and neither contains an entry 1_P. Define

$$f(u,k)_{\mathbf{E}} = \begin{cases} (r, \max(i+1, j)) & \text{if } (u_{\mathbf{E}}, k_{\mathbf{E}}) = ((q,i),(r,j)), \\ (q \wedge_{\mathbf{E}} \mu(k), i+1) & \text{if } (u_{\mathbf{E}}, k_{\mathbf{E}}) = ((q,i), 0_P), \\ (r,j) & \text{if } (u_{\mathbf{E}}, k_{\mathbf{E}}) = (0_P, (r,j)), \\ 0_P & \text{if } (u_{\mathbf{E}}, k_{\mathbf{E}}) = (0_P, 0_P). \end{cases}$$

There are many cases, but it is straightforward to check that conditions (C3a)–(C3f) hold in $\sum \mathbf{K_E}/\theta$.

For condition (C4), we choose B to be the set of all $x \in \sum \mathbf{K_E}$ such that $x_{\mathbf{E}}$ is nonzero for exactly one component $\mathbf{E}_0$, and $x_{\mathbf{E}_0} \neq 1_P$. For $b \in B$ with $b_{\mathbf{E}_0} \neq 0_P$ and $k \in K$, let

$$g(b,k)_{\mathbf{E}} = \begin{cases} g_{\mathbf{E}_0}(b_{\mathbf{E}_0}, k_{\mathbf{E}_0}) & \text{if } \mathbf{E} = \mathbf{E}_0, \\ k_{\mathbf{E}} & \text{otherwise} \end{cases}$$

and check that this works.

Now we turn to condition (C5). In the direct sum, it is possible to have $k < 1_K$ but $\mu(k) = 1_P$, *viz.*, when $\bigvee \mu(k_{\mathbf{E}}) = 1_P$. Suppose that such a k has nonzero entries $k_{\mathbf{E}_1}, \ldots, k_{\mathbf{E}_n}$. For each i, let u_i be the corresponding element from B, with u_i having its only nonzero entry $u_{\mathbf{E}_i} = k_{\mathbf{E}_i}$. (This is some (p, j) from $P \times \omega$, and $\mu(u_i) = p$.) Then $k = \bigvee u_i$ and $\mu(u_i) < 1_P$.

Let $\mathbf{F}$ be a finite distributive subsemilattice of $\mathbf{P}$ such that $F \supseteq E_1 \cup \cdots \cup E_n$. For each i, let v_i be the element with

$$(v_i)_{\mathbf{E}} = \begin{cases} k_{\mathbf{E}_i} & \text{if } \mathbf{E} = \mathbf{E}_i \text{ or } \mathbf{E} = \mathbf{F}, \\ 0_P & \text{otherwise.} \end{cases}$$

Surely $\mu(v_i) = \mu(u_i)$, but $\bigvee v_i = 1_K$ because $(\bigvee v_i)_{\mathbf{F}} = 1_P$.

Thus we have found $\mathbf{K}$ satisfying (C0)–(C5), which proves Theorem 9.1. Observe that the natural weak equaclosure operator Γ on $\mathrm{S_p}(\mathbf{S}, H)$ for this construction is the identity map.

Combining the theorem with Corollary 7.10 gives a particularly nice consequence.

Corollary 9.12 *Let* $\mathbf{D}$ *be a dually algebraic, distributive lattice. Then* $\mathbf{1} + \mathbf{D}$ *is isomorphic to* $\mathrm{L_q}(\mathcal{K})$ *for a quasivariety of structures with equality.*

9.2 Lattices of Order Ideals as $\mathrm{S_p}(\mathbf{S}, H)$

Lattices $\mathcal{O}(\mathbf{P})$ of order ideals of an ordered set have a well-known characterization [31, 59]; for the proof see Lemma 8.1.

Lemma 9.13 *A lattice is isomorphic to* $\mathcal{O}(\mathbf{P})$ *for some ordered set* $\mathbf{P}$ *if and only if it is distributive, algebraic, and dually algebraic.*

In the next section, we will show that these lattices are subquasivariety lattices. The result of the previous section yields the weaker result that the lattices $\mathcal{O}(\mathbf{P})$ are representable as lattices of algebraic subsets with operators. The simpler proof for this case may be instructive. (Earlier results for lattices of order ideals are in [64]. See also Example 2.61 and Sect. 8.2.4 above.)

Theorem 9.14 *For any ordered set* $\mathbf{P}$*, the lattice* $\mathcal{O}(\mathbf{P})$ *is isomorphic to a lattice* $\mathrm{S_p}(\mathbf{S}, H)$ *with* $\mathbf{S}$ *an algebraic lattice and* H *a monoid of operators.*

Proof In fact, we will represent $(\mathcal{O}(\mathbf{P}))^d \cong \mathcal{O}(\mathbf{P}^d)$.

Let $\mathbf{S} = \mathcal{O}(\mathbf{P})$. Thus $0_{\mathbf{S}} = \varnothing$ and $1_{\mathbf{S}} = P$.

Lemma 9.15 **S** *is an algebraic and dually algebraic lattice. Its compact members are the finitely generated order ideals.* □

The monoid H will contain operators ℓ_u for every $u \in P$. Define ℓ_u by $\ell_u(I) = I \vee \downarrow u$, which is in fact $I \cup \downarrow u$.

Lemma 9.16 *Each ℓ_u is an operator.* □

The crucial observation is that ℓ_u preserves meets because in **S**,

$$\ell_u(\bigwedge_j I_j) = (\bigwedge_j I_j) \vee \downarrow u = (\bigcap_j I_j) \cup \downarrow u = \bigcap_j (I_j \cup \downarrow u) = \bigwedge_j \ell_u(I_j).$$

Now for any ideal M of **P**, let

$$\begin{aligned} A_M &= \{I \in \mathbf{S} : I \supseteq M\} \\ &= \uparrow M. \end{aligned}$$

We claim that the map $\alpha : M \mapsto A_M$ is a dual isomorphism of $\mathcal{O}(\mathbf{P})$ onto $S_p(\mathbf{S}, H)$.

It is not hard to see that A_M is an H-closed algebraic subset of **S**. On the other hand, if A is any H-closed algebraic subset, then it has a least element N. Because of the maps ℓ_u we have $\uparrow N \subseteq A$, whence $\uparrow N = A$. Since α is containment-reversing, it is a dual isomorphism. □

9.3 Lattices of Order Ideals as $L_q(\mathcal{K})$

Recall that in $L_q(\mathcal{K})$, the quasivariety $\langle x \approx y \rangle$ is dually compact. In this section we will show that when **P** is an ordered set such that the least element $\varnothing$ of $\mathcal{O}(\mathbf{P})$ is dually compact, then $\mathcal{O}(\mathbf{P}) \cong L_q(\mathcal{K})$ for a quasivariety with equality. As a bonus, the natural equaclosure operator on $L_q(\mathcal{K})$ will be the identity map!

First, we need this lemma.

Lemma 9.17 *The following are equivalent for an ordered set* **P**.

(1) *$\varnothing$ is dually compact in $\mathcal{O}(\mathbf{P})$.*
(2) **P** *has only finitely many minimal elements, and for each $p \in P$, there exists a minimal element $m \in P$ with $p \geq m$.*

Proof Let M denote the set of minimal elements of **P**.

First assume that (2) fails. If there exists an element $p_0 \in P$ that is above no minimal element, choose a maximal chain C in $\downarrow p_0$. Then $\varnothing = \bigwedge_{c \in C} \downarrow c$ witnesses that $\varnothing$ is not dually compact in $\mathcal{O}(\mathbf{P})$. If on the other hand every element is above a minimal element, but there are infinitely many minimal elements, then for each $m \in M$ let $K_m = \{x \in P : x \ngeq m\}$. Each K_m contains all the other minimal elements, so no intersection of finitely many of them is empty, while $\bigcap_{m\in M} K_m = \varnothing$ since $p \geq m$ implies $p \notin K_m$. This again shows that $\varnothing$ is not dually compact.

Now assume that (2) holds, and that $\varnothing = \bigcap_{j\in J} I_j$ for some collection of order ideals. For each minimal element m there is an index j_m such that $m \notin I_{j_m}$. Since the finite intersection $\bigcap_{m\in M} I_{j_m}$ is an ideal containing no minimal elements, and every element is above a minimal element, it must be empty. Thus $\varnothing$ is dually compact. □

Theorem 9.18 *If the least element* $\varnothing$ *of* $\mathcal{O}(\mathbf{P})$ *is dually compact, then* $\mathcal{O}(\mathbf{P})$ *is isomorphic to* $\mathrm{L_q}(\mathcal{P})$ *for some quasivariety* $\mathcal{P}$ *with equality.*

Define the variety $\mathcal{P}_0$ to have a constant e and unary functions η_p for $p \in P$, satisfying the following laws.

(i) $\eta_p^2 x \approx \eta_p x$
(ii) $\eta_p \eta_q x \approx \eta_q \eta_p x$
(iii) $\eta_p \eta_q x \approx \eta_p x$ whenever $p \leq q$
(iv) $\eta_{m_1} \dots \eta_{m_k} x \approx e$ where $\{m_1, \dots, m_k\}$ is the set M of minimal elements of $\mathbf{P}$.

Note that (iii) implies (i).

The free $\mathcal{P}_0$-algebra generated by a singleton a is easy to construct. The elements have a canonical form: either a or $\eta_{p_1} \dots \eta_{p_k} a$ where $\{p_1, \dots, p_k\}$ is an antichain in $\mathbf{P}$. This is unique up to the order of the operations, and $e = \eta_{m_1} \dots \eta_{m_k} x$. The operations are such that $\eta_q(\eta_{p_1} \dots \eta_{p_k} a) = \eta_{r_1} \dots \eta_{r_\ell} a$ where $\{r_1, \dots, r_\ell\}$ is the set of minimal elements of $\{q, p_1, \dots, p_k\}$.

Example 9.19 Consider Fig. 9.1. In the upper left is a small distributive lattice $\mathbf{D}$ with $\mathrm{J}(\mathbf{D}) = \{p, q, r\}$, indicated by solid (red) vertices. These form the ordered set $\mathbf{P}$ such that $\mathbf{D} \cong \mathcal{O}(\mathbf{P})$. Thus the variety $\mathcal{P}_0$ will have a constant e and operations η_p, η_q, and η_r.

The free algebra $\mathbf{F}_{\mathcal{P}_0}(a)$ is in the lower middle. The solid (black) arrow indicates η_q, the dashed (blue) arrows are η_p, and the dotted (red) arrows are η_r; the loops for fixed points are not drawn (e.g., $\eta_p(\eta_p a) = \eta_p a$).

The upper right gives the representation of $\mathbf{D}$ as $\mathrm{L_q}(\mathcal{P}_0)$. There are 3 quasicritical algebras in $\mathcal{P}_0$, which are subalgebras of $\mathbf{F}_{\mathcal{P}_0}(a)$. The free algebra $\mathbf{F}_{\mathcal{P}_0}(a)$ generates $\mathcal{P}_0$, while $\mathrm{Sg}(\eta_p a)$ generates $\langle \eta_p x \approx x \rangle$, and $\mathrm{Sg}(\eta_r a)$ generates $\langle \eta_r x \approx x \rangle$.

Lemma 9.20 *The free* $\mathcal{P}_0$*-algebra* $\mathbf{F}_{\mathcal{P}_0}(a)$ *satisfies the quasi-equation*

(v) $\eta_p x \approx \eta_q y \rightarrow \eta_q x \approx x$ *whenever* $p \nleq q$.

Proof Let $p \nleq q$. Suppose $\eta_p x \approx \eta_q y$ in $\mathbf{F}_{\mathcal{P}_0}(a)$, where say $x = \eta_{r_1} \dots \eta_{r_k} a$ and $y = \eta_{s_1} \dots \eta_{s_\ell} a$. (We allow $k = 0$ or $\ell = 0$.) Then the sets of minimal elements satisfy $\min\{p, r_1, \dots, r_k\} = \min\{q, s_1, \dots, s_\ell\}$. Since $p \nleq q$, which means $r_i \leq q$ for some i. This in turn implies $\eta_q x = x$ by (iii). □

Now let $\mathcal{P}_1$ be the quasivariety generated by the free algebra $\mathbf{F}_{\mathcal{P}_0}(a)$. That is, $\mathcal{P}_1 = \mathbb{Q}(\mathbf{F}_{\mathcal{P}_0}(a))$, whence $\mathcal{P}_1 \leq \mathcal{P}_0$ with $\mathbf{F}_{\mathcal{P}_1}(a) = \mathbf{F}_{\mathcal{P}_0}(a)$. The laws (i)–(v) hold in $\mathcal{P}_1$; we do not know whether or not they completely describe it.

The next two lemmas and theorem show that subquasivarieties of $\mathcal{P}_1$ are determined (relative to $\mathcal{P}_1$) by 1-variable quasi-equations.

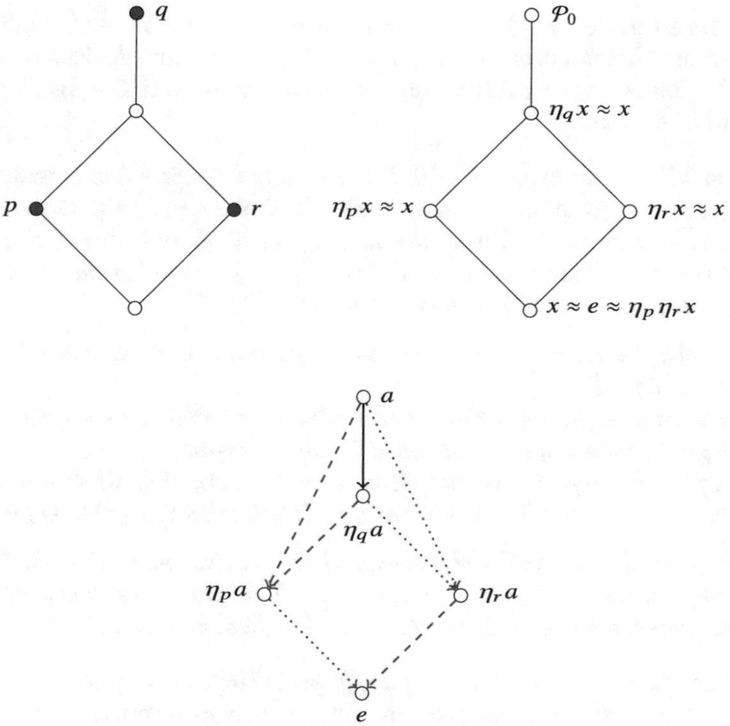

Fig. 9.1 Illustrating Example 9.19

Lemma 9.21 *Let* **S** *be a finitely generated subalgebra of* $\mathbf{F}_{\mathcal{P}_1}(a)$. *Then* **S** *is a subdirect product of its 1-generated subalgebras, which are retracts.*

Proof Consider a finitely generated subalgebra $\mathbf{S} = \mathrm{Sg}(x, y, \ldots, z)$ of $\mathbf{F}_{\mathcal{P}_1}(a)$. Then say $x = \eta_{p_1} \ldots \eta_{p_k} a$ and $y = \eta_{q_1} \ldots \eta_{q_\ell} a$, etc. The map $\eta_{p_1} \ldots \eta_{p_k}$ is a retraction of **S** onto $\mathrm{Sg}(x)$. □

Recall the following standard equivalence (see, e.g., Theorem 5.1 of [46]).

Lemma 9.22 *For any quasivariety* $\mathcal{K}$ *and structure* $\mathbf{T} \in \mathcal{K}$, *the following are equivalent.*

(1) **T** *is projective in* $\mathcal{K}$.
(2) **T** *is a retract of a* $\mathcal{K}$*-free structure.*
(3) *For any* $\mathbf{S} \in \mathcal{K}$, *any surjective homomorphism* $h : \mathbf{S} \twoheadrightarrow \mathbf{T}$ *is a retraction.*

For our purposes, a slightly weaker version of projectivity suffices. A structure **S** is *semi-projective* in a quasivariety $\mathcal{K}$ if whenever it is a homomorphic image of some $\mathbf{T} \in \mathcal{K}$, then it embeds into **T**, i.e., $\mathbf{S} \leq \mathbf{T}$.

By Lemma 9.21, the 1-generated subalgebras of $\mathbf{F}_{\mathcal{P}}(a)$ are projective. Let us state the next theorem somewhat more generally than needed for $\mathcal{P}_1$, in hopes that it can be used elsewhere as well. For $\mathcal{P}_1$, take $k = 1$. Since structures in $\mathcal{P}_1$ may be infinite, we cannot ignore ultraproducts; that fact necessitates a detour into the land of reducts.

In what follows, we say that a structure $\mathbf{T}$ is *m-generated* if $\mathbf{T} = \mathrm{Sg}(A)$ for some $A \subseteq T$ with $|A| \leq m$.

Theorem 9.23 *Fix an integer $k > 0$. Let Q be a quasivariety that is generated by a set X of structures. Assume that for each $m > 0$ and every finite subset F of the operation and relation symbols of the language of Q, there is a $G \supseteq F$ such that the following properties hold in the quasivariety Q' generated by the reducts of the structures $\mathbf{T} \in X$ to G. Let $\mathbf{T}'$ denote the reduct of $\mathbf{T}$ to G.*

(1) *For each $\mathbf{T} \in X$, every finitely generated substructure of the reduct $\mathbf{T}'$ is semi-projective in Q'.*
(2) *For each $\mathbf{T} \in X$, every finitely generated substructure of the reduct $\mathbf{T}'$ is a subdirect product of its k-generated G-substructures.*
(3) *There are finitely many isomorphism types $\mathbf{I}_1, \ldots, \mathbf{I}_B$ of G-structures such that for each $\mathbf{T} \in X$ and $A \subseteq T$ with $|A| \leq m$, it holds that $\mathrm{Sg}_G(A) \cong \mathbf{I}_j$ for some j.*

Then every quasi-equation in the language of Q is equivalent, modulo the theory of Q, to a set of quasi-equations in k or fewer variables. Thus every subquasivariety of Q is determined, relative to Q, by quasi-equations in at most k variables.

Proof Let β be a quasi-equation in the language of Q in say m variables. Without loss of generality $m > k$. Moreover, β involves only finitely many operation and relation symbols; call this set F. Let $G \supseteq F$ be such that the properties of the theorem hold.

Let $\mathbf{S}$ be a structure in Q that fails β. We may assume that $\mathbf{S}$ is generated by elements witnessing a failure of β, so that $\mathbf{S}$ is m-generated. Now $\mathbf{S} \in \mathbb{SPU}(X)$. Thus $\mathbf{S} \leq \mathbf{P} = \prod_{j \in J} \mathbf{R}_j$ with each $\mathbf{R}_j = \prod_{k \in K_j} \mathbf{T}_k / U_j$ for some appropriate index sets J, K_j ($j \in J$), structures $\mathbf{T}_k \in X$ ($k \in K_j$), and ultrafilters U_j on K_j.

This representation carries over to the reducts. That is, $\mathbf{S}' \leq \mathbf{P}' = \prod_{j \in J} \mathbf{R}'_j$ with each $\mathbf{R}'_j = \prod_{k \in K_j} \mathbf{T}'_k / U_j$ for some appropriate index sets and ultrafilters, where again $\mathbf{T}'$ denotes the reduct to G.

Now consider $\mathbf{S}''$, the G-substructure of $\mathbf{S}'$ generated by the chosen failure of β. Then, taking substructures, we get $\mathbf{S}'' \leq \mathbf{P}'' = \prod_{j \in J} \mathbf{R}''_j$ with each $\mathbf{R}''_j = \prod_{k \in K_j} \mathbf{W}_k / U_j$ and each $\mathbf{W}_k$ an m-generated Q'-substructure of $\mathbf{T}'_k$.

By assumption (3), there are only finitely many isomorphism types for an m-generated reduct of $\mathbf{T}'$ to G, say $\mathbf{I}_1, \ldots, \mathbf{I}_B$. Hence each $\mathbf{W}_j \cong \mathbf{I}_{\ell_j}$ for some ℓ_j. So each j-th ultraproduct $\mathbf{R}''_j = \prod_{k \in K_j} \mathbf{W}_k / U_j$ is isomorphic to some $\mathbf{I}_{\ell_j}$. Therefore, $\mathbf{S}'' \leq \prod_j \mathbf{I}_{\ell_j}$, and the ultraproduct is gone. Without loss of generality this last representation is subdirect (the projections are onto).

Since β fails in $\mathbf{S}''$, it fails in some $\mathbf{I}_{\ell_0}$. Now $\mathbf{I}_{\ell_0} \cong \mathbf{W}_0 \leq \mathbf{T}'_0$ for some $\mathbf{T}_0 \in X$. By (2), $\mathbf{I}_{\ell_0}$ is a subdirect product of some of its k-generated Q'-substructures. Therefore, β fails in some k-generated substructure $\mathbf{J}_0$ of $\mathbf{I}_{\ell_0}$. Note $\mathbf{J}_0$ is a subdirect factor of $\mathbf{I}_0$, which is a subdirect factor of $\mathbf{S}''$. By assumption (1), $\mathbf{J}_0$ embeds into $\mathbf{S}''$.

Now $\mathbf{J}_0$ is k-generated. Hence there is some substitution σ mapping $\{x_1, \ldots x_m\}$ to a set of terms in at most k variables such that $\mathbf{J}_0$ fails $\sigma(\beta)$. But $\mathbf{J}_0 \leq \mathbf{S}''$. Thus $\mathbf{S}''$, and hence $\mathbf{S}$, also fails $\sigma(\beta)$.

Clearly β implies any substitution $\sigma(\beta)$, and we have just shown that if β fails, then some k-variable $\sigma(\beta)$ fails. Therefore, under the assumptions of the theorem, each quasi-equation β is equivalent in Q to $\Sigma(\beta)$, the set of all substitutions $\sigma(\beta)$ of k-variable terms into β, which is the claim of the theorem. □

Lemma 9.21 shows that (1) and (2) of Theorem 9.23 hold for $\mathcal{P}_1$ with $k = 1$, while (3) holds because reducts of $\mathcal{P}_1$ to finitely many operations are locally finite. So we turn our attention to describing 1-variable quasi-equations in the language of $\mathcal{P}_1$.

Lemma 9.24 *The following are true in $\mathcal{P}_1$ and $L_q(\mathcal{P}_1)$.*

(1) *$\eta_p e \approx e$ for all $p \in P$.*
(2) *$\&_{m \in M} \eta_m x \approx x \rightarrow x \approx e$*
(3) *If $p \leq q$, then $\eta_p x \approx x$ implies $\eta_q x \approx x$.*
(4) *$\eta_{p_1} \ldots \eta_{p_k} x \approx x$ is equivalent to $\eta_{p_1} x \approx x \;\&\; \ldots \;\&\; \eta_{p_k} x \approx x$.*
(5) *$\eta_{p_1} \ldots \eta_{p_k} x \approx x \rightarrow \eta_q x \approx x$ is equivalent to $\eta_q x \approx x$ whenever $p_i \nleq q$ for all i.*
(6) *If say $p_1 \leq q$, then $\eta_{p_1} \ldots \eta_{p_k} x \approx x \rightarrow \eta_q x \approx x$ holds in $\mathcal{P}_1$.*
(7) *Let $A = \{p_1, \ldots, p_k\}$ and $B = \{q_1, \ldots, q_\ell\}$ be antichains in $\mathbf{P}$. The identity $\eta_{p_1} \ldots \eta_{p_k} x \approx \eta_{q_1} \ldots \eta_{q_\ell} x$ is equivalent to $\eta_{r_1} \ldots \eta_{r_n} x \approx x$, where $\{r_1, \ldots, r_n\}$ is the set of minimal elements of the symmetric difference $A \,\Delta\, B$.*

Proof Items (1) and (2) follow from laws (iii) and (iv).

For (3), assume $p \leq q$ and $\eta_p x = x$. Then $\eta_q x = \eta_q \eta_p x = \eta_p x = x$ where we have used laws (ii) and (iii).

Item (4) follows easily from laws (i) and (ii).

For (5), let $\mathbf{S}$ be an algebra in $\mathcal{P}_1$ that does not satisfy $\eta_q x \approx x$. Choose $s \in S$ such that $\eta_q s \neq s$, and consider $t = \eta_{p_1} \ldots \eta_{p_k} s$. Then $\eta_{p_1} \ldots \eta_{p_k} t = t$. Suppose $\eta_q t = t$. Then $\eta_{p_1} x = t = \eta_q t$ where $x = \eta_{p_2} \ldots \eta_{p_k} s$. By (v) we conclude that $\eta_q x = x$. Continuing in this manner, we eventually get $\eta_q s = s$, a contradiction. Therefore, $\eta_{p_1} \ldots \eta_{p_k} x \approx x \rightarrow \eta_q x \approx x$ fails in $\mathbf{S}$, as witnessed by t.

Item (6) follows from (3).

For (7), assume that $\eta_{p_1} \ldots \eta_{p_k} x \approx \eta_{q_1} \ldots \eta_{q_\ell} x$ holds in a subquasivariety of $\mathcal{P}_1$, and say q_1 is such that $p_i \nleq q_1$ for all i. Applying law (v) repeatedly, we obtain

$$\begin{aligned} \eta_{p_1} \ldots \eta_{p_k} x \approx \eta_{q_1} \ldots \eta_{q_\ell} x &\implies \eta_{q_1} \eta_{p_2} \ldots \eta_{p_k} x \approx \eta_{p_2} \ldots \eta_{p_k} x \\ &\implies \eta_{q_1} \eta_{p_3} \ldots \eta_{p_k} x \approx \eta_{p_3} \ldots \eta_{p_k} x \\ &\implies \ldots \implies \eta_{q_1} x \approx x. \end{aligned}$$

On the other hand, if $\eta_r x \approx x$ holds for all the minimal elements in $A \,\Delta\, B$, it is easy to derive the original identity, using (3). □

Note that it follows from (3) that for any subquasivariety $Q \leq \mathcal{P}_1$, the set of all $p \in P$ such that $\eta_p x \approx x$ holds in Q is an order filter. Also, by (2), any reference to e in a quasi-equation can be replaced by $\&_{m \in M} \eta_m x$.

Now we can handle the general case.

Lemma 9.25 *Every quasi-equation in $\mathcal{P}_1$ is equivalent to a collection of identities $\eta_p x \approx x$.*

Proof The general form of a quasi-equation is

$$\&_{R_i,S_i}\ \eta_{r_1}\dots\eta_{r_k}x \approx \eta_{s_1}\dots\eta_{s_\ell}x \to \eta_{u_1}\dots\eta_{u_m}x \approx \eta_{v_1}\dots\eta_{v_n}x \tag{£}$$

over some collections $R_1, S_1, \dots, R_t, S_t, U, V$ of finite subsets of P. Arguing as in the proof of (7), we first split the right hand side to show that (£) is equivalent to a set of quasi-equations of the form

$$\&_{R_i,S_i}\ \eta_{r_1}\dots\eta_{r_k}x \approx \eta_{s_1}\dots\eta_{s_\ell}x \to \eta_q x \approx x. \tag{\$}$$

Applying the same argument to the left-hand sides of each quasi-equation (\$), we obtain laws of the form

$$\&_j\ \eta_{p_j}x \approx x \to \eta_q x \approx x. \tag{¥}$$

But, *via* (4) and (5), these in turn are equivalent to a set of laws $\eta_t x \approx x$. □

Let $\mathcal{F}(\mathbf{P})$ denote the set of order filters of $\mathbf{P}$, *ordered by reverse set inclusion.* Then $\mathcal{O}(\mathbf{P}) \cong \mathcal{F}(\mathbf{P})$ *via* the complementation map.

The isomorphism $\varphi : \mathcal{F}(\mathbf{P}) \cong \mathrm{L_q}(\mathcal{P}_1)$ is

$$\varphi(F) = \bigcap_{p\in F} \langle \eta_p x \approx x\rangle.$$

The inverse map $\rho : \mathrm{L_q}(\mathcal{P}_1) \to \mathcal{F}(\mathbf{P})$ is

$$\rho(\mathcal{Q}) = \{p \in P : \mathcal{Q} \text{ satisfies } \eta_p x \approx x\}.$$

The crucial step is checking that φ is surjective, which is Lemma 9.25.

9.4 Ideals of Meet Semilattices as $\mathrm{L_q}(\mathcal{K})$

It is useful to view Theorem 9.18 in a general context, as the details of the construction can obscure an essentially simple idea.

Let $\mathbf{S}$ be a meet semilattice with 0. Let $\mathcal{Q}_0$ consist of algebras with a semilattice of operators $\mathbf{H} \cong \mathbf{S}$ and a constant e, such that the least element $0_\mathbf{S}$ corresponds to the constant operator with $\zeta x = e$. Then consider the 1-generated free $\mathcal{Q}_0$ algebra $\mathbf{F}$, and let $\mathcal{Q} = \mathbb{Q}(\mathbf{F})$ be the quasivariety generated by $\mathbf{F}$.

The elements of $\mathbf{F}$ are of the form ηa with $\eta \in \mathbf{H}$ and a the generator of $\mathbf{F}$. These inherit a natural order from $\mathbf{S}$: $a > \eta a \geq \kappa a \geq \zeta a = e$ whenever $\eta \geq \kappa$. Let $\mathcal{I}(\mathbf{F})$ denote the set of nonempty order ideals of $\mathbf{F}$. Since intersections and unions of order ideals are order ideals, $\mathcal{I}(\mathbf{F})$ is a completely distributive lattice.

The subalgebra $\mathrm{Sg}(\eta a) = \downarrow \eta a$ is an order ideal, and the map η retracts $\mathbf{F}$ onto $\downarrow \eta a$. Hence $\downarrow \eta a$ is projective in $\mathcal{Q}$. The reduct to any finite set of operators is of course locally finite. Thus $\mathcal{Q}$ satisfies exactly the setup for Theorem 9.23. The quasicritical algebras in $\mathcal{Q}$ are the ideals $\mathrm{Sg}(\eta a) = \downarrow \eta a$ of $\mathbf{F}$, and thus $\mathrm{L_q}(\mathcal{Q}) \cong \mathcal{I}(\mathbf{F})$.

Now $\mathcal{I}(\mathbf{F})$ is a special case of $\mathcal{O}(\mathbf{P})$. What Theorem 9.18 does is to start with an ordered set $\mathbf{P}$ that has finitely many strongly minimal elements, and constructs a semilattice $\mathbf{S}$ such that $\mathcal{I}(\mathbf{S}) \cong \mathcal{O}(\mathbf{P})$. Indeed, in the proof of that theorem, $\mathbf{S}$ is generated by $\{\eta_p : p \in P\}$. Laws (i) and (ii) make $\mathbf{S}$ a semilattice, while (iii) says that the order on $\mathbf{S}$ under composition reflects the order on $\mathbf{P}$. Law (iv) makes the constant e correspond to the empty order ideal of $\mathbf{P}$. The proof gives additional information about the subquasivarieties of $\mathcal{Q}$, which is interesting but not required to establish the isomorphism.

9.5 Recapitulation on Distributive, Dually Algebraic Lattices

Summarizing what we *do* know is easy. Let $\mathbf{D}$ be a distributive, dually algebraic lattice.

- $\mathbf{D}$ can be represented as $\mathrm{S_p}(\mathbf{S}, H)$ for some algebraic lattice $\mathbf{S}$ and monoid of operators H.
- Hence $\mathbf{1} + \mathbf{D}$ is isomorphic to $\mathrm{L_q}(\mathcal{K})$ for some quasivariety $\mathcal{K}$ of structures with equality.
- If $\mathbf{D}$ is distributive, algebraic, and dually algebraic, and $0_{\mathbf{D}}$ is dually compact, then $\mathbf{D} \cong \mathrm{L_q}(\mathcal{K})$ for some quasivariety $\mathcal{K}$ of structures with equality.

The conditions of the third bullet are equivalent to $\mathbf{D} \cong \mathcal{O}(\mathbf{P})$ for an ordered set $\mathbf{P}$ with only finitely many minimal elements, and every element above one of those.

This allows us to determine completely which chains are subquasivariety lattices. First, dually algebraic chains are included in the next lemma.

Lemma 9.26 *The following are equivalent for a chain* $\mathbf{S}$.

(1) $\mathbf{S}$ *is algebraic.*
(2) *Every element of* S *is a join of completely join irreducible elements.*
(3) $\mathbf{S} \cong \mathcal{O}(\mathbf{C})$ *for a chain* $\mathbf{C}$.
(4) $\mathbf{S}$ *is dually algebraic.*
(5) *Every element of* S *is a meet of completely meet irreducible elements.*

Proof In a chain, *nonzero compact* is the same as *completely join irreducible*. (In a complete lattice with 0, the least element is compact, but it is not completely join irreducible since $0 = \bigvee \varnothing$.) Thus (1) and (2) are equivalent.

Assuming (2), let $\mathbf{C}$ be the completely join irreducible elements of $\mathbf{S}$. Then the map $I \mapsto I \cup \{0\}$ yields $\mathcal{O}(\mathbf{C}) \cong \mathcal{I}(\mathbf{C} \cup \{0\}) = \mathbf{S}$. Also, (3) implies (1) because any $\mathcal{O}(\mathbf{P})$ is both algebraic and dually algebraic.

Now in general for ordered sets, using order filters, $\mathcal{O}(\mathbf{P}) \cong^d \mathcal{F}(\mathbf{P}) \cong \mathcal{O}(\mathbf{P}^d)$, which shows that (4) and (5) are also equivalent to (3). □

Note that there are no restrictions on the chain **C** in Lemma 9.26.

Since subquasivariety lattices with equality are always atomic, we have the following consequence.

Corollary 9.27 *The following are equivalent for a dually algebraic chain* **S**.

(1) $0_{\mathbf{S}}$ *is dually compact.*
(2) $0_{\mathbf{S}}$ *is completely meet irreducible.*
(3) **S** *has an atom.*
(4) $\mathbf{S} \cong \mathcal{O}(\mathbf{C})$ *for a chain* **C** *with a least element.*
(5) $\mathbf{S} \cong \mathrm{L_q}(\mathcal{K})$ *for some quasivariety* $\mathcal{K}$ *of structures with equality.*

Turning to distributive lattices in general, what we *do not* know remains immense, but it is pretty clear where to start looking. In trying to represent a general distributive, dually algebraic **D** as $\mathrm{L_q}(\mathcal{K})$, we need to locate the subquasivariety $\mathcal{E}$ of 1-element structures, determined by the law $x \approx y$. So generally speaking, we want to know

- what can the ideal $\downarrow\mathcal{E}$ look like in $\mathrm{L_q}(\mathcal{K})$? and
- how does $\downarrow\mathcal{E}$ fit into the whole lattice?

Of course $\mathcal{E}$ could be the least element, as in our representation of $\mathcal{O}(\mathbf{P})$ with 0 dually compact. But the place to begin the investigation is when it is the top element. Recall Theorem 2.64: $\mathbf{L} \cong \mathrm{L_q}(\mathcal{K})$ for a quasivariety of 1-element structures if and only if $\mathbf{L} \cong \mathrm{S_p}(\mathbf{S})$, with no operators, for some algebraic lattice **S**. That makes **L** atomistic. (But atomistic lattices may also be represented by quasivarieties not satisfying $x \approx y$.)

Thus we consider the problem: *Which distributive, dually algebraic lattices are isomorphic to* $\mathrm{S_p}(\mathbf{S})$ *for some algebraic lattice* **S**?

Lemma 9.28 *If* **D** *is distributive and* $\mathbf{D} \cong \mathrm{S_p}(\mathbf{S})$, *then* **S** *is a chain.*

Proof If **S** contains two incomparable elements, say a and b, then the algebraic subsets $\{a, 1\}$ and $\{b, 1\}$ generate a non-distributive sublattice of $\mathrm{S_p}(\mathbf{S})$, isomorphic to the lattice in Fig. 3.1. □

Recall also Theorem 1.12, from Gorbunov and Tumanov [53].

Theorem 9.29 *A Boolean lattice* **B** *is isomorphic to* $\mathrm{S_p}(\mathbf{S})$ *for an algebraic lattice* **S** *if and only if* $\mathbf{B} \cong \mathbf{2}^{\kappa}$ *for some* κ *with* $0 \leq \kappa \leq \aleph_0$.

Proof For k finite, $\mathbf{2}^k$ is isomorphic to $\mathrm{S_p}(\mathbf{k} + \mathbf{1})$ for a $k + 1$-element chain, while $\mathbf{2}^{\aleph_0}$ is isomorphic to $\mathrm{S_p}(\omega + 1)$.

Conversely, suppose $\mathbf{B} \cong \mathrm{S_p}(\mathbf{S})$. Then **B** is dually algebraic, and since complementation is a dual automorphism, it is also algebraic. Thus $\mathbf{B} \cong \mathbf{2}^A$ where A is the set of its atoms. In particular, those atoms are completely join prime.

But we have also seen that **S** is an algebraic chain. If in **S** there were a proper join $\bigvee B = a < 1$ or $\bigwedge B = a$ for some (infinite) $B \subseteq A$, then the atom $\{a, 1\}$ of $\mathrm{S_p}(\mathbf{S})$ would not be completely join prime, a contradiction. Thus **S** must be $\mathbf{k}+\mathbf{1}$ or $\omega + 1$. □

This sets us up to apply Theorem 2.65.

Corollary 9.30 *A Boolean lattice* $\mathbf{B} = \mathbf{2}^\kappa$ *can be represented as* $\mathrm{L_q}(\mathcal{K})$ *for a quasivariety with equality if and only if* $0 \leq \kappa \leq \aleph_0$.

Proof Boolean lattices are algebraic, so Theorem 2.65 indeed applies. If $\kappa \leq \aleph_0$, then $\mathbf{2}^\kappa \cong \mathrm{S_p}(\mathbf{S})$, so $\mathbf{2}^\kappa \cong \mathrm{L_q}(\mathcal{K})$ for a quasivariety of 1-element structures. On the other hand, if $\mathbf{2}^\kappa \cong \mathrm{L_q}(\mathcal{K})$, then the subquasivariety $\mathcal{E}$ corresponding to $\langle x \approx y \rangle$ must be dually compact and isomorphic to some $\mathrm{S_p}(\mathbf{S})$; this is property (I8) of equaclosure operators. The dually compact elements of $\mathbf{2}^\kappa$ correspond to cofinite subsets; if κ is infinite and e is cofinite in $\mathbf{2}^\kappa$, then $\downarrow e \cong \mathbf{2}^\kappa$. So in order for $\downarrow e$ to be isomorphic to $\mathrm{S_p}(\mathbf{S})$, we must have $\kappa \leq \aleph_0$ by Theorem 9.29. □

In appealing to Theorem 2.65, we have shortcut the representation of a lattice $\mathrm{S_p}(\mathbf{S})$ as $\mathrm{L_q}(\mathcal{K})$ for a quasivariety of 1-element structures, from Adaricheva, Dziobiak, and Gorbunov [14]. That construction is similar to the longstyle method of Theorem 7.7, except that for 1-element structures we no longer need the constant e nor the requirement that $1_\mathbf{S}$ be compact. Again we refer the reader to [14] for the details.

Note that for $\kappa > \aleph_0$, the lattice $\mathbf{2}^\kappa$ can still be represented as $\mathrm{S_p}(\mathbf{S}, H)$ using operators. This gives us more examples of lattices that are isomorphic to some $\mathrm{S_p}(\mathbf{S}, H)$ but not to $\mathrm{L_q}(\mathcal{K})$ for a quasivariety with equality, essentially because they fail property (I8) for an equaclosure operator.

Now in an arbitrary distributive $\mathbf{D} \cong \mathrm{S_p}(\mathbf{S})$, the lattice $\mathbf{D}$ is atomistic and its atoms are in one-to-one correspondence with $S \setminus \{1\}$. The thing that keeps $\mathbf{D}$ from being Boolean is that in $\mathbf{S}$, some elements are infinite joins or meets of others. (As $\mathbf{S}$ is a chain, finite joins and meets are trivial.)

We can think of $\mathbf{D}$ as a closure system. Put $A = S \setminus \{1\}$. Since $\mathbf{D}$ is atomistic, the elements of $\mathbf{D}$ can be regarded as subsets of A. The closure rules for a set X to be in $\mathbf{D}$ are

$$B \subseteq X \to \bigvee B \in X$$
$$B \subseteq X \to \bigwedge B \in X$$

running over all infinite subsets $B \subseteq A$ such that $\bigvee B \neq 1$. The sets $X \subseteq A$ satisfying all those rules are closed under joins and meets, making $\mathbf{D}$ a sublattice of $\mathbf{2}^A$. Note that $\mathbf{D}$ includes all finite subsets of A. If $a \leq \bigvee X$ properly in $\mathbf{D}$, then a is a limit point of X, i.e., $a = \bigvee Y$ or $a = \bigwedge Y$ for some infinite $Y \subseteq X$.

The foregoing description falls short of being a characterization, until we can determine what sort of rule collections can actually occur. But it does help to construct examples. Remember that $\mathbf{C}$ is an arbitrary chain, and $\mathbf{S} = \mathcal{O}(\mathbf{C})$. Taking $\mathbf{C}$ to be an ordinal, or $\omega + \omega^d$, or the rational numbers $\mathbb{Q}$ are some possibilities.

As an extreme case, consider the distributive, dually algebraic, atomistic lattice $\mathbf{Y}_X$ consisting of an infinite set X and all its finite subsets, ordered by inclusion. Now $\mathbf{Y}_X$ is not isomorphic to any $\mathrm{S_p}(\mathbf{S})$: in an infinite chain, not every point can be either in or a limit point of every infinite subset. But we cannot eliminate the possibility that $\mathbf{Y}_X$ is a subquasivariety lattice $\mathrm{L_q}(\mathcal{K})$!

Chapter 10
Problems and an Advertisement

Solvitur ambulando. – St. Jerome

Покой нам только снится
– Александр Блок, На поле Куликовом

In this chapter we propose some open problems. The first ten problems concern the construction project of Chap. 7. Problem (11) asks about representations of atomistic lattices, where some loose ends remain. Problems (12)–(14) are about possible strengthening of the conditions for an equaclosure operator. Problems (15)–(18) are about locally finite quasivarieties. Problems (19)–(20) concern varieties and relative varieties, while (21) asks about quasivarieties of groups.

We conclude the chapter with a short review of our monograph [63] on lattices of subquasivarieties of a locally finite quasivariety.

10.1 Problems

(1) Which finite lattices admit a preclop? In other words, when does the algorithm of Sect. 4.2 return a positive answer? (See Sects. 4.1–4.2.)

(2) Can every pair $(\mathbf{L}, \gamma)$ with $\mathbf{L}$ a finite, lower bounded lattice and γ an equaclosure operator, such that $\mathrm{J}(\mathbf{L}) \subseteq \tau(\mathbf{L})$, be represented as the lattice of subsemilattices of a semilattice with operators, $\mathbf{L} \cong \mathrm{Sub}(\mathbf{S}, \wedge, 1, H)$ with γ corresponding to the natural equaclosure operator? There is an algorithm in Sect. 5.2 to decide this for any given pair $(\mathbf{L}, \gamma)$, but we do not know when the answer is YES! If it is not always possible, what additional restrictions should be imposed on equaclosure operators?

(3) If such a pair $(\mathbf{L}, \gamma)$ with $\mathrm{J}(\mathbf{L}) \subseteq \tau(\mathbf{L})$ can be represented as $(\mathrm{Sub}(\mathbf{S}, \wedge, 1, H), \Gamma)$, when can this be extended to a representation $(\mathrm{L_q}(\mathcal{K}), \Gamma)$ where $\mathcal{K}$ is a quasivariety with equality?

K. Adaricheva et al., *A Primer of Subquasivariety Lattices*, CMS/CAIMS Books in Mathematics 3, https://doi.org/10.1007/978-3-030-98088-7_10

(4) Find a systematic construction to handle the case $\mathrm{J}(\mathbf{L}) \not\subseteq \tau(\mathbf{L})$, and decide which of those can be represented as $\mathrm{Sub}(\mathbf{S}, \wedge, 1, H)$. This is the topic of Sect. 4.3 and Chap. 6; see also Running Example 5 in Chap. 7, and Figs. A.1 and A.2 in Sect. A.1.

(5) Find more classes C of lattices such that, for a lattice $\mathbf{L} \in C$, it is straightforward to determine whether $\mathbf{L} \cong \mathrm{L_q}(\mathcal{K})$ for a quasivariety $\mathcal{K}$. So far we have satisfactory answers only for distributive lattices (Theorem 9.18) and atomistic lattices (Theorem 2.65), and even these answers are not complete in the infinite case.

(6) In particular, is there a class of lattices C such that a longstyle representation as a subquasivariety lattice is guaranteed for each $\mathbf{L} \in C$? One might count the lattices $\mathbf{1} + \mathrm{S_p}(\mathbf{S}, H)$ as such a class (Corollary 7.10), but there may well be something more general.

(7) Often in the course of this book we have represented a finite pair $(\mathbf{L}, \gamma)$ as $(\mathrm{Sub}(\mathbf{S}, \wedge, 1, H), \Gamma)$ but then, unless there was an obvious longstyle representation, not determined whether the pair could be represented as $(\mathrm{L_q}(\mathcal{K}), \Gamma)$. It might prove instructive to resolve some of these cases. In particular, one could look at the pairs $(\mathbf{W}, \gamma_j)$ for $1 \leq j \leq 5$ from Chap. 6, the unresolved pairs in Appendix A.1, and those in Figs. 10.1, 10.2, and 10.3 below.

(8) We have not used join semidistributivity nor (except in Sect. 4.3) lower boundedness? What gives?

(9) Describe $\mathrm{L_q}(\mathcal{K})$ for a quasivariety generated by a finite semilattice with operators, perhaps with simplifying conditions. (This could help in making step 6 systematic.)

(10) Can the construction project be adapted to the infinite case?

(11) Can every dually algebraic, atomistic lattice that admits an equaclosure operator be represented as $\mathrm{S_p}(\mathbf{S}, H)$? as $\mathrm{L_q}(\mathcal{K})$? (See Theorems 2.64, 2.65, 5.13, and 9.30.) In particular, what about the distributive, dually algebraic, atomistic lattice $\mathbf{Y}_X$ consisting of an infinite set X and all its finite subsets, ordered by inclusion?

(12) Find a common generalization of conditions (K9) and (K10).

(13) The condition (K9) is a strengthening of (‡) to include infinite index sets. Is there a similar strengthening of (K10)? (See [20, 92].)

(14) If the condition of Theorem 3.29 for equaclosure operators strictly stronger than (K10)?

(15) Do the pairs $(\mathbf{L}, \gamma)$ in Figs. 7.19 and 7.20 have representations as $(\mathrm{L_q}(\mathcal{K}), \Gamma)$ where $\mathcal{K}$ is a locally finite quasivariety with equality?

(16) Some properties special to subquasivariety lattices for locally finite quasivarieties of algebras are found in Freese, Kearnes, and Nation [47]. *Do these properties hold for locally finite quasivarieties of structures, or just algebras?* (Tame Congruence Theory was used in [47].)

(17) More generally, what special properties pertain to lattices of subquasivarieties of *algebras*, as opposed to more general structures?

(18) What is the role of finite critical structures in locally finite varieties? (Start with groups, as in H. Neumann [93].)

(19) How does the Zipper Condition affect the natural equaclosure operator on $L_q(\mathcal{K})$? Can we better describe the connection between $ETh(\mathcal{K})$ and $L_q(\mathcal{K})$? Perhaps the quasivariety in Chapter 5 of the monograph [63] shows this is impossible?

(20) Characterize lattices $L_v(\mathcal{V})$ where $\mathcal{V}$ is a variety of *structures* in a language that may not include equality. For *algebras with equality as the only relation* the Zipper Condition and its generalizations apply [41, 78, 79]. What about general structures with equality? or structures that may or may not contain equality? For general structures *without* equality, see Sect. A.2.

(21) Find some Q-universal quasivarieties of groups. Is the variety of all nilpotent groups of rank 2 Q-universal?

We conjecture that if $\mathbf{L}$ is a finite lower bounded lattice and $(\mathbf{L}, \gamma)$ has $J(\mathbf{L}) \subseteq \tau(\mathbf{L})$ and satisfies a few necessary conditions, like (K9) and (K10), then the pair can be represented as $\mathrm{Sub}(\mathbf{S}, \wedge, 1, H)$ with $\mathbf{S} = \gamma^d(\mathbf{L})$. This is a worthwhile enterprise, because failure to represent will likely lead to more necessary conditions. The examples in Figs. 10.1, 10.2, and 10.3 indicate some of the complexities, but each of these can be represented using the algorithm in Sect. 5.2.

Recall there were also two conjectures in Chap. 8:

(A) For any finite join semidistributive lattice, the linear sum $\mathbf{1} + \mathbf{L}$ has a longstyle representation as a subquasivariety lattice.

(B) More generally, if $\mathbf{L}$ is join semidistributive and dually algebraic, the linear sum $\mathbf{1} + \mathbf{L}$ has a longstyle representation as $L_q(\mathcal{K})$.

While these may be difficult, if true they would tell us a lot about what restrictions hold for subquasivariety lattices (in a negative way).

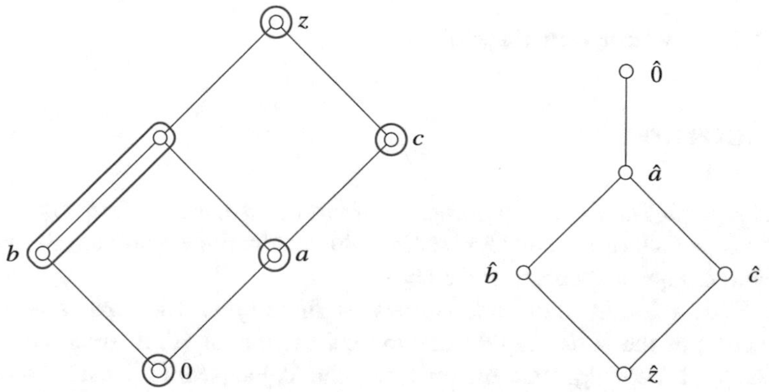

Fig. 10.1 Tricky to represent as $\mathrm{Sub}(\mathbf{S}, H)$

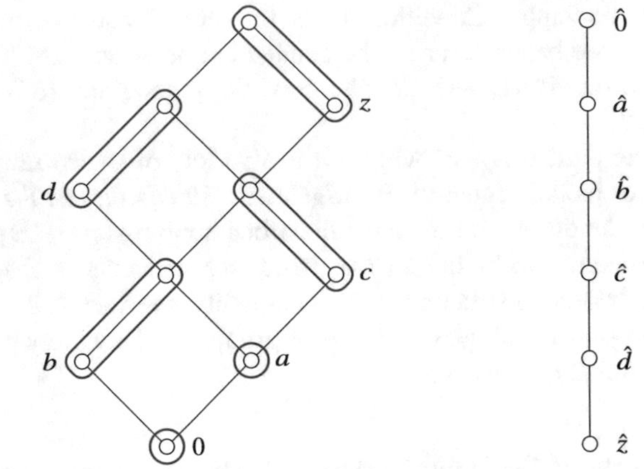

Fig. 10.2 Trickier to represent

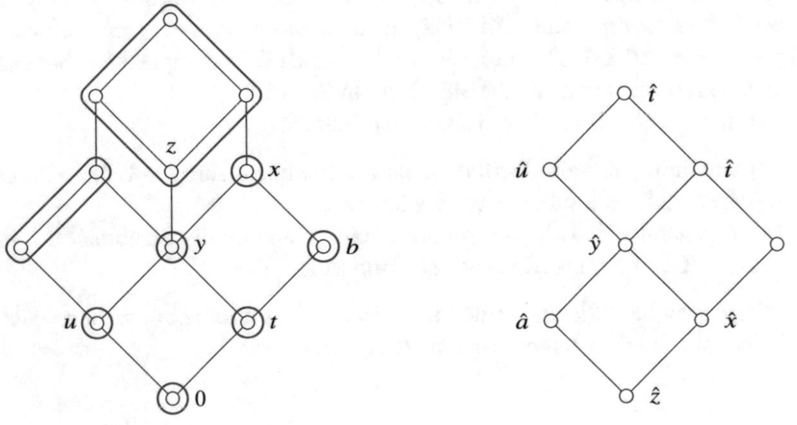

Fig. 10.3 Harder yet to represent, but doable

10.2 Advertisement

The monograph *The Lattice of Subquasivarieties of a Locally Finite Quasivariety* [63] develops a set of methods for dealing with locally finite quasivarieties of finite type. Here is a quick synopsis of the book.

Let $\mathcal{K}$ be a locally finite quasivariety of finite type. The goal is to analyze the structure of the lattice $L_q(\mathcal{K})$ of subquasivarieties of $\mathcal{K}$. It turns out that the lattice $L_q(\mathcal{K})$ is both algebraic and dually algebraic, join semidistributive, and it is a fermentable lattice. The completely join irreducible quasivarieties in $L_q(\mathcal{K})$ are $\langle \mathbf{T} \rangle$ with $\mathbf{T}$ a finite quasicritical structure in $\mathcal{K}$.

For any finite algebra $\mathbf{T}$ in a locally finite quasivariety $\mathcal{K}$ of finite type, there is a finite set $\mathcal{E}(\mathbf{T})$ of quasi-equations such that, for any subquasivariety $Q \leq \mathcal{K}$, it is the case that $\mathbf{T} \notin \mathcal{Q}$ if and only if $\mathcal{Q}$ satisfies ε for some $\varepsilon \in \mathcal{E}(\mathbf{T})$. Moreover, given $\mathbf{T}$ and its $\mathcal{K}$-congruence lattice $\mathrm{Con}_{\mathcal{K}}\, \mathbf{T}$, it is straightforward to find the quasi-equations in $\mathcal{E}(\mathbf{T})$. For each quasi-equation ε in $\mathcal{E}(\mathbf{T})$, there is a finite list $\mathbf{U}_1, \ldots, \mathbf{U}_k$ of finite algebras in $\mathcal{K}$ such that an algebra $\mathbf{S} \in \mathcal{K}$ satisfies ε if and only if $\mathbf{S}$ contains no $\mathbf{U}_i$ as a subalgebra.

Since $\mathrm{L_q}(\mathcal{K})$ is algebraic, every subquasivariety of $\mathcal{K}$ is a meet of completely meet irreducible quasivarieties. For every completely meet irreducible quasivariety $\mathcal{M}$ in $\mathrm{L_q}(\mathcal{K})$, there is a finite quasicritical algebra $\mathbf{T}$, not in $\mathcal{M}$, such that $\mathcal{M} = \langle \varepsilon \rangle$ for a quasi-equation ε in $\mathcal{E}(\mathbf{T})$. Quasivarieties $\mathcal{Q}$ that are finitely based relative to $\mathcal{K}$ are then the intersection of finitely many such quasivarieties $\langle \varepsilon_i \rangle$, with each ε_i in $\mathcal{E}(\mathbf{T}_i)$ for some $\mathbf{T}_i$ not in $\mathcal{Q}$. Thus finitely based quasivarieties can also be characterized in terms of omitting finitely many subalgebras.

Quasivarieties that are completely join prime or completely meet prime in $\mathrm{L_q}(\mathcal{K})$ are also characterized.

The methods are illustrated in the monograph by applying them to quasivarieties of 1-unary and 2-unary algebras, lattices, abelian groups, and pure unary relational structures.

Appendices

A number of Americans, driven mad by the Interstates, awakened. Knowing the frontier trails were gone but for broken traces and ruts, travelers began seeking highway adventures in a broadly settled and thoroughly roaded land now rarely wilderness but still similarly full of challenges and unpredictabilities to sharpen one's senses, whittle perceptions into keenness, and carve a lasting mark into memory. ... Ike's Interstates don't much disturb me any longer because their 76,000 miles did not just open up three million miles of two-lanes; they also reminded us how to travel in ways that give a chance to enter the American landscape and to inhabit our heritage of history and place. I think that may be why, the week after next, I'll be headed for Tennessee. There's this place I remember not far out of Tullahoma. – William Least Heat-Moon in *Here, There, Everywhere*

Если ждать минуты, когда всё, решительно всё будет готово, никогда не придётся начинать. – Иван Тургенев, Новь

These appendices contain topics that we thought were interesting and relevant, but which did not seem to fit into the text anywhere without interrupting the flow of the narrative. Each can be read independently.

Appendix A.1 contains additional representations of finite pairs $(\mathbf{L}, \gamma)$ as either $(\mathrm{S_p}(\mathbf{S}, H), \Gamma)$ or $(\mathrm{L_q}(\mathcal{K}), \Gamma)$. These pairs are generally more challenging than those in Chap. 7 and illustrate some techniques that are useful in practice.

Appendix A.2 describes lattices of atomic theories in languages that do not contain equality. These lattices are isomorphic to the lattice of ideals of an ordered set. Hence they are distributive, algebraic, and dually algebraic.

Appendix A.3, inspired by a paper of Adams and Dziobiak [7], proposes a notion of quasicriticality for quasivarieties that may not be locally finite.

K. Adaricheva et al., *A Primer of Subquasivariety Lattices*, CMS/CAIMS Books in Mathematics 3, https://doi.org/10.1007/978-3-030-98088-7

A.1 Additional Examples

In this section, we give a few more examples of the six-step program.

Example A.1 The lattice **N** in Fig. A.1 is the same lattice as the top right lattice in Fig. 3.3 and Example 6.4. As mentioned in Sect. 3.1, this lattice has four preclops. The operator $\gamma_1 = \mu$ of Fig. 3.3 fails (‡), so the pair $(\mathbf{N}, \gamma_1)$ is not representable. The operator γ_2, with $\gamma_2(p) = 1$ and $\gamma_2(z) = z \vee a$, has an easy representation as $\mathrm{Sub}(\hat{\mathbf{S}}, \wedge, \hat{0}, h)$, which we leave to the reader. The operator γ_3 of Fig. A.1 is representable as indicated in Example 6.4. Since the representations of both γ_2 and γ_3 satisfy the condition ϖ, completing the construction project through step 6, using a longstyle representation, yields $(\mathbf{N}, \gamma_j)$ as $(\mathrm{L_q}(\mathcal{K}_j), \Gamma)$ for $j = 2, 3$.

The operator γ_4 on **N** has $\gamma_4(z) = 1$ and $\gamma_4(a) = b$. This can be represented as $\mathrm{Sub}(\hat{\mathbf{S}}, \wedge, \hat{0}, h, k)$ where $\hat{\mathbf{S}} \cong \mathbf{3} \times \mathbf{2}$, with $z_1 > p > \hat{z}$ down the lower edge and $\hat{0} > \hat{b} > \hat{a}$ down the upper edge. This is another exercise for the reader. It does not lead to a longstyle representation as a subquasivariety lattice; it remains to be determined whether a shortstyle representation can be found.

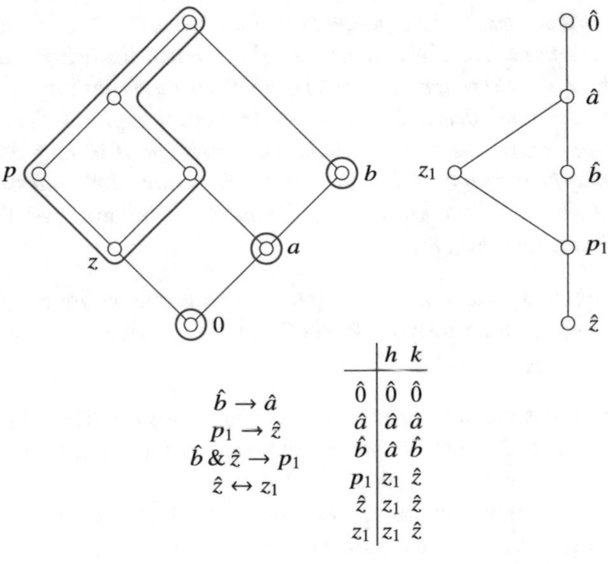

$\hat{b} \to \hat{a}$
$p_1 \to \hat{z}$
$\hat{b} \,\&\, \hat{z} \to p_1$
$\hat{z} \leftrightarrow z_1$

	h	k
$\hat{0}$	$\hat{0}$	$\hat{0}$
$\hat{a}$	$\hat{a}$	$\hat{a}$
$\hat{b}$	$\hat{a}$	$\hat{b}$
p_1	z_1	$\hat{z}$
$\hat{z}$	z_1	$\hat{z}$
z_1	z_1	$\hat{z}$

Fig. A.1 Example A.1: $(\mathbf{N}, \gamma_3)$

Example A.2 Let us briefly consider Day's doubling construction [35]. Let $\mathbf{B}_3$ denote the Boolean lattice $\mathbf{2}^3$. We have already seen that with the "wrong" closure operator, $\mathbf{B}_3$ fails (‡); this is the first example in Fig. 3.3. Let $\mathbf{B}_3[c]$ denote the lattice obtained by doubling a coatom. The least closure operator on $\mathbf{B}_3[c]$ satisfying (I1)–(I8), which collapses the new coatom to the top, still fails (K9). Figure A.2 shows that $\mathbf{B}_3[c]$ with a different closure operator CAN be represented as Sub($\mathbf{S}$, $\wedge$, 1, H). (We do not know whether this pair can be represented as $\mathrm{L_q}(\mathcal{K})$ for a quasivariety with equality.)

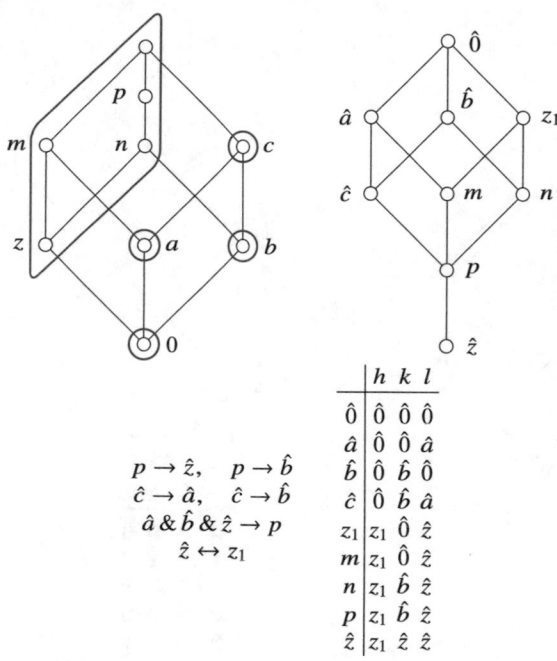

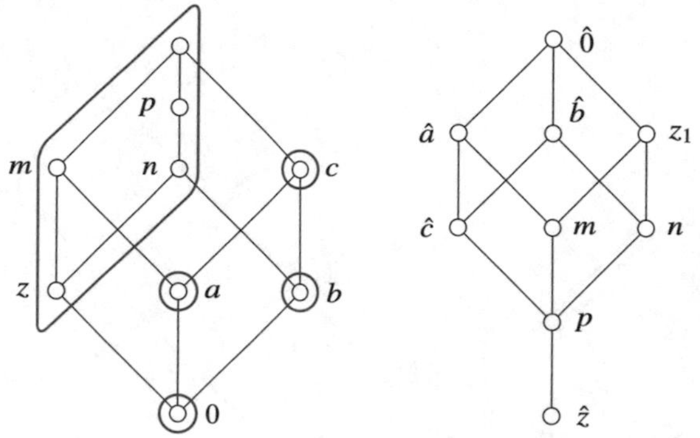

$p \to \hat{z}, \quad p \to \hat{b}$
$\hat{c} \to \hat{a}, \quad \hat{c} \to \hat{b}$
$\hat{a} \,\&\, \hat{b} \,\&\, \hat{z} \to p$
$\hat{z} \leftrightarrow z_1$

	h	k	l
$\hat{0}$	$\hat{0}$	$\hat{0}$	$\hat{0}$
$\hat{a}$	$\hat{0}$	$\hat{0}$	$\hat{a}$
$\hat{b}$	$\hat{0}$	$\hat{b}$	$\hat{0}$
$\hat{c}$	$\hat{0}$	$\hat{b}$	$\hat{a}$
z_1	z_1	$\hat{0}$	$\hat{z}$
m	z_1	$\hat{0}$	$\hat{z}$
n	z_1	$\hat{b}$	$\hat{z}$
p	z_1	$\hat{b}$	$\hat{z}$
$\hat{z}$	z_1	$\hat{z}$	$\hat{z}$

Fig. A.2 A representation of $\mathbf{B}_3[c]$ from Example A.2

Likewise, if we double an atom, then $\mathbf{B}_3[a]$ with its least preclop μ fails (K10); this is the example in Fig. 3.6. But as Example 7.21, we saw that $\mathbf{B}_3[a]$ with a different closure operator has a representation as $\mathrm{L_q}(\mathcal{K}_1)$.

Thus both $\mathbf{B}_3[a]$ and $\mathbf{B}_3[c]$ have representations as Sub($\mathbf{S}$, $\wedge$, 1, H), but you must choose γ carefully. Extensions to $\mathbf{B}_n[a]$ and beyond are in Examples 7.22 and 7.23.

It seems that if $\mathbf{P} = \mathbf{m} \times \mathbf{n}$ is a direct product of two finite chains, and you double *any* single point x, then $\mathbf{P}[x]$ with its least equaclosure operator μ has an easy representation as Sub($\mathbf{S}$, $\wedge$, 1, H). We do not have a formal proof of that, though; cf. Example 7.31 and Theorem 8.7.

On the other hand, the lattice $\mathbf{K}$ in Fig. 3.7 is obtained by doubling a convex subset with a least element in a distributive lattice, so it is lower bounded of rank 1.

That lattice has only one preclop, and it fails both conditions (K9) and (K10) for an equaclosure operator. Thus **K** is not representable as Sub(**S**, ∧, 1, H) or $\mathrm{S_p}(\mathbf{S}, H)$.

Example A.3 The lattice **K** in Fig. A.3 admits only one preclop. It took considerable effort to find the representation of **K** as Sub(**S**, ∧, 1, H) indicated in the figure. Again, we do not know whether **K** can be represented as $\mathrm{L_q}(\mathcal{K})$ for a quasivariety with equality.

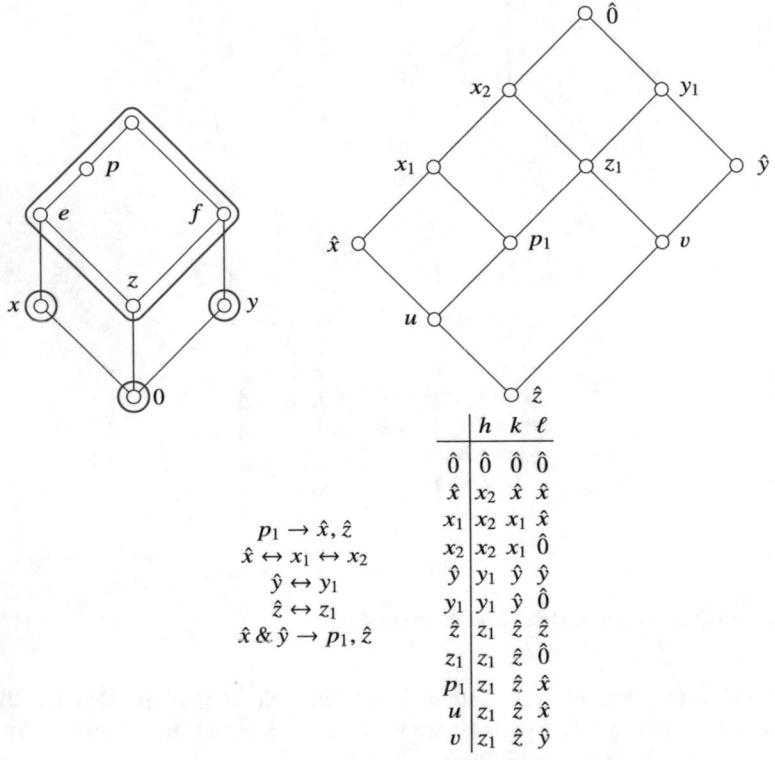

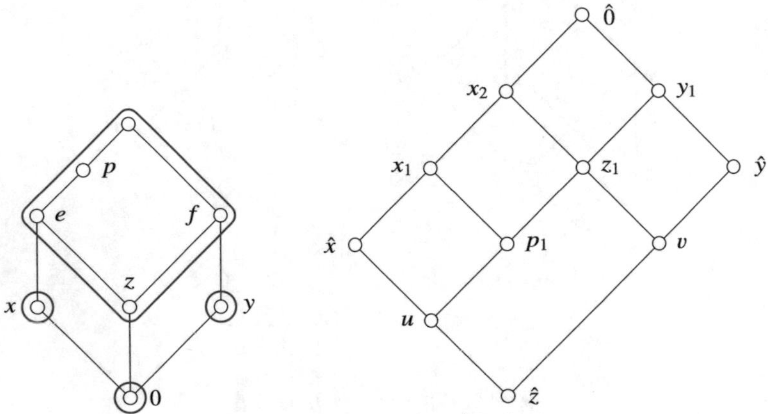

$$p_1 \to \hat{x}, \hat{z}$$
$$\hat{x} \leftrightarrow x_1 \leftrightarrow x_2$$
$$\hat{y} \leftrightarrow y_1$$
$$\hat{z} \leftrightarrow z_1$$
$$\hat{x} \,\&\, \hat{y} \to p_1, \hat{z}$$

	h	k	ℓ
$\hat{0}$	$\hat{0}$	$\hat{0}$	$\hat{0}$
$\hat{x}$	x_2	$\hat{x}$	$\hat{x}$
x_1	x_2	x_1	$\hat{x}$
x_2	x_2	x_1	$\hat{0}$
$\hat{y}$	y_1	$\hat{y}$	$\hat{y}$
y_1	y_1	$\hat{y}$	$\hat{0}$
$\hat{z}$	z_1	$\hat{z}$	$\hat{z}$
z_1	z_1	$\hat{z}$	$\hat{0}$
p_1	z_1	$\hat{z}$	$\hat{x}$
u	z_1	$\hat{z}$	$\hat{x}$
v	z_1	$\hat{z}$	$\hat{y}$

Fig. A.3 The pair (**K**, μ) of Example A.3. Note $\lambda u = e$ and $\lambda v = f$

Example A.4 The pair (**G**, μ) from Example 4.7, reproduced here in Fig. A.4, is an interesting challenge. We want to represent (**G**, μ) as (Sub(**T**, ∧, 1, H), Γ).

From the figure, we need

$$v \longrightarrow s$$
$$y \longrightarrow x$$
$$x \longrightarrow s$$
$$s \,\&\, t \longrightarrow y$$
$$v \,\&\, x \longrightarrow y.$$

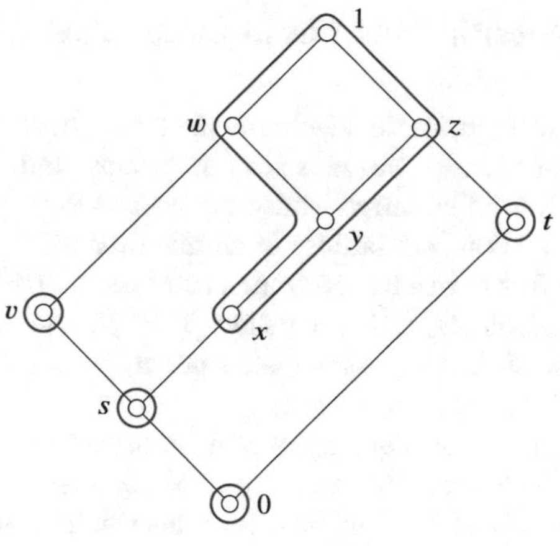

Fig. A.4 $(\mathbf{G}, \mu)$ from Examples 4.7 and A.4

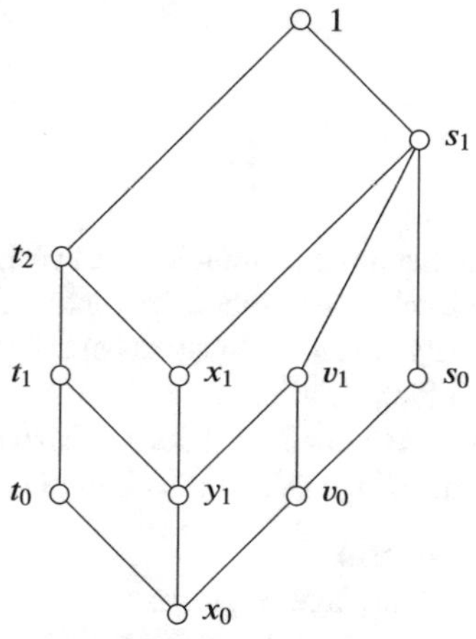

Fig. A.5 The semilattice **T**

The semilattice **T** is drawn in Fig. A.5. We have $s_1 \wedge t_1 = y_1$ and $x_1 \wedge v_1 = y_1$. Operators will be needed to enforce the remaining arrows, plus $x_0 \leftrightarrow x_1$, etc. Here we list one set of operators that works, omitting the fixed points of each operation.

$h_1 : t_0, t_1 \mapsto t_2, \quad x_0, y_1 \mapsto x_1, \quad v_0, v_1, s_0 \mapsto s_1$
$h_2 : t_1, t_2 \mapsto t_0, \quad x_1, y_1 \mapsto x_0, \quad v_1 \mapsto v_0, \quad s_1 \mapsto s_0$
$h_3 : t_2 \mapsto t_1, \quad x_1, y_1 \mapsto x_0, \quad v_1 \mapsto v_0, \quad s_1 \mapsto s_0$
$h_4 : t_1, t_2 \mapsto t_0, \quad x_1, y_1 \mapsto x_0, \quad v_0 \mapsto v_1, \quad s_0 \mapsto s_1$
$h_5 : p \mapsto 1$ for all $p \geq t_0, \quad q \mapsto s_0$ for all $q \leq s_1$.

We leave to the reader the task of checking that these maps are admissible and enforce all the arrows.

A.2 Lattices of Atomic Theories in Languages Without Equality

Lattices of equational theories of algebraic structures have been a rich topic of investigation, both for specific varieties such as groups, and for universal algebra. Section 1.3 contains a brief summary and some references.

In this appendix, we consider lattices of atomic theories in a language that does not contain equality. These results are from a seminar in 2010 at the University of Hawai'i with Tristan Holmes, Dayna Kitsuwa, J. B. Nation, and Sheri Tamagawa. The participants were at the time a graduate student, undergraduate, professor, and high school student, respectively.

Let $\mathcal{L}$ be a language without equality. We will construct an ordered set **P** such that the lattice of atomic theories in the language $\mathcal{L}$ is isomorphic to the lattice of order ideals of **P**. Thus the lattice of atomic theories is distributive, and both algebraic and dually algebraic. We will be concerned with describing the properties of the ordered sets **P** that can occur in this way.

A.2.1 Atomic Theories

Let us work in a language $\mathcal{L}$ that has a countable set of variables $X = \{x_0, x_1, x_2, \dots\}$, constants, function symbols, relation symbols, parentheses and commas for punctuation, but no primitive equality relation. Constants are regarded as nullary functions, but assume that $\mathcal{L}$ has no nullary relations.

Implicitly, the logic used is Boolean, with its standard truth values and functions.

Recall that *terms* in $\mathcal{L}$ are strings of symbols defined thusly.

(1) Each variable $x \in X$ is a term.
(2) Every constant c of the language is a term.
(3) If $t_1, \dots, t_k$ are terms and f is a k-ary function symbol of the language, then $f(t_1, \dots, t_k)$ is a term.
(4) Only strings obtained by (1)–(3) are terms.

Two terms are equal only if they are identical.

The set of terms, with the operations of $\mathcal{L}$ acting in the obvious way, form the *term algebra* or *absolutely free algebra* $\mathbf{F}_{\mathcal{L}}(X)$. The term algebra is used often, and will be denoted simply **F**.

No relations hold on **F** as an algebraic structure. An *atomic formula* of $\mathcal{L}$ is an expression $R(t_1, \dots, t_k)$ where R is a k-ary relation symbol of $\mathcal{L}$ and $t_1, \dots, t_k$ are terms. The set of atomic formulas of $\mathcal{L}$ will be denoted by $\mathcal{P}_{\mathcal{L}}$, or just $\mathcal{P}$.

Since **F** is an absolutely free algebra, any map $\sigma : X \to \mathbf{F}$ can be extended to a homomorphism. That is, given $\sigma(x)$ for all $x \in X$, we define $\sigma(c) = c$ for constants c, and recursively

$$\sigma(f(t_1, \dots, t_k)) = f(\sigma(t_1), \dots, \sigma(t_k))$$

for a k-ary function symbol f. Because each term has a unique expression, this uniquely defines the extension $\sigma : \mathbf{F} \to \mathbf{F}$. We refer to these endomorphisms as *substitutions*, and use Sbn($\mathbf{F}$) to denote the monoid of all substitutions.

Generally speaking, a *theory* is a set of sentences closed under deduction. In this appendix, we consider only universally quantified atomic formulas, i.e., sentences of the form

$$\forall x_1 \ldots \forall x_m\, R(t_1(x_1, \ldots, x_m), \ldots, t_k(x_1, \ldots, x_m))$$

with R a relation symbol and $t_1, \ldots, t_k$ terms. Under the circumstances, we can suppress the quantification symbols.

The only relevant rule of deduction is then substitution. That is, inference should be determined solely by the scheme of rules that when R is a relation symbol, $t_1, \ldots, t_k$ terms and σ a substitution, then

$$R(t_1, \ldots, t_k) \vdash R(\sigma(t_1), \ldots, \sigma(t_k)).$$

Note that the relation $\vdash$ defined thusly on the set of atomic formulas $\mathcal{P}$ is a quasi-order, i.e., reflexive and transitive. As usual, we then define two atomic formulas Φ, Ψ to be *equivalent*, denoted $\Phi \equiv \Psi$, if $\Phi \vdash \Psi$ and $\Psi \vdash \Phi$. Then $\equiv$ is an equivalence relation on $\mathcal{P}$, and $\vdash$ is a partial order on $\mathcal{P}/\equiv$. Let $\mathbf{P}$ be the ordered set $\langle \mathcal{P}/\equiv, \vdash \rangle$.

It is not hard to characterize equivalence: $R(s_1, \ldots, s_k) \equiv R'(t_1, \ldots, t_\ell)$ if and only if $R = R'$ and there is a permutation π of the variables such that the induced substitution has $\pi(s_i) = t_i$ for $1 \le i \le k = \ell$.

Note in passing that the presence of a primitive equality relation, that is, a binary relation $\approx$ that is assumed to be a congruence relation, would require a more complicated deduction scheme, albeit a familiar one.

A set of atomic formulas $\Sigma \subseteq \mathcal{P}$ is an *atomic theory* if, for all relation symbols R and substitutions $\sigma \in$ Sbn($\mathbf{F}$), whenever $R(t_1, \ldots, t_k) \in \Sigma$ then $R(\sigma(t_1), \ldots, \sigma(t_k)) \in \Sigma$. That is, atomic theories are just sets of atomic formulas that are closed under substitution, or equivalently, deduction.

Models for atomic theories in languages without equality are discussed in Blok and Pigozzi [26], Czelakowski [32], Elgueta [40], and Nation [91]. These would be a digression at this point.

A.2.2 Lattices of Atomic Theories

By general principles, the lattice of all atomic theories of $\mathcal{L}$, ordered by set inclusion, forms an algebraic lattice ATh($\mathcal{L}$). Likewise, given an atomic theory $\mathcal{B}$ of $\mathcal{L}$, the set of all theories containing $\mathcal{B}$ forms an algebraic lattice ATh($\mathcal{B}$). The latter is naturally the principal filter $\uparrow \mathcal{B}$ in ATh($\mathcal{L}$), and we obtain ATh($\mathcal{L}$) by taking $\mathcal{B} = \varnothing$.

If Φ and Ψ are atomic formulas with $\Phi \vdash \Psi$, then for atomic theories $\mathcal{T}$ we have $\Phi \in \mathcal{T}$ implies $\Psi \in \mathcal{T}$. Thus it makes sense to interpret $\vdash$ as $\ge$ on $\mathbf{P}$. With that convention, we have these characterizations.

Theorem A.5 *Let $\mathcal{L}$ be a language without equality, and consider* $\mathbf{P} = \langle \mathcal{P}/\equiv, \vdash \rangle$ *as above. Then* ATh($\mathcal{L}$) *is isomorphic to the lattice of order ideals* $O(\mathbf{P})$.

Theorem A.6 *Let $\mathcal{B}$ be an atomic theory in* ATh($\mathcal{L}$), *corresponding to an order ideal I. Then* $\mathrm{ATh}(\mathcal{B}) = \uparrow \mathcal{B} \cong O(\mathbf{P} - I)$.

It follows that lattices of atomic theories are distributive, algebraic and dually algebraic.

Now each atomic formula involves only one relation. So if $\mathcal{R}$ denotes the set of relation symbols of $\mathcal{L}$, then we can write $\mathcal{P}$ as a disjoint union $\mathcal{P} = \dot{\bigcup}_{R \in \mathcal{R}} \mathcal{P}_R$, where $\mathcal{P}_R$ denotes all atomic formulas of the form $R(t_1, \ldots, t_k)$. Correspondingly, $\mathbf{P} = \dot{\bigcup}_{R \in \mathcal{R}} \mathbf{P}_R$.

Corollary A.7 *Let $\mathcal{L}$ be a language without equality, and let $\mathcal{B}$ be an atomic theory in $\mathcal{L}$. Each lattice* ATh($\mathcal{L}$) *and* ATh($\mathcal{B}$) *is isomorphic to a direct product of the corresponding lattices for a language with only one relation symbol.*

For a fixed R, the greatest element of $\mathbf{P}_R$ is the equivalence class of $R(x_1, \ldots, x_k)$, each formula in $\mathcal{P}_R$ being obtained from that one by a substitution. Moreover, up to a permutation of the variables, each atomic formula $R(t_1, \ldots, t_k)$ can be obtained from $R(x_1, \ldots, x_k)$ by a sequence of substitutions of the following basic types, perhaps in more than one way.

(1) For variables x, y both appearing in Φ, $\beta(x) = \beta(y) = x$ and $\beta(z) = z$ for all other $z \in X$.
(2) For a variable x appearing in Φ and a constant c, $\gamma(x) = c$ and $\gamma(z) = z$ for all other $z \in X$.
(3) For a variable x appearing in Φ, a function symbol f, and variables $y_1, \ldots, y_m$ not appearing in Φ, $\delta(x) = f(y_1, \ldots, y_m)$ and $\delta(z) = z$ for all other $z \in X$.

Type (2) is of course technically a special case of type (3). These basic types of substitutions give the covering relations in $\mathbf{P}_R$.

This analysis yields the first simple properties of $\mathbf{P}_R$.

Theorem A.8 *For each relation symbol R, the ordered set $\mathbf{P}_R$ has these properties.*

(1) *$\mathbf{P}_R$ has a greatest element.*
(2) *For each $p \in \mathbf{P}_R$, the filter $\uparrow p$ is finite.*

Example A.9 An example would serve us well at this point. Let $\mathcal{L}$ be the language with one binary predicate R, one unary function symbol f, and one constant c. The atomic formulas that can be obtained by 0, 1, or 2 basic substitutions are

(A_0) $R(x, y)$
(A_1) $R(f(x), y)$
(A_2) $R(c, y)$
(A_3) $R(x, x)$
(A_4) $R(x, c)$
(A_5) $R(x, f(y))$
(A_6) $R(f^2(x), y)$
(A_7) $R(f(c), y)$
(A_8) $R(f(x), x)$
(A_9) $R(c, f(y))$
(A_{10}) $R(c, c)$
(A_{11}) $R(f(x), f(x))$
(A_{12}) $R(f(x), f(y))$
(A_{13}) $R(f(x), c)$
(A_{14}) $R(y, f(y))$
(A_{15}) $R(x, f(c))$
(A_{16}) $R(x, f^2(y))$

The filter of $\mathbf{P}_R$ consisting of these atomic formulas, ordered by $\vdash$, is drawn in Fig. A.6.

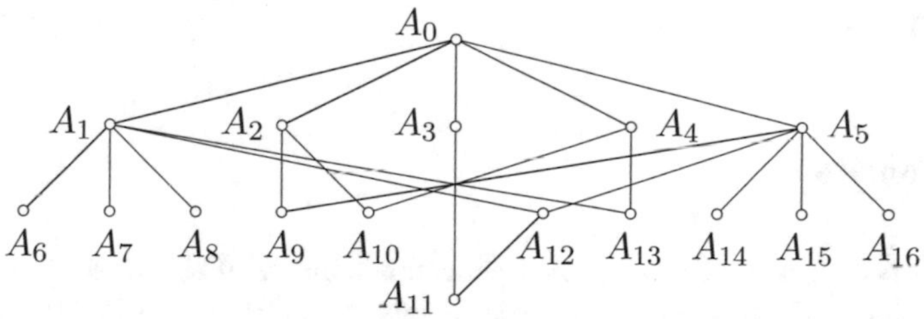

Fig. A.6 Top of $\mathbf{P}_R$ in Example A.9

In general, the structure at the top of $\mathbf{P}_R$ is more restricted than indicated by Theorem A.8. Define the *depth* of an element b in an ordered set with 1 to be the shortest length of a maximal chain from b to 1.

Theorem A.10 *Let R be a relation symbol in a language $\mathcal{L}$. If b is an element of depth two in $\mathbf{P}_R$, then the interval $[b, 1]$ is isomorphic to one of the ordered sets in* Fig. A.7.

Note that each of these possibilities occurs in our example. The proof is by considering all the ways we can construct an atomic formula of depth 2.

A similar analysis would yield all upper intervals of depth k for other small values of k. The point, though, is that only a few types of intervals can occur at the top of $\mathbf{P}_R$.

Another restriction on the structure of $\mathbf{P}_R$ is a consequence of a well-known result from computer science.

Theorem A.11 *For each relation symbol R, the ordered set $\mathbf{P}_R$ is a join semilattice.*

Suppose that $R(\mathbf{s}_i)$ for $1 \leq i \leq k$ are expressions that have a common lower bound in $\mathbf{P}_R$. A common lower bound is called a *unifier* for $R(\mathbf{s}_1), \ldots, R(\mathbf{s}_k)$, and

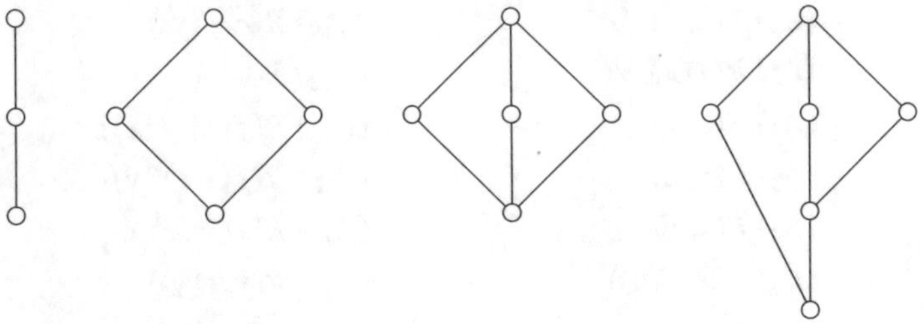

Fig. A.7 Intervals at the top of $\mathbf{P}_R$

a greatest lower bound is called a *most general unifier*. The unification algorithm of J. A. Robinson [103] (see e.g. Theorem 4.3 of Lloyd [80]) gives a most general unifier for a set of expressions that can be unified, i.e., have a common lower bound.

So, to find the join of two expressions $R(\mathbf{s})$ and $R(\mathbf{t})$, we may apply the unification algorithm to the set of all common upper bounds of $R(\mathbf{s})$ and $R(\mathbf{t})$. Note that this set is finite by Theorem A.8(2).

A.2.3 Conclusion

Theorems A.5, A.6 and Corollary A.7 show that a lattice of atomic theories $\mathrm{ATh}(\mathcal{B})$ is isomorphic to the lattice $\mathcal{O}(\mathbf{P})$ of order ideals of an ordered set $\mathbf{P}$. Theorems A.8–A.11 then give some restrictions on the ordered sets $\mathbf{P}$ that occur in this way. These results tend to support the idea that the complexities of lattices of universal theories in general arise, not from the atomic sentences in the axiom system (e.g., reflexivity for equality), but rather from the quasi-identities (e.g., symmetry and transitivity). This is hardly surprising!

A.3 In Search of Quasicriticality

A.3.1 Quasicriticality

A paper by Adams and Dziobiak [7] induced us to revisit the old problem, *How should quasicriticality be defined?* for structures that are not necessarily finite. Adams and Dziobiak proposed the definition given below, which we have used in this book. This section explores the advantages of their definition, while the next section suggests a strengthening, followed by examples.

Definition: A structure $\mathbf{T}$ is *quasicritical* if $\mathbf{T}$ is finitely generated and $\mathbb{Q}(\mathbf{T})$-subdirectly irreducible.

Let us show that their definition has six desirable properties. Moreover, the arguments are more natural with this general definition.

1. The old definition is that a *finite* structure $\mathbf{T}$ is quasicritical if $\mathbf{T}$ is not in the quasivariety generated by its proper substructures. Since $\mathbb{Q}(\mathcal{X}) = \mathbb{SP}(\mathcal{X})$ when $\mathcal{X}$ is a collection of finitely many finite structures, the definitions are equivalent for finite structures. (This is true even when the type is not finite. A finite structure still has only finitely many substructures, so an ultraproduct of those is isomorphic to one of them.)
2. Quasicriticality is an inherent property of an algebra: whether $\mathbf{T}$ is quasicritical depends only on $\mathbb{Q}(\mathbf{T})$.
3. Every finitely generated structure is either quasicritical or a subdirect product of quasicritical factors. To see this, let $\mathbf{T}$ be a finitely generated structure, and let $Q = \mathbb{Q}(\mathbf{T})$. The lattice $\mathrm{Con}_Q \mathbf{T}$ is algebraic, and every element of an algebraic lattice is a meet of completely meet irreducible elements. If $\mathbf{T}$ is not quasicritical, i.e., not subdirectly irreducible in Q, then it has a subdirect decomposition $\mathbf{T} \leq \prod_i \mathbf{S}_i$ based on the decomposition of the least element Δ of $\mathrm{Con}_Q \mathbf{T}$ as a meet of completely meet irreducible Q-congruences. Each factor $\mathbf{S}_i$ is finitely generated, in Q, and Q-subdirectly irreducible. Since $\mathbb{Q}(\mathbf{S}_i) \leq Q$, each $\mathbf{S}_i$ is *a fortiori* subdirectly irreducible in $\mathbb{Q}(\mathbf{S}_i)$.

Recall that every quasivariety is generated by its finitely generated members. Indeed, every structure embeds into an ultraproduct of its finitely generated substructures (Mal'cev [84]). Combining this with property 3 yields:

4. Every quasivariety is determined by its quasicritical members.
5. If a quasivariety Q is completely join irreducible in $\mathrm{L_q}(\mathcal{K})$, then Q is generated by any $\mathbf{T}$ in $Q \setminus Q_*$. Moreover, $\mathbf{T}$ can be chosen to be finitely generated, and by property 3, quasicritical.

Now for any quasivariety $\mathcal{K}$, the lattice $\mathrm{L_q}(\mathcal{K})$ is dually algebraic, so that every subquasivariety is a join of completely join irreducible quasivarieties. Property 5 says that each completely join irreducible quasivariety is generated by a quasicritical structure.

When $\mathcal{K}$ is locally finite of finite type, then conversely each quasicritical structure generates a quasivariety that is completely join irreducible in $\mathrm{L_q}(\mathcal{K})$. The next property says that, without the finiteness conditions, we still get finite join irreducibility.

6. The quasivariety $\mathbb{Q}(\mathbf{T})$ generated by a quasicritical structure is finitely join irreducible. If $\mathbf{T}$ is finite and of finite type, then $\mathbb{Q}(\mathbf{T})$ is completely join irreducible.

For assume that $\mathbf{T}$ is quasicritical and $\mathbb{Q}(\mathbf{T}) = Q_1 \vee \cdots \vee Q_m$. Then $\mathbf{T} \in \mathbb{SP}(Q_1 \cup \cdots \cup Q_m)$, that is, there exist structures $\mathbf{S}_i \in Q_i$ such that $\mathbf{T} \leq \mathbf{S}_1 \times \cdots \times \mathbf{S}_m$, and we may take this to be a subdirect product. But each $\mathbf{S}_i$ is in $\mathbb{Q}(\mathbf{T})$, whence $\mathbf{T} \cong \mathbf{S}_i$ for some i, and $\mathbb{Q}(\mathbf{T}) = Q_i$.

If $\mathbf{T}$ is finite and of finite type, then $\mathbb{Q}(\mathbf{T})$ is compact in $\mathrm{L_q}(\mathcal{K})$ for any quasivariety $\mathcal{K}$ containing $\mathbf{T}$. (This is Lemma 2.4 of [63].) For compact elements, join irreducible is the same as completely join irreducible.

A.3.2 Strong Quasicriticality

Now we strengthen the definition in a way that allows us to characterize those structures $\mathbf{T}$ such that $\mathbb{Q}(\mathbf{T})$ is completely join irreducible.

Definition: A structure $\mathbf{T}$ is *strongly quasicritical* if $\mathbf{T}$ is finitely generated and $\mathbb{Q}(\mathbf{T})$-subdirectly irreducible and satisfies the condition

(¥) If $\mathbf{S}_i \in \mathbb{Q}(\mathbf{T})$ for all $i \in I$ and $\mathbf{T} \leq (\prod_i \mathbf{S}_i)/U$ for an ultrafilter U on I, then there exist $i_0 \in I$, a set J, and an ultrafilter V on J such that $\mathbf{T} \leq \mathbf{S}_{i_0}^J/V$.

Let $\mathbb{P}_s$ denote the class operator of taking subdirect products.

Lemma A.12 $\mathbb{SP}(\mathcal{X}) \subseteq \mathbb{P}_s\mathbb{S}(\mathcal{X})$, *whence* $\mathbb{Q}(\mathcal{X}) = \mathbb{P}_s\mathbb{SU}(\mathcal{X})$.

Theorem A.13 *A quasivariety Q is completely join irreducible in* $\mathrm{L_q}(\mathcal{K})$ *for any* $\mathcal{K} \supseteq Q$ *if and only if* $Q = \mathbb{Q}(\mathbf{T})$ *for a strongly quasicritical algebra.*

Proof Suppose Q is completely join irreducible. Let $\mathbf{T}$ be a quasicritical algebra in $Q \setminus Q_*$. By property 5, $Q = \mathbb{Q}(\mathbf{T})$ so $\mathbf{T}$ is Q-subdirectly irreducible and finitely generated. Suppose $\mathbf{T} \leq (\prod_i \mathbf{S}_i)/U$ for a collection $\mathbf{S}_i$ $(i \in I)$ with each $\mathbf{S}_i \in Q$. Then $Q = \bigvee_i \mathbb{Q}(\mathbf{S}_i)$, whence $Q = \mathbb{Q}(\mathbf{S}_{i_0})$ for some i_0. Therefore, $\mathbf{T} \in \mathbb{P}_s\mathbb{SU}(\mathbf{S}_{i_0})$, and since $\mathbf{T}$ is Q-subdirectly irreducible, $\mathbf{T} \in \mathbb{SU}(\mathbf{S}_{i_0})$, as desired.

Conversely, assume that $\mathbf{T}$ is strongly quasicritical and that $\mathbb{Q}(\mathbf{T}) = \bigvee_j \mathcal{R}_j$ in $\mathrm{L_q}(\mathbb{Q}(\mathbf{T}))$. Then $\mathbb{Q}(\mathbf{T}) = \mathbb{SPU}(\bigcup_j \mathcal{R}_j)$, so there exist structures $\mathbf{S}_i$ $(i \in I)$ in $\bigcup_j \mathcal{R}_j$ such that $\mathbb{Q}(\mathbf{T}) = \mathbb{P}_s\mathbb{SU}(\bigcup_i \mathbf{S}_i)$. Because $\mathbf{T}$ is quasicritical and the $\mathbf{S}_i$ are in $\mathbb{Q}(\mathbf{T})$, we get $\mathbf{T} \in \mathbb{SU}(\bigcup_i \mathbf{S}_i)$. As $\mathbf{T}$ satisfies (¥), this implies $\mathbf{T} \in \mathbb{SU}(\mathbf{S}_{i_0})$ for some i_0. Thus $\mathbf{T} \in \mathbb{Q}(\mathbf{S}_{i_0}) \leq \mathcal{R}_{j_0} \leq \mathbb{Q}(\mathbf{T})$ for the appropriate j_0, so that $\mathbb{Q}(\mathbf{T}) = R_{j_0}$, and $\mathbb{Q}(\mathbf{T})$ is completely join irreducible. □

It must be admitted that the condition (¥) is not very practical.

A.3.3 Three Examples

Example A.14 Let $\mathbf{L}_0$ be the lattice of subspaces of a 3-dimensional vector space over the rational numbers. Note that

- $\mathbf{L}_0$ is 4-generated, modular, and simple.
- Every proper sublattice of $\mathbf{L}_0$ is 2-distributive, i.e., satisfies

$$x \wedge (y \vee z \vee t) = (x \wedge (y \vee z)) \vee (x \wedge (y \vee t)) \vee (x \wedge (z \vee t)).$$

Indeed, the maximal proper sublattices are isomorphic to either $\mathbf{2}^3$ or $\mathbf{M}_\omega \times \mathbf{2}$ or $\mathbf{M}_{\omega,\omega}$.

We claim that the quasivariety $\mathbb{Q}(\mathbf{L}_0)$ is completely join irreducible.

Consider a finitely generated lattice **T** in $\mathbb{Q}(\mathbf{L}_0) = \mathbb{P}_s\mathbb{S}\mathbb{U}(\mathbf{L}_0)$. A sublattice of an ultrapower of $\mathbf{L}_0$ has height at most 3, and a finitely generated subdirect product of lattices of bounded height itself has finite height. By a result of Huhn [62], if **T** does not contain a projective plane (necessarily of characteristic 0 by Freese [45]) as a sublattice, then it is 2-distributive. Thus every finitely generated lattice in $\mathbb{Q}(\mathbf{L}_0)$ that does not contain $\mathbf{L}_0$ is 2-distributive. We conclude that $\mathbb{Q}(\mathbf{L}_0)_* = \mathbb{Q}(\mathbf{L}_0) \cap \mathcal{D}_2$.

Example A.15 We give an example of a lattice **K** that is quasicritical but not strongly quasicritical, so that $\mathbb{Q}(\mathbf{K})$ is join irreducible but not completely join irreducible.

Consider the lattices $\mathbf{K}_n$ $(n \in \omega)$ where $\mathbf{K}_0 = \mathbf{2} \times \mathbf{2}$ and $\mathbf{K}_1$, $\mathbf{K}_2$ are drawn in Fig. A.8. Each $\mathbf{K}_n$ is generated by $\{a_0, b_0, c_n\}$. The lattice **K**, as drawn in Fig. A.9, contains each $\mathbf{K}_n$ as a sublattice (0 and all elements with subscripts $\leq n$), and **K** is generated by $\{a_0, b_0, \overline{c}, \overline{d}\}$.

Moreover, **K** embeds into a reduced product of the lattices $\mathbf{K}_n$, as follows. On the index set ω, let F be the filter consisting of all sets $\uparrow k = \{j : k \leq j < \omega\}$. We

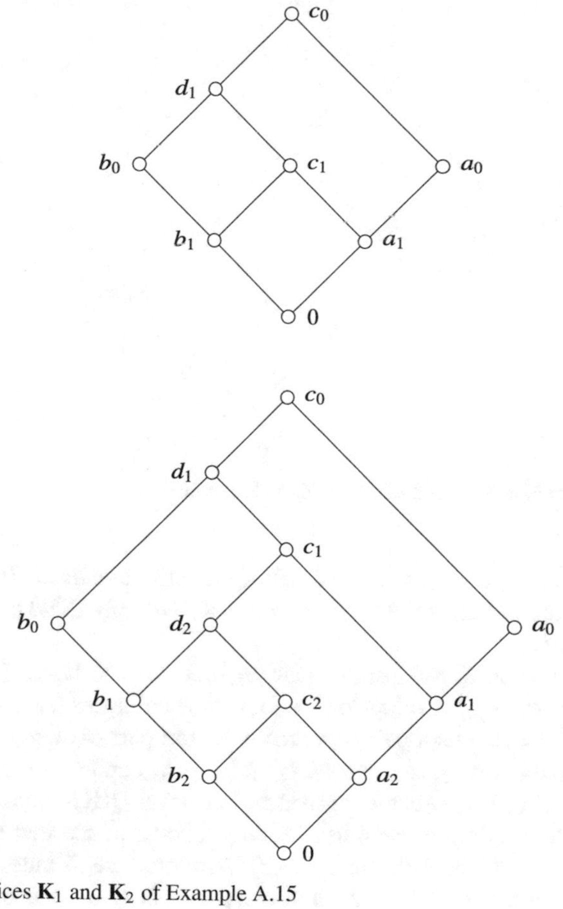

Fig. A.8 The lattices $\mathbf{K}_1$ and $\mathbf{K}_2$ of Example A.15

can embed $\mathbf{K}$ into $\prod \mathbf{K}_n/{\approx_F}$ as follows, where $[\mathbf{x}]$ denotes the equivalence class of $\mathbf{x}$ modulo F.

- $a_n \mapsto [a_0, \dots, a_{n-1}, a_n, a_n, \dots]$ and similarly for b_n, c_n, and d_n,
- $\overline{a} \mapsto [a_0, \dots, a_{k-1}, a_k, a_{k+1}, \dots]$ and similarly for $\overline{b}$ and $\overline{c}$,
- $\overline{d} \mapsto [d_1, \dots, d_k, d_{k+1}, d_{k+2}, \dots]$,
- $\overline{\overline{a}} \mapsto [a_1, \dots, a_k, a_{k+1}, a_{k+2}, \dots]$,
- $0 \mapsto [0, 0, 0, \dots]$.

We conclude that $\mathbb{Q}(\mathbf{K}) = \bigvee_n \mathbb{Q}(\mathbf{K}_n)$. This is a proper join, since each $\mathbf{K}_n$ is a splitting lattice, that is, a finite subdirectly irreducible projective lattice.

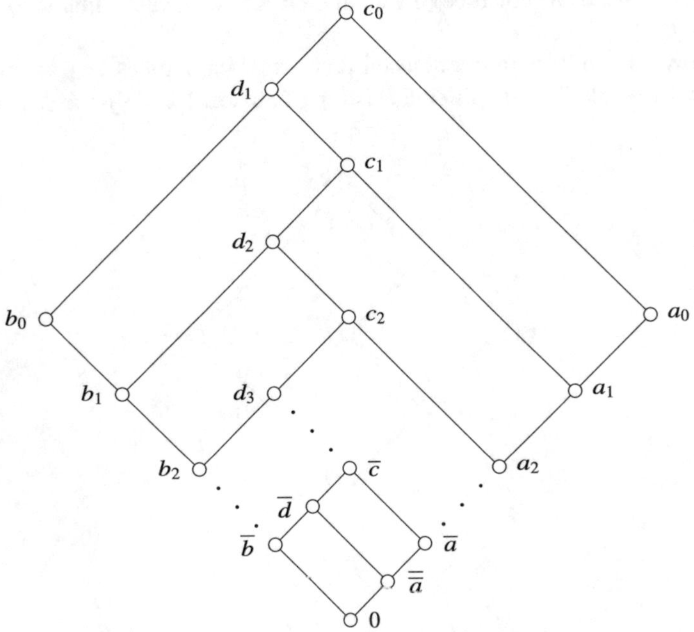

Fig. A.9 The lattice $\mathbf{K}$, a limit of the lattices $\mathbf{K}_n$ in Example A.15

On the other hand, $\mathbf{K}$ is quasicritical with $\mathbb{Q}(\mathbf{K})$-critical pair $(\overline{a}, \overline{\overline{a}})$. First note that $\mathbf{K}$ is semidistributive. Hence the quotient lattice $\mathbf{K}/\theta$ for any $\mathbb{Q}(\mathbf{K})$ congruence must be semidistributive.

Any congruence on $\mathbf{K}$ collapsing a pair (a_i, a_{i+1}) also collapses (c_i, c_{i+1}), which in turn collapses (b_i, b_{i+1}). Collapsing (b_i, b_{i+1}) then collapses (a_{i+1}, a_{i+2}). Thus any congruence θ that contains a pair (x, y) from the top part of $\mathbf{K}$ must collapse each of the chains $a_k, a_{k+1}, a_{k+2}, \dots$, and $b_k, b_{k+1}, b_{k+2}, \dots$, and $c_k, d_{k+1}, c_{k+1}, \dots$ from some point on. This is a lattice congruence, but not a $\mathbb{Q}(\mathbf{K})$-congruence, because the quotient lattice fails join semidistributivity. (You can see this with the dotted lines at the bottom of Fig. A.9.) To be a $\mathbb{Q}(\mathbf{K})$-congruence, $\overline{a}$ must be in the class of $a_k, a_{k+1}, \dots$, and likewise for b_k and $\overline{b}$, c_k and $\overline{c}$. (Any one of these implies the others.) It follows that $\overline{d} = (\overline{\overline{a}} \vee \overline{b})\,\theta\,(\overline{\overline{a}} \vee b_k) = d_{k+1}$; meeting with $\overline{a}$ yields $\overline{\overline{a}}\,\theta\,\overline{a}$. It remains to observe that collapsing $(\overline{b}, 0)$ or $(\overline{\overline{a}}, 0)$ forces $a_0\,\theta\,c_0$ or $b_0\,\theta\,d_1$, respectively, both in the top part. Thus $(\overline{a}, \overline{\overline{a}})$ is a $\mathbb{Q}(\mathbf{K})$-critical pair, whence $\mathbf{K}$ is $\mathbb{Q}(\mathbf{K})$-subdirectly irreducible.

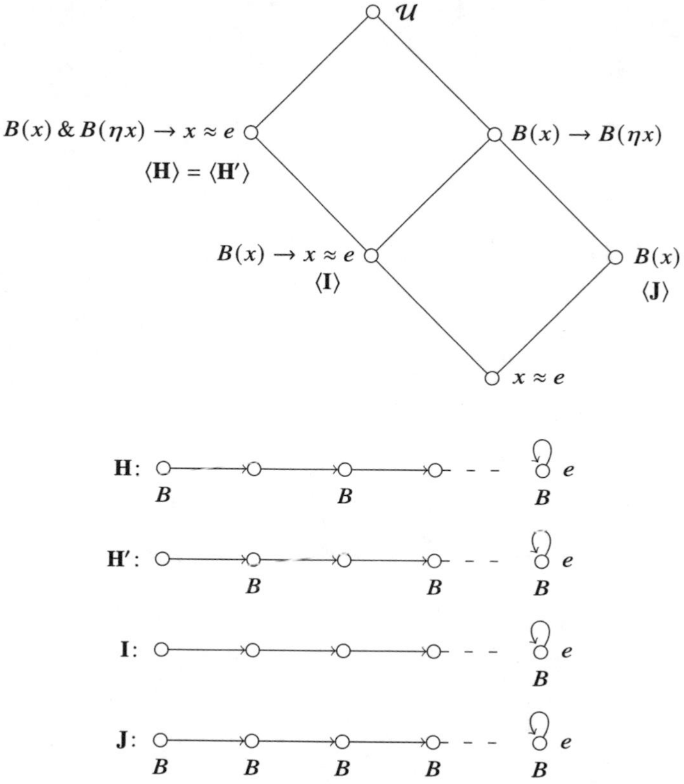

Fig. A.10 $\mathrm{L_q}(\mathcal{U})$ and the quasicritical structures in $\mathcal{U}$ from Example A.16

Example A.16 Let $\mathcal{U}$ be the longstyle quasivariety with operations η, e, and a unary predicate B satisfying these laws:

$$B(e) \qquad \eta e \approx e$$

$$B(x) \leftrightarrow B(\eta^2 x)$$

$$\eta x \approx \eta y \rightarrow x \approx y$$

$$\eta^k x \approx x \rightarrow x \approx e \quad \text{for } k \geq 1.$$

The lattice of subquasivarieties $\mathrm{L_q}(\mathcal{U})$ is given in Fig. A.10, along with its (strongly) quasicritical structures. Note that $\mathbf{I} \leq \mathbf{H}^2$ and $\mathbf{H} \leq \mathbf{H}' \leq \mathbf{H}$. In particular,

$\mathbb{Q}(\mathbf{H}) = \mathbb{Q}(\mathbf{H}')$ is completely join irreducible with two non-isomorphic quasicritical generators, which cannot happen for locally finite quasivarieties of finite type (Theorem 2.11 of [63]).

Bibliography

1. Adams, M., Adaricheva, K., Dziobiak, W., Kravchenko, A.: Open questions related to the problem of Birkhoff and Maltsev. Studia Logica **78**, 357–378 (2004)
2. Adams, M., Dziobiak, W.: Lattices of quasivarieties of 3-element algebras. Proc. Amer. Math. Soc **120**, 1053–1059 (1994)
3. Adams, M., Dziobiak, W.: Q-universal quasivarieties of algebras. Proc. Amer. Math. Soc. **120**, 1053–1059 (1994)
4. Adams, M., Dziobiak, W.: Finite-to-finite universal quasivarieties are q-universal. Algebra Univers. **46**, 253–283 (2001)
5. Adams, M., Dziobiak, W.: Quasivarieties of idempotent semigroups. Int. J. Algebra Comput. **13**, 733–752 (2003)
6. Adams, M., Dziobiak, W.: Universal quasivarieties of algebras. In: Proceedings of the 9th "Dr. Antonio A. R. Monteiro" Congress, pp. 11–21. Univ. Nac. del Sur, Baía Blanca (2008)
7. Adams, M., Dziobiak, W.: Two remarks about the q-lattice of the variety of lattices. Algebra Univers. **82**, 5 (2021)
8. Adaricheva, K.: The structure of finite lattices of subsemilattices. Algebra Logic **30**, 249–264 (1990)
9. Adaricheva, K.: Characterization of lattices of subsemilattices. Algebra Logic **30**, 385–404 (1991)
10. Adaricheva, K.: Two embedding theorems for lower bounded lattices. Algebra Univers. **36**, 425–430 (1996)
11. Adaricheva, K.: Lattices of algebraic subsets. Algebra Univers. **52**, 167–183 (2004)
12. Adaricheva, K.: On the prevariety of perfect lattices. Algebra Univers. **65**, 21–39 (2011)
13. Adaricheva, K., Dziobiak, W., Gorbunov, V.: Finite atomistic lattices that can be represented as lattices of quasivarieties. Fund. Math. **142**, 19–43 (1993)
14. Adaricheva, K., Dziobiak, W., Gorbunov, V.: Algebraic point lattices of quasivarieties. Algebra Logic **36**, 213–225 (1997)
15. Adaricheva, K., Freese, R., Nation, J.: Semidistributive semilattices (2020). Manuscript, available at /math.hawaii.edu/~jb/
16. Adaricheva, K., Gorbunov, V.: Equational closure operator and forbidden semidistributive lattices. Sib. Math. J. **30**, 831–849 (1989)
17. Adaricheva, K., Gorbunov, V.: On lower bounded lattices. Algebra Univers. **46**, 203–213 (2001)
18. Adaricheva, K., Gorbunov, V., Tumanov, V.: Join-semidistributive lattices and convex geometries. Adv. Math. J. **173**, 1–49 (2003)
19. Adaricheva, K., Maroti, M., Mckenzie, R., Nation, J., Zenk, E.: The Jónsson-Kiefer property. Studia Logica **83**, 111–131 (2006)
20. Adaricheva, K., Nation, J.: Lattices of quasi-equational theories as congruence lattices of semilattices with operators, Parts I and II. Int. J. Algebra Comput. **22**, N7 (2012)

K. Adaricheva et al., *A Primer of Subquasivariety Lattices*, CMS/CAIMS Books in Mathematics 3, https://doi.org/10.1007/978-3-030-98088-7

21. Adaricheva, K., Nation, J.: Classes of semidistributive lattices (Chapter 3). In: G. Grätzer, F. Wehrung (eds.) Lattice Theory: Special Topics and Applications, vol. 2, pp. 59–101. Birkhäuser, Cham (2016)
22. Adaricheva, K., Nation, J.: Lattices of algebraic subsets and implicational classes (Chapter 4). In: G. Grätzer, F. Wehrung (eds.) Lattice Theory: Special Topics and Applications, vol. 2, pp. 103–151. Birkhäuser, Cham (2016)
23. Bergman, C.: Universal Algebra: Fundamentals and Selected Topics. CRC Press, Boca Raton (2012)
24. Birkhoff, G.: On the structure of abstract algebras. Proc. Cambridge Phil. Soc. **31**, 432–454 (1935)
25. Blanco, J., Campercholi, M., Vaggione, D.: The subquasivariety lattice of a discriminator variety. Adv. Math. **159**, 18–50 (2001)
26. Blok, W., Pigozzi, D.: Algebraic semantics for universal Horn logic without equality. In: Universal Algebra and Quasigroup Theory (Jadwisin, 1989), Res. Exp. Math., vol. 19, pp. 1–56. Heldermann, Berlin (1992)
27. Bulman-Fleming, S., Werner, H.: Equational compactness in quasiprimal varieties. Algebra Univers. **7**, 33–46 (1977)
28. Burris, S., Sankappanavar, H.P.: A Course in Universal Algebra. Springer-Verlag, New York (1981). Available from http://www.math.uwaterloo.ca/~snburris/htdocs/ualg.html
29. Chang, C.C., Kiesler, J.: Model Theory. North Holland, Amsterdam (1990)
30. Chang, C.C., Morel, A.: On closure under direct products. J. Symbolic Logic **23**, 149–154 (1959)
31. Crawley, P., Dilworth, R.P.: Algebraic Theory of Lattices. Prentice-Hall, Englewood Cliffs (1973)
32. Czelakowski, J.: Protoalgebraic Logics, *Trends in Logic, Studia Logica Library*, vol. 10. Kluwer, Dordrecht (2001)
33. Davey, B., Priestley, H.: Introduction to Lattices and Order, 2nd ed. Cambridge University Press, New York (2002)
34. Davey, B.A., Sands, B.: An application of Whitman's condition to lattices with no infinite chains. Algebra Univers. **7**, 171–178 (1977)
35. Day, A.: Characterizations of finite lattices that are bounded-homomorphic images of sublattices of free lattices. Canad. J. Math. **31**, 69–78 (1979)
36. Dellunde, P., Jansana, R.: Some characterization theorems for infinitary universal Horn logic without equality. J. Symbolic Logic **61**, 1242–1260 (1996)
37. Dziobiak, W.: On subquasivariety lattices of semiprimal varieties. Algebra Univers. **20**, 127–129 (1985)
38. Dziobiak, W.: On lattice identities satisfied in subquasivariety of modular lattices. Algebra Univers. **22**, 205–214 (1986)
39. Dziobiak, W.: On atoms in the lattice of quasivarieties. Algebra Univers. **24**, 32–35 (1987)
40. Elgueta, R.: Characterizing classes defined without equality. Studia Logica **58**, 357–394 (1997)
41. Erné, M.: Weak distributive laws and their role in lattices of equational theories. Algebra Univers. **25**, 290–321 (1988)
42. Fajtlowicz, S., Schmidt, J.: Bézout families, join congruences and meet-irreducible ideals. In: Lattice Theory (Szeged, 1974), Colloq. Math. Soc. János Bolyai, vol. 14, pp. 51–76. North Holland, Amsterdam (1976)
43. Frayne, T., Morel, A., Scott, D.: Reduced direct products. Fund. Math. **51**, 195–228 (1962)
44. Frayne, T., Morel, A., Scott, D.: Correction to the paper "Reduced direct products". Fund. Math. **53**, 117 (1963)
45. Freese, R.: The variety of modular lattices is not generated by its finite members. Trans. Amer. Math. Soc. **255**, 277–300 (1979)
46. Freese, R., Ježek, J., Nation, J.: Free Lattices, *Mathematical Surveys and Monographs*, vol. 42. Amer. Math. Soc., Providence (1995)

47. Freese, R., Kearnes, K., Nation, J.: Congruence lattices of congruence semidistributive algebras. In: Lattice Theory and its Applications (Darmstadt, 1991), Res. Exp. Math., vol. 23, pp. 63–78. Heldermann, Lemgo (1995)
48. Freese, R., Nation, J.: Congruence lattices of semilattices. Pac. J. Math. **49**, 51–58 (1973)
49. Freese, R., Nation, J.: A simple semidistributive lattice. Int. J. Algebra Comput. (2021)
50. Gorbunov, V.: Covers in lattices of quasivarieties and independent axiomatizability. Algebra Logic **16**, 340–369 (1977)
51. Gorbunov, V.: The structure of lattices of quasivarieties. Algebra Univers. **32**, 493–530 (1994)
52. Gorbunov, V.: Algebraic Theory of Quasivarieties. Plenum, New York (1998)
53. Gorbunov, V., Tumanov, V.: A class of lattices of quasivarieties. Algebra Logic **19**, 38–52 (1980)
54. Gorbunov, V., Tumanov, V.: Construction of lattices of quasivarieties. In: Mathematical logic and the theory of algorithms, Trudy Inst. Mat., vol. 2, pp. 12–44. Nauka Sibirsk. Otdel., Novosibirsk (1982)
55. Grätzer, G.: Equational classes of lattices. Duke Math. J. **33**, 613–622 (1966)
56. Grätzer, G.: Universal Algebra, 2nd ed. Springer-Verlag, New York (1979)
57. Grätzer, G.: Lattice Theory: Foundations. Springer, New York (2011)
58. Grätzer, G., Lakser, H.: The lattice of quasivarieties of lattices. Algebra Univers. **9**, 102–115 (1979)
59. Grätzer, G., Schmidt, E.T.: On congruence lattices of lattices. Acta Math. Acad. Sci. Hung. **13**, 179–185 (1962)
60. Hoehnke, H.J.: Fully invariant algebraic closure systems of congruences and quasivarieties of algebras. In: Lectures in Universal Algebra (Szeged, 1983), Colloq. Math. Soc. János Bolyai, vol. 43, pp. 189–207. North Holland, Amsterdam (1986)
61. Horn, A.: On sentences which are true of direct unions of algebras. J. Symbolic Logic **16**, 14–21 (1951)
62. Huhn, A.: Two notes on n-distributive lattices. In: Lattice Theory (Szeged, 1974), Colloq. Math. Soc. János Bolyai, vol. 14, pp. 137–147. North Holland, Amsterdam (1974)
63. Hyndman, J., Nation, J.: The Lattice of Subquasivarieties of a Locally Finite Quasivariety. Canadian Math. Soc. Books in Mathematics. Springer, New York (2018)
64. Hyndman, J., Nation, J., Nishida, J.: Congruence lattices of semilattices with operators. Studia Logica **104**, 305–316 (2016)
65. Ježek, J.: Intervals in the lattice of varieties. Algebra Univers. **6**, 147–158 (1976)
66. Jipsen, P., Nation, J.: Primitive lattice varieties (2020). Manuscript
67. Jipsen, P., Rose, H.: Varieties of lattices (chapter 1). In: G. Grätzer, F. Wehrung (eds.) Lattice Theory: Special Topics and Applications, vol. 2, pp. 1–26. Birkhäuser, Cham (2016)
68. Jónsson, B.: Equational classes of lattices. Math. Scand. **22**, 187–196 (1968)
69. Jónsson, B., Kiefer, J.: Finite sublattices of a free lattice. Canad. J. Math **14**, 487–497 (1962)
70. Jónsson, B., Nation, J.: A report on sublattices of a free lattice. In: Universal Algebra and Lattice Theory, Contributions to Universal Algebra, *Lecture Notes in Mathematics, Coll. Math. Soc. János Bolyai*, vol. 17, pp. 223–257. North-Holland (1977)
71. Kearnes, K., Nation, J.: Axiomatizable and nonaxiomatizable congruence prevarieties. Algebra Univers. **59**, 323–335 (2008)
72. Keimel, K., Werner, H.: Stone duality for varieties generated by quasiprimal algebras. In: Recent advances in the representation theory of rings and C^*-algebras by continuous sections (Sem., Tulane Univ., New Orleans, 1973), *Mem. Amer. Math. Soc.*, vol. 148, pp. 59–85. Amer. Math. Soc., Providence (1974)
73. Kravchenko, A., Nurakunov, A., Schwidefsky, M.: On the structure of quasivariety lattices I: Independent axiomatizability. Algebra Logic **57**, 445–462 (2019)
74. Kravchenko, A., Nurakunov, A., Schwidefsky, M.: On the structure of quasivariety lattices II: Undecidable problems. Algebra Logic **58**, 123–136 (2019)
75. Kravchenko, A., Nurakunov, A., Schwidefsky, M.: On the complexity of the lattices of subvarieties and congruences. Int. J. Algebra Comput. **30**, 1609–1624 (2020)
76. Kravchenko, A., Nurakunov, A., Schwidefsky, M.: On the structure of quasivariety lattices III: Finitely partitionable bases. Algebra Logic **59**, 222–229 (2020)

77. Kravchenko, A., Schwidefsky, M.: On the complexity of the lattices of subvarieties and congruences II: Differential groupoids and unary algebras. Sib. Elektron. Mat. Izv. **17**, 753–768 (2020)
78. Lampe, W.: A property of the lattice of equational theories. Algebra Univers. **23**, 61–69 (1986)
79. Lampe, W.: Further properties of lattices of equational theories. Algebra Univers. **28**, 459–486 (1991)
80. Lloyd, J.: Foundations of Logic Programming, 2nd. ed. Springer-Verlag, New York (1987)
81. Lutsak, S.M.: On the complexity of quasivariety lattices. Sib. Èlektron. Mat. Izv. **14**, 92–97 (2017)
82. Mal'cev, A.I.: Untersuchungen aus dem Gebiete mathematischen Logik. Rec. Mat. N. S. **1**, 323–335 (1936)
83. Mal'cev, A.I.: Several remarks on quasivarieties of algebraic systems. Algebra i Logika Sem. **5**, 3–9 (1966). (Russian)
84. Mal'cev, A.I.: Algebraic Systems. Springer-Verlag, New York (1973)
85. McKenzie, R.: Equational bases and non-modular lattice varieties. Trans. Amer. Math. Soc. **174**, 1–43 (1972)
86. McKenzie, R.: Finite forbidden lattices. In: Universal Algebra and Lattice Theory, Lecture Notes in Mathematics, vol. 1004, pp. 176–205. Springer-Verlag, Berlin (1983)
87. McKenzie, R., McNulty, G., Taylor, W.: Algebras, Lattices, Varieties, volume I. Wadsworth & Brooks/Cole, Belmont, CA (1987)
88. Murskiĭ, V.L.: The existence of finite bases of identities, and other properties of "almost all" finite algebras (Russian). Problemy Kibernet **30**, 43–56 (1975)
89. Nation, J.: An approach to lattice varieties of finite height. Algebra Univers. **27**, 521–543 (1990)
90. Nation, J.: Notes on lattice theory (1990). Available at /math.hawaii.edu/~jb/
91. Nation, J.: Lattices of theories in languages without equality. Notre Dame J. of Formal Logic **54**, 167–175 (2013)
92. Nation, J., Nishida, J.: A refinement of the equaclosure operator. Algebra Univers. **79**, 46 (2018)
93. Neumann, H.: Varieties of Groups. Springer-Verlag, New York (1967)
94. Newrly, N.: Lattices of equational theories are congruence lattices of monoids with one additional unary operation. Algebra Univers. **30**, 217–220 (1993)
95. Nurakunov, A.: Equational theories as congruences of enriched monoids. Algebra Univers. **58**, 357–372 (2008)
96. Nurakunov, A.: Unreasonable lattices of quasivarieties. Int. J. Algebra Comput. **22**, 125006 (2012)
97. Nurakunov, A.: Lattices of quasivarieties of pointed abelian groups. Algebra Logic **53**, 238–257 (2014)
98. Papert, D.: Congruence relations in semilattices. J. London Math. Soc. **39**, 723–729 (1964)
99. Pigozzi, D., Tardos, G.: The representation of certain abstract lattices as lattices of subvarieties (1999). Manuscript
100. Pudlák, P., Tůma, J.: Yeast graphs and fermentation of algebraic lattices. In: Lattice Theory (Szeged, 1974), Colloq. Math. Soc. János Bolyai, vol. 14, pp. 301–341. North Holland, Amsterdam (1976)
101. Reinhold, J.: Weak distributive laws and their role in free lattices. Algebra Univers. **33**, 209–215 (1995)
102. Repnitskiĭ, V.: On finite lattices which are embeddable in subsemigroup lattices. Semigroup Forum **46**, 388–397 (1993)
103. Robinson, J.: A machine-oriented logic based on the resolution principle. J. ACM **12**, 23–41 (1965)
104. Sapir, M.: The lattice of quasivarieties of semigroups. Algebra Univers. **21**, 172–180 (1985)
105. Schmidt, E.T.: Zur charakterisierung der kongruenzverbände der verbände. Mat. Casopis Sloven. Akad. Vied **18**, 3–20 (1968)

106. Schmidt, E.T.: Kongruenzrelationen Algebraischer Strukturen, vol. 25. Dt. Verlag d. Wiss. (1969)
107. Schwidefsky, M.: Complexity of quasivariety lattices. Algebra Logic **54**, 245–257 (2015)
108. Semenova, M.: On lattices that are embeddable into subsemigroup lattices. I. Semilattices. Algebra Logic **45**, 124–133 (2006)
109. Shafaat, A.: On the structure of certain idempotent semigroups. Trans. Amer. Math. Soc. **149**, 371–378 (1970)
110. Shafaat, A.: On implicational completeness. Canad. J. Math. **26**, 761–768 (1974)
111. Slavík, V.: A note on subquasivarieties of some varieties of lattices. Comment. Math. Univ. Carolinae **16**, 173–181 (1975)
112. Tumanov, V.: Finite distributive lattices of quasivarieties. Algebra Logic **22**, 119–129 (1983)
113. Vinogradov, A.A.: Quasivarieties of abelian groups (Russian). Algebra i Logika **4**, 15–19 (1965)
114. Wehrung, F.: Sublattices of complete lattices with continuity conditions. Algebra Univers. **53**, 149–173 (2005)
115. Whitman, P.M.: Free lattices. Ann. of Math. (2) **42**, 325–330 (1941)

Symbol Index

K. Adaricheva et al., *A Primer of Subquasivariety Lattices*, CMS/CAIMS Books in Mathematics 3, https://doi.org/10.1007/978-3-030-98088-7

Author Index

K. Adaricheva et al., *A Primer of Subquasivariety Lattices*, CMS/CAIMS Books in Mathematics 3, https://doi.org/10.1007/978-3-030-98088-7

Subject Index

K. Adaricheva et al., *A Primer of Subquasivariety Lattices*, CMS/CAIMS Books in Mathematics 3, https://doi.org/10.1007/978-3-030-98088-7

Printed in the United States
by Baker & Taylor Publisher Services